# The Box

# The Box

## How the Shipping Container Made the World Smaller and the World Economy Bigger

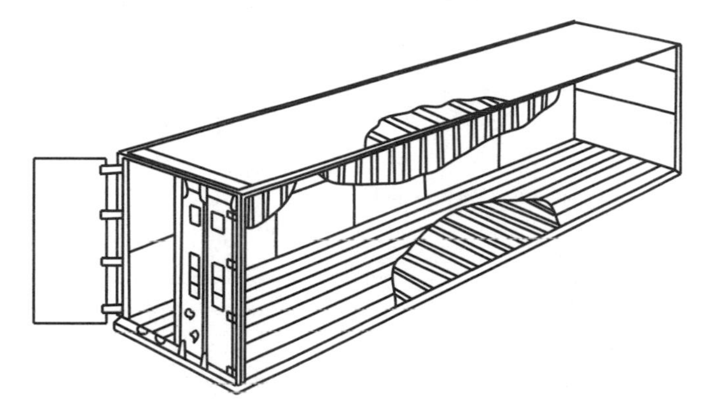

With a new preface by the author

**Marc Levinson**

PRINCETON UNIVERSITY PRESS
PRINCETON AND OXFORD

Published by Princeton University Press, 41 William Street,
Princeton, New Jersey 08540
In the United Kingdom: Princeton University Press,
3 Market Place, Woodstock, Oxfordshire OX20 1SY



Ninth printing, and first paperback printing, with a new preface by the author, 2008
Paperback ISBN: 978-0-691-13640-0

The Library of Congress has cataloged the cloth edition of this book as follows

Levinson, Marc.
The box : how the shipping container made the world smaller and
the world economy bigger / Marc Levinson.
p. cm.
Includes bibliographical references and index.
ISBN-13: 978-0-691-12324-0 (hardcover : alk. paper)
ISBN-10: 0-691-12324-1 (hardcover)
1. Containerization—History. 2. McLean, Malcolm Purcell, 1913–2001. I. Title.
TA1215.L47 2006
387.5′442—dc22 2005030021

British Library Cataloging-in-Publication Data is available

This book has been composed in Janson text with Clarendon Family Display

Printed on acid-free paper. ∞

press.princeton.edu

Printed in the United States of America

20

To **Aaron, Rebecca, and Deborah**

# Contents

## Preface to the Paperback Edition

**Writing a book** is usually a solitary venture, but *The Box* was a more private project than most. This was not entirely my choice. Early on in my work, when acquaintances would ask what I'd been doing, I would proudly tell them I was writing a history of the shipping container. The result of this disclosure was invariably stunned silence, as my interlocutors tried to think of something to say about a boring metal box. Eventually, I stopped talking about thc book altogether, simply to avoid the embarrassment that every mention of the topic would bring.

The response to the book's publication in the spring of 2006, then, caught me by surprise. I knew that the history of containerization would show itself to be a far more absorbing topic than readers could imagine, and I figured that economists and logistics specialists might be intrigued by my argument that tumbling transport costs were critical in opening the way to what we now call globalization. I had not the slightest clue, however, that the container was on its way to becoming something trendy. Then the invitations began to arrive. In New York, I shared a platform with architects using containers to design office buildings and apartments. In Genoa, I spoke alongside an entrepreneur who turned containers into temporary art galleries, while in Santa Barbara, California, the local museum joined forces with a university to promote a series of public events on ramifications of the container that I had never considered. The ugliness of stacks of abandoned containers; the security threat posed by millions of boxes with undetermined contents; the environmental damage caused by massive movement of cargo: all of these issues came to the fore in reviews and critical articles.

Then business executives weighed in. A leading computer manufacturer embraced *The Box* as a metaphor for modular products, announcing a "data center in a box." A major oil company drew insights from the container that helped it cut the cost of exploration in the Canadian Arctic. Several consulting firms applied lessons from containerization to a variety of business problems having nothing to do with freight transportation. A software house developed the notion of a computer system that passed "containerized" pieces of data from one location to the next, an extension of containerization well beyond any I had imagined.

Academics drew on *The Box* to open some new lines of intellectual inquiry. There has been, as I discovered in the course of my research, very little serious study of the container and its consequences, except in the area of labor relations. That neglect is partly due to lack of data: chapter 13 lays out the obstacles to developing reliable estimates of how the container has changed transport costs since the 1950s. The reluctance of academics to cross traditional boundaries also has inhibited exploration of the container's impact. One expert on logistics, for example, was quite familiar with containerships but told me he had never considered the container's impact on shore. Mainly, however, I think academics have ignored the container because it seems so prosaic. A noted economic historian, a reader informed me, had long proclaimed to students that while the container was an important development, it was too simple to be worth much study. The book may at least have knocked down that last objection. It seems to have provided fodder for a variety of conferences and symposia, stimulating a new intellectual dialogue about the role of transportation in economic change.

The media have undertaken a similar reconsideration. Since the late 1980s, commentators have filled columns and airwaves with glib chatter about globalization, as if it were merely a matter of bits and bytes and corporate cost-cutting. Since *The Box* appeared, however, many news stories and articles have acknowledged that, digital communication notwithstanding, the integration of the world economy depends less on call centers and trans-Pacific exports of technical services than on the ability to move goods cheaply from here to there. *The Box*, I hope, has contributed to public understanding that inadequate port, road, and rail infrastructure can cause economic harm by raising the cost of moving freight.

Many aspects of the response to *The Box* were startling, but perhaps the most unexpected concerns a widespread stereotype about innovation. In his later years, Malcom McLean, the former trucker whose audacious scheme to create the first containership line is recounted in chapter 3, was frequently asked how he came up with the idea of the container. He responded with a tale about how, after spending hours in late 1937 queuing at a Jersey City pier to unload his truck, he realized that it would be quicker simply to hoist the entire truck body on board. From this incident, we are meant to believe, came his decision eighteen years later to buy a war-surplus tanker and equip it to carry 33-foot-long containers.

The story of this "Aha!" moment does not appear in *The Box*, because I believe the event never occurred. There is certainly no contemporaneous evidence for it. I suspect that the story of McLean's stroke of genius took on a life of its own as, decades later, well-meaning people asked McLean where the container came from. As I show in chapter 2, ship lines and railroads had been experimenting with containers for half a century before Malcom McLean's trip to Jersey City, and containers were already in wide use in North America and Europe when McLean's first ship set sail in 1956.

Malcom McLean's real contribution to the development of containerization, in my view, had to do not with a metal box or a ship, but with a managerial insight. McLean understood that transport companies' true business was moving freight rather than operating ships or trains. That understanding helped his version of containerization succeed where so many others had failed. To my consternation, though, I quickly learned that many people quite fancy the tale of McLean's dockside epiphany. The idea of a single moment of inspiration, of the apple landing on young Isaac Newton's head, stirs the soul, even if it turns out to be apocryphal. In contrast, the idea that innovation occurs in fits and starts, with one person adapting a concept already in use and another figuring out how to make a profit from it, has little appeal. The world likes heroes, even if the worshipful story of one person's heroic effort is rarely an accurate representation of the complex path of technological advance.

How innovation really works is certainly one of the lessons of *The Box*, but for me there is another that looms even larger: the role of unintended consequences. Economists, myself included, are in the business of predicting events; we like to think that we can analyze what has happened and draw insight into what will occur in the days to come. Business school students take a similar approach, learning to apply quantitative analysis to historical data in order to draw conclusions about the future. In the business world, this way of looking at the world through a spreadsheet is treated as modern management thinking. It's the bread and butter of some of the world's most famous, and expensive, consulting firms.

The story of containerization attests to the limits of this sort of rational analysis, for the developments recounted in *The Box* turned out not at all as expected. Containerization, after all, began as a means of shaving a few dollars off the cost of sending Malcom McLean's trucks between New York and North Carolina. At best, it was regarded as a minor innovation, "an expedient," as one leading naval architect opined in 1958. Perhaps, the experts thought, containers might capture a small share of America's declining coastal shipping business. They were deemed impractical for most types of cargo and for shipments to distant places, such as Asia.

Absolutely no one anticipated that containerization would open the way to vast changes in where and how goods are manufactured, that it would provide a major impetus to transport deregulation, or that it would help integrate East Asia into a world economy that previously had centered on the North Atlantic. That containerization would eliminate the jobs of dockworkers was clear from the start, but no one imagined that it would cause massive job loss among workers in manufacturing and wholesaling whose employment had long been tied to the presence of nearby docks. Political leaders, trade unionists, and corporate executives made costly mistakes because they failed to apprehend the container's influence. U.S. railroads fought containerization tooth and nail in the 1960s and 1970s, convinced that it would destroy their traditional boxcar business, never imagining that early in the twenty-first century they would be carrying 12 million containers every year. Many shipping magnates—including, eventually, McLean himself—led their ship lines to failure by misjudging how the container business would develop. And certainly, no one in the early days of container shipping foresaw that this American-born industry would come to be dominated by European and Asian firms, as the U.S.-flag ship lines, burdened by a legacy of protected markets and heavy regulation, proved unable to compete in a fast-changing world.

And, of course, no one involved with the container's development imagined that metal boxes would come to be regarded as a major security threat. Improved security, ironically, was originally one of the container's big selling points: cargo packed inside a locked container was far less susceptible to theft and damage than cargo handled loose. Ship lines and border-control officials were taken by surprise in the 1980s, when smugglers figured out that the relative secrecy, anonymity, and reliability of container shipping made it ideal for transporting drugs and undocumented migrants as well. In those days, enclosing container yards with fences and locked gates was thought adequate to solve the problem.

Two decades later, evaluating potential threats after a series of devastating attacks, antiterrorism experts hit upon the possibility that terrorists might cripple world trade by exploding radioactive weapons secreted in containers. The seriousness of that threat was impossible to evaluate, although experience has made clear that terrorists bent on wreaking large-scale devastation can do so with readily available materials—ammonium nitrate fertilizer, propane, explosives embedded with nails—without going to the trouble of building a "dirty bomb." Nonetheless, containers suddenly came into public consciousness as an urgent threat, one that no government anywhere was equipped to confront. A large-scale spending program inevitably followed, with radiation detectors appearing at port gates and port workers mandated to wear supposedly tamper-proof identity cards. Whether these efforts have improved security remains unclear; tests using satellites to track container shipments from origin to destination have not been promising. The frenetic attempt to bolster security at the ports, however, may have created a risk that may be even more difficult to address: the risk that precipitous government orders to detain vessels or close ports in the face of a real or imagined terrorist action could cause grave damage to economies all around the world.

The history of the shipping container is humbling. Careful planning and thorough analysis have their place, but they provide little guidance in the face of abrupt changes that alter an industry's very fundamentals. Flexibility is a virtue in such a situation. Resistance can be a vice, but so can a rush to action. In this kind of situation, "expect the unexpected" may be as good a motto as any.

Just as no one in the container's early years dreamed that the world's ports would soon be handling the arrival of one-and-a-half million 40-foot containers every week, so, too, did no one conceive that steel shipping containers could be turned into houses and sculptures or that abandoned containers would become a serious nuisance. That simple metal box was what we today label a disruptive technology. Even now, more than half a century after it came into use, it continues to affect our world in unexpected ways.

*October 2007*

# Acknowledgments

**Container shipping** is not ancient history, but much relatively recent source material proved surprisingly difficult to locate. Many relevant corporate records have been destroyed. The early growth of containerization was nurtured by the Port of New York Authority (now the Port Authority of New York and New Jersey), but many of that agency's records were destroyed in the terrorist attacks on the World Trade Center on September 11, 2001. That this book came to be is a tribute to the work of the many dedicated archivists and librarians who helped me identify extant materials in collections that researchers rarely look at, as well as to private individuals who combed their own files for important records.

Back in the early 1990s, when I first thought of writing about Malcom McLean, George Stevenson of the North Carolina State Archives came up with hard-to-find material about the McLean family. When I decided to revisit containerization more recently, Kenneth Cobb of the New York Municipal Archives, Doug DiCarlo of the LaGuardia and Wagner Archives at LaGuardia Community College in New York, and Bette M. Epstein of the New Jersey State Archives in Trenton helped me piece together the story of how the container decimated New York's port.

The lack of historical material on the International Longshoremen's Association is a serious impediment to historical work on longshore labor relations. Gail Malmgreen of the Robert F. Wagner Labor Archives at New York University helped me locate documents and oral histories in that remarkable collection. Patrizia Sione and Melissa Holland of the Kheel Center, Catherwood Library, at the Cornell University School of Industrial and Labor Relations, guided me through the papers of Vernon Jensen, which contain a wealth of detail on the ILA.

Military history is not my field, but my efforts to learn about the role of container shipping in the Vietnam War benefited from much expert guidance. Gina Akers and Wade Wyckoff of the Operational Archives Branch of the Naval Historical Center, in Washington, helped me with the records of the Military Sea Transportation Service and with the U.S. Navy's extensive collection of oral histories. Jeannine Swift and Rich Boylan, of the Modern Military Records division of the National Archives in College Park, Maryland, went to great lengths to locate little-used material on Vietnam-era logistics. William Moye, of the U.S. Army Materiel Command Historical Office at Fort Belvoir, Virginia, furnished important information on General Frank S. Besson Jr., who persuaded the U.S. armed forces to embrace containerization.

Roger Horowitz and Christopher T. Baer, of the Hagley Museum and Library in Wilmington, Delaware, suggested files I would never have thought to investigate in the archives of the Penn Central railroad. Beth Posner of the City University of New York Graduate Center located much obscure material for me. I also drew on resources from the Bancroft Library of the University of California at Berkeley, the Library of Congress, the Cornell University library system, the New York Public Library, and the Seattle Public Library, and I wish to record my appreciation for their assistance.

The oral histories prepared for the Smithsonian by Arthur Donovan, professor emeritus at the U.S. Merchant Marine Academy, and the late Andrew Gibson are an important source for any researcher on this subject, and Professor Donovan also pointed me to records on container standards. Marilyn Sandifur, Midori Tabata, Jerome Battle, and Mike Beritzhoff of the Port of Oakland were kind enough to guide me around the port and bring my knowledge of terminal management up to date. I owe a particular debt to Jim Doig, who allowed me to use material (now in the New Jersey State Archives) that he compiled in preparing his masterful book on the Port of New York Authority, and to Les Harlander, whose files on the negotiation of container standards are the major source for chapter 7.

A number of people read portions of the manuscript, caught embarrassing errors, pointed me to additional sources, and provided valuable comments. I especially wish to thank Jim Doig, Joshua Freeman, Vincent Grey, Les Harlander, Thomas Kessner, Nelson Lichtenstein, Kathleen McCarthy, Bruce Nelson, and Judith Stein. The material in chapter 5 was presented to the Business History Conference, several of whose members provided insights and suggestions. Portions of chapter 5 appeared in *Business History Review*, whose anonymous referees made extremely helpful suggestions, and the referees who reviewed the manuscript for Princeton University Press did much to improve it. I would also like to thank my editors at Princeton University Press, Lauren Lepow, who did a superb job of copyediting, and Tim Sullivan, who enthusiastically shared my vision of this book and my belief that the container really did change the world.

*August 2005*

# The Box

Chapter 1

# The World the Box Made

**n April 26, 1956,** a crane lifted fifty-eight aluminum truck bodies aboard an aging tanker ship moored in Newark, New Jersey. Five days later, the *Ideal-X* sailed into Houston, where fifty-eight trucks waited to take on the metal boxes and haul them to their destinations. Such was the beginning of a revolution.

Decades later, when enormous trailer trucks rule the highways and trains hauling nothing but stacks of boxes rumble through the night, it is hard to fathom just how much the container has changed the world. In 1956, China was not the world's workshop. It was not routine for shoppers to find Brazilian shoes and Mexican vacuum cleaners in stores in the middle of Kansas. Japanese families did not eat beef from cattle raised in Wyoming, and French clothing designers did not have their exclusive apparel cut and sewn in Turkey or Vietnam. Before the container, transporting goods was expensive—so expensive that it did not pay to ship many things halfway across the country, much less halfway around the world.

What is it about the container that is so important? Surely not the thing itself. A soulless aluminum or steel box held together with welds and rivets, with a wooden floor and two enormous doors at one end: the standard container has all the romance of a tin can. The value of this utilitarian object lies not in what it is, but in how it is used. The container is at the core of a highly automated system for moving goods from anywhere, to anywhere, with a minimum of cost and complication on the way.

The container made shipping cheap, and by doing so changed the shape of the world economy. The armies of ill-paid, ill-treated workers who once made their livings loading and unloading ships in every port are no more, their tight-knit waterfront communities now just memories. Cities that had been centers of maritime commerce for centuries, such as New York and Liverpool, saw their waterfronts decline with startling speed, unsuited to the container trade or simply unneeded, and the manufacturers that endured high costs and antiquated urban plants in order to be near their suppliers and their customers have long since moved away. Venerable ship lines with century-old pedigrees were crushed by the enormous cost of adapting to container shipping. Merchant mariners, who had shipped out to see the world, had their traditional days-long shore leave in exotic harbors replaced by a few hours ashore at a remote parking lot for containers, their vessel ready to weigh anchor the instant the high-speed cranes finished putting huge metal boxes off and on the ship.

Even as it helped destroy the old economy, the container helped build a new one. Sleepy harbors such as Busan and Seattle moved into the front ranks of the world's ports, and massive new ports were built in places like Felixstowe, in England, and Tanjung Pelepas, in Malaysia, where none had been before. Small towns, distant from the great population centers, could take advantage of their cheap land and low wages to entice factories freed from the need to be near a port to enjoy cheap transportation. Sprawling industrial complexes where armies of thousands manufactured products from start to finish gave way to smaller, more specialized plants that shipped components and half-finished goods to one another in ever lengthening supply chains. Poor countries, desperate to climb the rungs of the ladder of economic development, could realistically dream of becoming suppliers to wealthy countries far away. Huge industrial complexes mushroomed in places like Los Angeles and Hong Kong, only because the cost of bringing raw materials in and sending finished goods out had dropped like a stone.[1]

This new economic geography allowed firms whose ambitions had been purely domestic to become international companies, exporting their products almost as effortlessly as selling them nearby. If they did, though, they soon discovered that cheaper shipping benefited manufacturers in Thailand or Italy just as much. Those who had no wish to go international, who sought only to serve their local clientele, learned that they had no choice: like it or not, they were competing globally because the global market was coming to them. Shipping costs no longer offered shelter to high-cost producers whose great advantage was physical proximity to their customers; even with customs duties and time delays, factories in Malaysia could deliver blouses to Macy's in Herald Square more cheaply than could blouse manufacturers in the nearby lofts of New York's garment district. Multinational manufacturers—companies with plants in different countries—transformed themselves into international manufacturers, integrating once isolated factories into networks so that they could choose the cheapest location in which to make a particular item, yet still shift production from one place to another as costs or exchange rates might dictate. In 1956, the world was full of small manufacturers selling locally; by the end of the twentieth century, purely local markets for goods of any sort were few and far between.

For workers, of course, this has all been a mixed blessing. As consumers, they enjoy infinitely more choices thanks to the global trade the container has stimulated. By one careful study, the United States imported four times as many varieties of goods in 2002 as in 1972, generating a consumer benefit—not counted in official statistics—equal to nearly 3 percent of the entire economy. The competition that came with increased trade has diffused new products with remarkable speed and has held down prices so that average households can partake. The ready availability of inexpensive imported consumer goods has boosted living standards around the world.[2]

As wage earners, on the other hand, workers have every reason to be ambivalent. In the decades after World War II, wartime devastation created vast demand while low levels of international trade kept competitive forces under control. In this exceptional environment, workers and trade unions in North America, Western Europe, and Japan were able to negotiate nearly continuous improvements in wages and benefits, while government programs provided ever stronger safety nets. The workweek grew shorter, disability pay was made more generous, and retirement at sixty or sixty-two became the norm. The container helped bring an end to that unprecedented advance. Low shipping costs helped make capital even more mobile, increasing the bargaining power of employers against their far less mobile workers. In this highly integrated world economy, the pay of workers in Shenzhen sets limits on wages in South Carolina, and when the French government ordered a shorter workweek with no cut in pay, it discovered that nearly frictionless, nearly costless shipping made it easy for manufacturers to avoid the higher cost by moving abroad.[3]

A modern containerport is a factory whose scale strains the limits of imagination. At each berth—the world's biggest ports have dozens—rides a mammoth oceangoing vessel, up to 1,100 feet long and 140 feet across, carrying nothing but metal containers. The deck is crowded with row after row of them, red and blue and green and silver, stacked 15 or 20 abreast and 6 or 7 high. Beneath the deck are yet more containers, stacked 6 or 8 deep in the holds. The structure that houses the crew quarters, topped by the navigation bridge, is toward the stern, barely visible above the stacks of boxes. The crew accommodations are small, but so is the crew. A ship carrying 3,000 40-foot containers, filled with 100,000 tons of shoes and clothes and electronics, may make the three-week transit from Hong Kong around the Cape of Good Hope to Germany with only twenty people on board.[4]

On the wharf, a row of enormous cranes goes into action almost as soon as the ship ties up. The cranes are huge steel structures, rising 200 feet into the air and weighing more than two million pounds. Their legs stretch 50 feet apart, easily wide enough for several truck lanes or even train tracks to pass beneath. The cranes rest on rails running parallel to the ship's side, so that they can move forward or aft as required. Each crane extends a boom 115 feet above the dock and long enough to span the width of a ship broader than the Panama Canal.

High up in each crane, an operator controls a trolley able to travel the length of the boom, and from each trolley hangs a spreader, a steel frame designed to lock onto all four top corners of a 40-ton box. As unloading begins, each operator moves his trolley out the boom to a precise location above the ship, lowers the spreader to engage a container, raises the container up toward the trolley, and pulls trolley and container quickly toward the wharf. The trolley stops above a rubber-tired transporter waiting between the crane's legs, the container is lowered onto the transporter, and the spreader releases its grip. The transporter then moves the container to the adjacent storage yard, while the trolley moves back out over the ship to pick up another box. The process is repeated every two minutes, or even every ninety seconds, each crane moving 30 or 40 boxes an hour from ship to dock. As parts of the ship are cleared of incoming containers, reloading begins, and dockside activity becomes even more frenzied. Each time the crane places an incoming container on one vehicle, it picks up an outbound container from another, simultaneously emptying and filling the ship.

In the yard, a mile-long strip paved with asphalt, the incoming container is driven beneath a stacking crane. The stacker has rubber-tired wheels 50 feet apart, wide enough to span a truck lane and four adjacent stacks of containers. The wheels are linked by a metal structure 70 feet in the air, so that the entire machine can move back and forth above the rows of containers stacked six high. The crane engages the container, lifts it from the transporter, and moves it across the stacks of other containers to its storage location. A few hours later, the process will be reversed, as the stacking crane lifts the container onto a steel chassis pulled by an over-the-road truck. The truck may take the cargo hundreds of miles to its destination or may haul it to a nearby rail yard, where low-slung cars specially designed for containers await loading.

The colorful chaos of the old-time pier is nowhere in evidence at a major container terminal, the brawny longshoremen carrying bags of coffee on their shoulders nowhere to be seen. Terry Malloy, the muscular hero played by Marlon Brando in *On the Waterfront*, would not be at home. Almost every one of the intricate movements required to service a vessel is choreographed by a computer long before the ship arrives. Computers, and the vessel planners who use them, determine the order in which the containers are to be discharged, to speed the process without destabilizing the ship. The actions of the container cranes and the equipment in the yard all are programmed in advance. The longshoreman who drives each machine faces a screen telling him which container is to be handled next and where it is to be moved—unless the terminal dispenses with longshoremen by using driverless transporters to pick up the containers at shipside and centrally controlled stacker cranes to handle container storage. The computers have determined that the truck picking up incoming container ABLQ 998435 should be summoned to the terminal at 10:45 a.m., and that outgoing container JKFC 119395, a 40-foot box bound for Newark, carrying 56,800 pounds of machinery and currently stacked at yard location A-52-G-6, will be loaded third from the bottom in the fourth slot in the second row of the forward hold. They have ensured that the refrigerated containers are placed in bays with electrical hookups, and that containers with hazardous contents are apart from containers that could increase the risk of explosion. The entire operation runs like clockwork, with no tolerance for error or human foibles. Within twenty-four hours, the ship discharges its thousands of containers, takes on thousands more, and steams on its way.

Every day at every major port, thousands of containers arrive and depart by truck and train. Loaded trucks stream through the gates, where scanners read the unique number on each container and computers compare it against ships' manifests before the trucker is told where to drop his load. Tractor units arrive to hook up chassis and haul away containers that have just come off the ship. Trains carrying nothing but double-stacked containers roll into an intermodal terminal close to the dock, where giant cranes straddle the entire train, working their way along as they remove one container after another. Outbound container trains, destined for a rail yard two thousand miles away with only the briefest of stops en route, are assembled on the same tracks and loaded by the same cranes.

The result of all this hectic activity is a nearly seamless system for shipping freight around the world. A 25-ton container of coffeemakers can leave a factory in Malaysia, be loaded aboard a ship, and cover the 9,000 miles to Los Angeles in 16 days. A day later, the container is on a unit train to Chicago, where it is transferred immediately to a truck headed for Cincinnati. The 11,000-mile trip from the factory gate to the Ohio warehouse can take as little as 22 days, a rate of 500 miles per day, at a cost lower than that of a single first-class air ticket. More than likely, no one has touched the contents, or even opened the container, along the way.

This high-efficiency transportation machine is a blessing for exporters and importers, but it has become a curse for customs inspectors and security officials. Each container is accompanied by a manifest listing its contents, but neither ship lines nor ports can vouch that what is on the manifest corresponds to what is inside. Nor is there any easy way to check: opening the doors at the end of the box normally reveals only a wall of paperboard cartons. With a single ship able to disgorge 3,000 40-foot-long containers in a matter of hours, and with a port such as Long Beach or Tokyo handling perhaps 10,000 loaded containers on the average workday, and with each container itself holding row after row of boxes stacked floor to ceiling, not even the most careful examiners have a remote prospect of inspecting it all. Containers can be just as efficient for smuggling undeclared merchandise, illegal drugs, undocumented immigrants, and terrorist bombs as for moving legitimate cargo.[5]

Getting from the *Ideal-X* to a system that moves tens of millions of boxes each year was not an easy voyage. Both the container's promoters and its opponents sensed from the very beginning that this was an invention that could change the way the world works. That first container voyage of 1956, an idea turned into reality by the ceaseless drive of an entrepreneur who knew nothing about ships, unleashed more than a decade of battle around the world. Many titans of the transportation industry sought to stifle the container.

Powerful labor leaders pulled out all the stops to block its ascent, triggering strikes in dozens of harbors. Some ports spent heavily to promote it, while others spent huge sums for traditional piers and warehouses in the vain hope that the container would prove a passing fad. Governments reacted with confusion, trying to figure out how to capture its benefits without disturbing the profits, jobs, and social arrangements that were tied to the status quo. Even seemingly simple matters, such as the design of the steel fitting that allows almost any crane in any port to lift almost any container, were settled only after years of contention. In the end, it took a major war, the United States' painful campaign in Vietnam, to prove the merit of this revolutionary approach to moving freight.

How much the container matters to the world economy is impossible to quantify. In the ideal world, we would like to know how much it cost to send one thousand men's shirts from Bangkok to Geneva in 1955, and to track how that cost changed as containerization came into use. Such data do not exist, but it seems clear that the container brought sweeping reductions in the cost of moving freight. From a tiny tanker laden with a few dozen containers that would not fit on any other vessel, container shipping matured into a highly automated, highly standardized industry on a global scale. An enormous containership can be loaded with a minute fraction of the labor and time required to handle a small conventional ship half a century ago. A few crew members can manage an oceangoing vessel longer than three football fields. A trucker can deposit a trailer at a customer's loading dock, hook up another trailer, and drive on immediately, rather than watching his expensive rig stand idle while the contents are removed. All of those changes are consequences of the container revolution. Transportation has become so efficient that for many purposes, freight costs do not much effect economic decisions. As economists Edward L. Glaeser and Janet E. Kohlhase suggest, "It is better to assume that moving goods is essentially costless than to assume that moving goods is an important component of the production process." Before the container, such a statement was unimaginable.[6]

In 1961, before the container was in international use, ocean freight costs alone accounted for 12 percent of the value of U.S.

TABLE 1
Cost of Shipping One Truckload of Medicine from Chicago to Nancy, France (estimate ca. 1960)

| | *Cash Outlay* | *Percent of Cost* |
|---|---|---|
| Freight to U.S. port city | $341 | 14.3% |
| Local freight in port vicinity | $95 | 4.0% |
| Total port cost | $1,163 | 48.7% |
| Ocean shipping | $581 | 24.4% |
| European inland freight | $206 | 8.6% |
| Total | $2,386 | |

*Source*: American Association of Port Authority data reported by John L. Eyre. See n.7.

exports and 10 percent of the value of U.S. imports. "These costs are more significant in many cases than governmental trade barriers," the staff of the Joint Economic Committee of Congress advised, noting that the average U.S. import tariff was 7 percent. And ocean freight, dear as it was, represented only a fraction of the total cost of moving goods from one country to another. A pharmaceutical company would have paid approximately $2,400 to ship a truckload of medicines from the U.S. Midwest to an interior city in Europe in 1960. This might have included payments to a dozen different vendors: a local trucker in Chicago, the railroad that carried the truck trailer on a flatcar to New York or Baltimore, a local trucker in the port city, a port warehouse, a steamship company, a warehouse and a trucking company in Europe, an insurer, a European customs service, and the freight forwarder who put all the pieces of this complicated journey together. Half the total outlay went for port costs.[7]

This process was so expensive that in many cases selling internationally was not worthwhile. "For some commodities, the freight may be as much as 25 per cent of the cost of the product," two engineers concluded after a careful study of data from 1959. Shipping steel pipe from New York to Brazil cost an average of $57 per ton in 1962, or 13 percent of the average cost of the pipe being exported—a figure that did not include the cost of getting the pipe from the steel mill to the dock. Shipping refrigerators from London to Capetown cost the equivalent of 68 U.S. cents per cubic foot, adding $20 to the wholesale price of a midsize unit. No wonder that, relative to the size of the economy, U.S. international trade was smaller in 1960 than it had been in 1950, or even in the Depression year of 1930. The cost of conducting trade had gotten so high that in many cases trading made no sense.[8]

By far the biggest expense in this process was shifting the cargo from land transport to ship at the port of departure and moving it back to truck or train at the other end of the ocean voyage. As one expert explained, "a four thousand mile voyage for a shipment might consume 50 percent of its costs in covering just the two ten-mile movements through two ports." These were the costs that the container affected first, as the elimination of piece-by-piece freight handling brought lower expenses for longshore labor, insurance, pier rental, and the like. Containers were quickly adopted for land transportation, and the reduction in loading time and transshipment cost lowered rates for goods that moved entirely by land. As ship lines built huge vessels specially designed to handle containers, ocean freight rates plummeted. And as container shipping became intermodal, with a seamless shifting of containers among ships and trucks and trains, goods could move in a never-ending stream from Asian factories directly to the stockrooms of retail stores in North America or Europe, making the overall cost of transporting goods little more than a footnote in a company's cost analysis.[9]

Transport efficiencies, though, hardly begin to capture the economic impact of containerization. The container not only lowered freight bills, it saved time. Quicker handling and less time in storage translated to faster transit from manufacturer to customer, reducing the cost of financing inventories sitting unproductively on railway sidings or in pierside warehouses awaiting a ship. The container, combined with the computer, made it practical for companies like Toyota and Honda to develop just-in-time manufacturing, in which a supplier makes the goods its customer wants only as the customer needs them and then ships them, in containers, to arrive at a specified time. Such precision, unimaginable before the container, has led to massive reductions in manufacturers' inventories and correspondingly huge cost savings. Retailers have applied those same lessons, using careful logistics management to squeeze out billions of dollars of costs.

These savings in freight costs, in inventory costs, and in time to market have encouraged ever longer supply chains, allowing buyers in one country to purchase from sellers halfway around the globe with little fear that the gaskets will not arrive when needed or that the dolls will not be on the toy store shelf before Christmas. The more reliable these supply chains become, the further retailers, wholesalers, and manufacturers are willing to reach in search of lower production costs—and the more likely it becomes that workers will feel the sting of dislocation as their employers find distant sources of supply.

Some scholars have argued that reductions in transport costs are at best marginal improvements that have had negligible effects on trade flows. This book disputes that view. In the decade after the container first came into international use, in 1966, the volume of international trade in manufactured goods grew more than twice as fast as the volume of global manufacturing production, and two and a half times as fast as global economic output. Something was accelerating the growth of trade even though the economic expansion that normally stimulates trade was weak. Something was driving a vast increase in international commerce in manufactured goods even though oil shocks were making the world economy sluggish. While attributing the vast changes in the world economy to a single cause would be foolhardy, we should not dismiss out of hand the possibility that the extremely sharp drop in freight costs played a major role in increasing the integration of the global economy. [10]

The subject of this book lies at the confluence of several major streams of research. One delves into the impact of changes in transportation technology, a venerable subject for both historians and economists. The steamship, invented in the 1780s and put to regular use by 1807, strengthened New York's prominence as a port, and the Erie Canal, an undertaking of unprecedented size, had an even greater impact. The radical decline in ocean freight rates during the nineteenth century, the result of technological change and improved navigation techniques, encouraged a huge increase in world trade and added to Europe's eagerness to found colonies. The connection between railroad development and U.S. economic growth has been debated strenuously, but there is little dispute that lower rail freight rates increased agricultural productivity, knitted the North together before the Civil War, and eventually made Chicago the hub of a region stretching a thousand miles to the west. A transport innovation of the 1880s, the refrigerated railcar, made meat affordable for average households by allowing meat companies to ship carcasses rather than live animals across the country. The truck and the passenger car reshaped urban development starting in the 1920s, and more recently commercial aviation redrew the economic map by bringing formerly isolated communities within a few hours of major cities. This book will argue that container shipping has had a similarly large effect in stimulating trade and economic development—and that, as with steamships, railroads, and airplanes, government intervention both encouraged and deterred its growth.[11]

The importance of innovation is at the center of a second, and rapidly growing, body of research. Capital, labor, and land, the basic factors of production, have lost much of their fascination for those looking to understand why economies grow and prosper. The key question asked today is no longer how much capital and labor an economy can amass, but how innovation helps employ those resources more effectively to produce more goods and services. This line of research makes clear that new technology, by itself, has little economic benefit. As economist Nathan Rosenberg observed, "innovations in their early stages are usually exceedingly ill-adapted to the wide range of more specialised uses to which they are eventually put." Resistance to new methods can impede their adoption. Potential users may avoid commitments until the future is more certain; as early buyers of Betamax video players can attest, it is risky to bet on a technology that turns out to be a dead end. Even after a new technology is proven, its spread must often wait until prior investments have been recouped; although Thomas Edison invented the incandescent lightbulb by 1879, only 3 percent of U.S. homes had electric lighting twenty years later. The economic benefits arise not from innovation itself, but from the entrepreneurs who eventually discover ways to put innovations to practical use—and most critically, as economists Erik Brynjolfsson and Lorin M. Hitt have pointed out, from the organizational changes through which businesses reshape themselves to take advantage of the new technology.[12]

This book contends that, just as decades elapsed between the taming of electricity in the 1870s and the widespread use of electrical power, so too did the embrace of containerization take time. Big savings in the cost of handling cargo on the docks did not translate immediately into big savings in the total cost of transportation. Transportation companies were generally ill-equipped to exploit the container's advantages, and their customers had designed their operations around different assumptions about costs. Only with time, as container shipping developed into an entirely new system of moving goods by land and sea, did it begin to affect trade patterns and industrial location. Not until firms learned to take advantage of the opportunities the container created did it change the world. Once the world began to change, it changed very rapidly: the more organizations that adopted the container, the more costs fell, and the cheaper and more ubiquitous container transportation became.[13]

The third intellectual stream feeding into this book is the connection between transportation costs and economic geography, the question of who makes what where. This connection might seem self-evident, but it is not. When David Ricardo showed in 1817 that both Portugal and England could gain by specializing in making products in which they had a comparative advantage, he assumed that only production costs mattered; the costs of shipping Portuguese wine to England and English cloth to Portugal did not enter his analysis. Ricardo's assumption that transportation costs were zero has been incorporated into economists' models ever since, despite ample real-world evidence that transportation costs matter a great deal.[14]

Economists have devoted serious effort to studying the geographic implications of transport costs only since the early 1990s. This new stream of work shows formally what common sense suggests. When transport costs are high, manufacturers' main concern is to locate near their customers, even if this requires undesirably small plants or high operating costs. As transportation costs decline relative to other costs, manufacturers can relocate first domestically, and then internationally, to reduce other costs, which come to loom larger. Globalization, the diffusion of economic activity without regard for national boundaries, is the logical end point of this process. As transport costs fall to extremely low levels, producers move from high-wage to low-wage countries, eventually causing wage levels in all countries to converge. These geographic shifts can occur quickly and suddenly, leaving long-standing industrial infrastructure underutilized or abandoned as economic activity moves on.[15]

Have declines in the cost of shipping really caused such significant economic shifts? Some scholars doubt that ocean freight costs have fallen very much since the middle of the twentieth century. Others, pointing to the undeniable fact that countries trade much more with neighbors than with distant lands, argue that transportation costs still matter a great deal. The present work intentionally takes a non-quantitative approach in addressing these questions. The data on freight costs from the mid-1950s through the 1970s are so severely deficient that they will never provide conclusive proof, but the undisputed fact that the transportation world raced to embrace containerization is very strong evidence that this new shipping technology significantly reduced costs. Nor does this book employ economic models to prove the container's impact. Given the vast changes in the world economy over a span that saw the breakdown of the exchange-rate system, repeated oil crises, the end of colonialism, the invention of jet travel, the spread of computers, the construction of hundreds of thousands of miles of expressways, and many other developments, no model is likely to be conclusive in distinguishing the impact of containerization from that of the many other forces. Nonetheless, dramatic shifts in trade patterns and in the location of economic activity over the past half century suggest that the connection between containerization and changes in economic geography is extremely strong.[16]

Mysteriously, the container has escaped all three of these very lively fields of research. It has no engine, no wheels, no sails: it does not fascinate those captivated by ships and trains and planes, or by sailors and pilots. It lacks the flash to draw attention from those who study technological innovation. And so many forces have combined to alter economic geography since the middle of the twentieth century that the container is easily overlooked. There is, half a century after its arrival, no general history of the container.[17]

In telling the remarkable story of containerization, this book represents an attempt to fill that historical void. It treats containerization not as shipping news, but as a development that has sweeping consequences for workers and consumers all around the globe. Without it, the world would be a very different place.

Chapter 2

# Gridlock on the Docks

**In the early 1950s,** before container shipping was even a concept, most of the world's great centers of commerce had docks at their heart. Freight transportation was an urban industry, employing millions of people who drove, dragged, or pushed cargo through city streets to or from the piers. On the waterfront itself, swarms of workers clambered up gangplanks with loads on their backs or toiled deep in the holds of ships, stowing boxes and barrels in every available corner. Warehouses stood at the heads of many of the wharves, and where there were no warehouses, there were factories. As they had for centuries, manufacturers still clustered near the docks for easier delivery of raw materials and faster shipment of finished goods. Whether in San Francisco or Montreal, Hamburg or London, Rio or Buenos Aires, the surrounding neighborhoods were filled with households that made their livings from the port, bound together by the special nature of waterfront work and the unique culture that developed from it.

Though ships had been plying the seas for thousands of years, using them to move goods was still a hugely complicated project in the 1950s. At the shipper's factory or warehouse, the freight would be loaded piece by piece on a truck or railcar. The truck or train would deliver hundreds or thousands of such items to the waterfront. Each had to be unloaded separately, recorded on a tally sheet, and carried to storage in a transit shed, a warehouse stretching alongside the dock. When a ship was ready to load, each item was removed from the transit shed, counted once more, and hauled or dragged to shipside. The dock would be covered with a jumble of paperboard cartons and wooden crates and casks. There might be steel drums of cleaning compound and beef tallow alongside 440-pound bales of cotton and animal skins. Borax in sacks so heavy it took two men to lift them, loose pieces of lumber, baskets of freshly picked oranges, barrels of olives, and coils of steel wire might all be part of the same load of "mixed cargo," waiting on the dock amid a tangle of ropes and cables, as lift trucks and handcarts darted back and forth.

Getting all of this loaded was the job of the longshoremen. On the dock or in the pierside warehouse, a gang of longshore workers would assemble various boxes and barrels into a "draft" of cargo atop a wooden pallet, the sling board. Some sling loads were wrapped in rope or netting, but pallets often held stacks of loose cartons or bags. When the draft was ready, the longshoremen on the dock would slip cables beneath the sling board and tie the ends together. On the ship's deck, the winch driver, or "deck man," waited for his signal. When it came, he positioned the hook of the shipboard crane over the sling. The dockside men placed the cables on the hook, and the winch hoisted the draft from the dock, maneuvered it over an open hatch, and lowered it into the hold. The hook was released quickly and lifted out to grab another load, lest the foreman complain that "the hook is hanging." Meanwhile, in the dimness of the ship's interior, another gang of longshoremen removed each item from the sling board and found a secure place to stow it, maneuvering it into position with a four-wheeled cart, a forklift, or brute force. Every longshoreman carried a steel hook with a wooden handle, designed to grab a recalcitrant piece of cargo and jerk it into place under power of nothing but human muscle.

Unloading could be just as difficult. An arriving ship might be carrying 100-kilo bags of sugar or 20-pound cheeses nestled next to 2-ton steel coils. Simply moving one without damaging the other was hard enough. A winch could lift the coiled steel out of the hold, but the sugar and cheese needed men to lift them. Unloading bananas required the longshoremen to walk down a gangplank carrying 80-pound stems of hard fruit on their shoulders. Moving coffee meant carrying fifteen 60-kilo bags to a wooden pallet placed in the hold, letting a winch lift the pallet to the dock, and then removing each bag from the pallet and stacking it atop a massive pile. The work could be brutally physical. In Edinburgh, unloading a hold full of bagged cement meant digging through a thirty-foot-high pile of dusty bags, tightly packed together, and lifting them into a sling, one by one. Copper came from Peru to New York in the form of bars too big for a man to handle. Longshoremen had to move these enormous hunks of metal across the dock, from the incoming ship to a lighter, or barge, which would transport them to a plant in New Jersey. "Because they had to bend over to do that, you'd see these fellows going home at the end of the day kind of like orangutans," a former pier superintendent remembered. "I mean, they were just kind of all bent, and they'd eventually straighten up for the next day."[1]

Automation had arrived during World War II, but in a very limited way. Forklifts, used in industry since the 1920s, were widely used by the 1950s to move pallets from the warehouse to the side of the ship, and some ports installed conveyors to unload bags of coffee and potatoes. Even with machinery at hand, though, muscle was often the ultimate solution. Longshoremen had to be prepared to handle small cartons of delicate tropical fruits one day, tons of filthy carbon black the next. They labored sometimes in daylight, sometimes at night, in all weather conditions. Sweltering holds, icy docks, and rain-slicked gangways were part of the job. The risk of tripping over a load of pipe or being knocked down by a draft on the hook was ever present. In Marseilles, forty-seven dockworkers were killed on the job between 1947 and 1957, while in Manchester, where dockers serviced oceangoing vessels that ascended a canal from the Irish Sea, one out of two longshoremen suffered an injury in 1950, and one out of six landed in the hospital. New York, with a lesser injury rate, reported 2,208 serious accidents in 1950. Government safety rules and inspections were almost nonexistent. Outsiders may have found romance and working-class solidarity in dock labor, but for the men on the docks it was an unpleasant and often dangerous job, with an injury rate three times that of construction work and eight times that in manufacturing.[2]

The ships of the era were breakbulk vessels, built with several levels of open space below deck to handle almost any kind of dry cargo.* Much of the world's commercial fleet had been destroyed during the war, but nearly 3,000 U.S. merchant ships survived and were available for merchant service by 1946. Among them were more than 2,400 of the Liberty Ships that U.S. shipyards had turned out between 1941 and 1945. Designed as convoy vessels and built in fewer than 70 days from prefabricated parts, the Liberty Ships were very slow and cheap enough to be expendable. The vessels were deliberately built small so that little cargo would be lost if a ship were sunk by German submarines; Liberty Ships were just 441 feet long. In 1944, U.S. shipyards started to make Victory Ships, which were much faster than the 11-knot** Liberty Ships but only a few feet longer and wider. The U.S. Navy sold 450 Liberty Ships to U.S. merchant lines after the war, and sold another 450 or so for commercial use in Europe and China. More than 540 Victory Ships outlasted the war, and the navy began selling them off in late 1945 as well.[3]

Neither type of vessel was designed for commercial efficiency. The interiors were cramped. The curvature of the ships' sides meant that the five small holds on each vessel were wider near the top and narrower at the bottom, and more spacious toward the middle of the vessel than forward or aft. Longshoremen had to know how to fill these odd dimensions: for the shipowner, wasted space meant money lost. Each hold was covered by its own hatch, a watertight metal cover secured to the deck; cargo for the first port of call had

* "Bulk" cargo usually refers to commodities such as coal or grain, which can be loaded on a ship in a continuous process without packaging or sorting. "Breakbulk" cargo, by contrast, consists of discrete items that must be handled individually.

** A nautical mile is equal to approximately 6,080 feet, 1.15 statute miles, and 1.85 kilometers. A speed of 11 knots, or nautical miles per hour, is equivalent to 12.7 statute miles per hour, or 20.7 kilometers per hour.

to be loaded last so it would be near a hatch, available for easy unloading, while cargo for the final port on the ship's itinerary was shoved to the distant corners of the hold. At the same time, every single piece of freight had to be stowed tightly so that it would not shift as the ship rolled at sea; a loose box or barrel could break, damaging the contents and other cargo as well. Experienced longshoremen knew which items to push into the irregular spaces along the outside walls and which to weave into interior bulkheads, intermingling cartons and sacks and lumber into temporary walls to keep the cargo wedged in place while still having it available for discharge when the ship reached port. Mistakes could be fatal. If a load shifted in an ocean swell, the ship could capsize.[4]

At journey's end, loading for the next voyage could not begin until every bit of incoming cargo had been removed. Cargo in the hold was too tightly packed to be sorted, so longshoremen often piled things on the dock and then picked through them, checking labels and tags to figure out what should be moved to the transit shed and what was being picked up on the spot. If the ship was arriving from abroad, customs inspectors walked the pier prying open crates to assess duties. Buyers' representatives came onto the dock to make sure their orders had arrived in good shape, and meat and produce dealers sent agents around to sample the new merchandise. The longshore workforce included a small army of carpenters and coopers, whose job was to repair broken crates and barrels once these various inspectors were done. At that point, noisy diesel trucks might back onto the dock to pick up their loads, while forklifts would move other cargo off to the transit sheds. Moving an incoming shipload of mixed cargo from ship to transit shed and then taking on an outbound load could keep a vessel tied up at the dock for a week or more.[5]

These waterfront realities meant that shipping was a highly labor-intensive industry in the postwar era. Depression and war had sharply curtailed the construction of privately built merchant vessels since the 1920s, so ship operators had little capital invested in the business. In the United States, total private outlays for ships and barges from 1930 through 1951 amounted to only $2.5 billion, which was less than shipowners had invested during the decade of the 1920s. Ship lines could buy surplus Liberty Ships, Victory Ships, and tankers for as little as $300,000 apiece, so the carrying cost of ships that were sitting in port rather than earning revenue was not a major expense. Outlays for shoreside facilities were negligible. The big cost item was the wages of longshore gangs, which could eat up half the total expense of an ocean voyage. Add in the tonnage fees paid to pier owners and "60 to 75 percent of the cost of transporting cargo by sea is accounted for by what takes place while the ship is at the dock and not by steaming time," two analysts concluded in 1959. There was little sense investing in fancier docks or bigger vessels when the need to handle cargo by hand made it hard to cut turnaround times and use docks and ships more efficiently.[6]

One fact above all had traditionally defined life along the waterfront: employment was highly irregular. One day, the urgent need to unload perishable cargo could create jobs for all comers. The next day, there might be no work at all. A port needed a big labor supply to handle the peaks, but on an average day the demand for workers was much smaller. Longshoremen, truckers, and warehouse workers were caught up in a world of contingent labor that shaped the communities built around the docks.[7]

Almost everywhere, longshoremen had been forced to compete for work each morning in an age-old ritual. In America, it was known as shape-up. The Australians called it the pick-up. The British had a more descriptive name: the scramble. In most places, the process involved begging, flattery, and kickbacks to get a day's work. In 1930s Edinburgh, "[t]he foremen got up on the platform about five tae eight in the mornin' and it wis jus a mad scramble for a damned job," remembered Scottish longshoreman George Baxter. The same had been true in Portland, Oregon: "They would hire their gangs and maybe you would be on that dock at seven o'clock Tuesday morning. And maybe that ship would get in at nine o'clock Tuesday night. But you didn't dare leave. You were hired, but you weren't getting paid." In Marseilles, the workday in 1947 began at 6:30 in Place de la Juliette, where workers milled on the sidewalks in the winter darkness until a foreman made a sign to the workers he wanted; the chosen could proceed to a nearby café to await the start of work, while the others went looking for another foreman. In San Francisco, men shaped on the sidewalk near the Ferry Building. In Liverpool, they congregated beneath the concrete structure of the "dockers' umbrella," more formally known as the Liverpool Overhead Railway, and waited for a foreman to come and tap them on the shoulder.[8]

The shape-up was more than just a ritual. It was an invitation to corruption. *On the Waterfront* was a dramatization, but payments to pier foremen were often the price of getting work. Newark longshoreman Morris Mullman testified that he could no longer get hired after declining to contribute to a union official's "vacation fund" in 1953. In New Orleans, a weekly payoff of two or three dollars was the norm to secure work the following week. Compulsory bets were another means of extracting money from the men; workers who failed to bet might find it difficult to get selected for work. In many ports, foremen commonly had a side business in moneylending. Liverpool dock foremen specializing in forced lending were called "gombeen men," a term derived from "gaimbin," an Irish word meaning usury. By taking a loan to be repaid with a threepenny premium on every shilling—25 percent interest for just a brief period of borrowing—a docker could be assured of being hired, because he knew that the gombeen man would take repayment from his wages.[9]

Pressure from labor unions and governments gradually eliminated some of the worst excesses of the shape-up. On the U.S. Pacific coast, employers lost control of the hiring process after a bitter strike in 1934; thereafter, the order of hiring was determined by the public drawing of longshoremen's badge numbers each morning in the shelter of a union-controlled hiring hall. The Australian Stevedoring Board took over longshore work assignments after World War II, and the creation of Britain's National Dock Labour Board in 1947 did away with the scramble. In Rotterdam, violent strikes over working conditions in 1945 and 1946 persuaded employers that they were better off with full-time staff than with occasional labor;

by 1952, more than half the port's longshoremen worked regularly for a single company. New Zealand and France started government agencies to regulate longshore hiring. The Waterfront Commission of New York Harbor, created by the states of New York and New Jersey to fight corruption on the docks, took charge of hiring in the Port of New York in 1953.[10]

These reforms led to a major change in the nature of waterfront employment. Although the longshore labor force was vast in the years after World War II—more than 51,000 men worked as dockers in New York in 1951, and there were 50,000 registered dockers in London—very few of these men had full-time jobs. With the end of the shape-up, governments and unions sought to raise longshoremen's incomes by restricting the supply of labor, especially "casual labor," the men who shaped only when their off-dock work fell through. New rules limited or blocked entry into the dockworker profession. Authorized longshoremen were required to obtain registration books, and ship lines and stevedore companies were barred from hiring anyone other than a registered longshoreman assigned by the hiring hall. The men who registered were assigned hiring categories based on their seniority. Hiring began with men in the highest category—the "A" men in New York, the *professionnels* in Marseilles—being selected in random order, and less senior workers could not get on until all higher-category men who wanted to work on a given day had been offered jobs. The expectation was that those who did not work frequently would find other careers, leaving a cadre of better-paid workers with fairly regular incomes.[11]

Thanks to the new hiring halls, longshoremen no longer needed to endure the daily humiliation of literally fighting for a job. But their incomes remained most uncertain, because the demand for their services varied hugely. In the most extreme case, Liverpool, stevedoring firms needed twice as many workers on busy days as on quiet ones. In London, where dockworkers did not win a pension scheme until 1960, men over the age of seventy commonly showed up in hopes of winning a light assignment. Even where government schemes provided payments to dockers who were unable to find work, the payments were far lower than regular wages, and many dockers were ineligible. Of the non-Communist world's major ports, only in Rotterdam and in Hamburg, where semicasual workers were guaranteed income equal to five shifts per week in 1948, could most dockers look forward to earning steady incomes.[12]

The peculiarities of dockworker life had long since given rise to a distinct waterfront culture. Longshoremen rarely worked for a single employer for long; their loyalty was to their colleagues, not to "the company." Many believed that no one knew or cared how well they did their work. Their labor was arduous and often dangerous in ways that outsiders could not appreciate, contributing to an unusual esprit de corps. Lack of control over their own time interfered with dockers' involvement in off-the-job activities scheduled around workers with regular shifts. "A longshoreman's wife seldom knows when her husband will be working, and owing to the uncertain length of the workshift, she is seldom certain when he will be home for supper," wrote Oregon longshoreman William Pilcher. And, of course, income was highly irregular. Most dockers earned hourly wages above the local average for manual labor—when they worked. Frequent episodes of part-day work or unemployment could lead to days or weeks with little income. On the other hand, many dockers cherished the fact that their work was inherently casual. If a longshoreman chose not to work on any particular day, if he decided to go fishing rather than shaping, he was entirely within his rights.[13]

Thanks to these particularities, one sociologist observed, "More than in any other industry in a big city, it appears that waterfront jobs belong to particular working class communities." Longshoremen often spent their entire lives near the waterfront. In Manchester, England, 54 percent of the dockworkers hired on in the years after World War II lived within one mile of the docks; although the houses were small and dilapidated and neighborhood amenities were few, sociologists found that "few of the dock workers living there want to move away." In Fremantle, Western Australia, half the dockworkers in the 1950s lived within two miles of the docks. In South Brooklyn, a heavily Italian neighborhood adjoining the Brooklyn docks, one in five workers in 1960 was either a trucker or a longshoreman.[14]

As often as not, dockworkers had fathers, sons, brothers, uncles, and cousins on the docks as well, and they frequently lived nearby. Strangers, including men of different ethnic groups, were unwelcome. In London and Liverpool, the Irish ruled the docks, and nonwhite immigrants from the West Indies or Africa had no chance of finding employment. In the American South, where about three-quarters of all longshoremen were black, white and black dockworkers belonged to separate union locals and often worked separate ships; the main exception, an unusual alliance in New Orleans that had an equal number of black and white longshoremen working every hatch of every ship, had collapsed under intense employer pressure in 1923. In Boston, the Irish-controlled Longshoremen's Union made no effort to sign up blacks even after many were hired as strikebreakers in 1929. The International Longshoremen's Association (ILA) in New York had locals that were identifiably Irish, Italian, and black in practice, if not by rule, and Baltimore had separate locals for black longshoremen and whites. Although the International Longshoremen's and Warehousemen's Union (ILWU) in the West barred discrimination on the basis of race, its locals in Portland and Los Angeles were almost lily-white into the early 1960s; the Portland local even called off its efforts to represent a group of grain handlers when it was discovered that some of them were black.[15]

Even where race and ethnicity were not major issues, longshore unions openly discriminated against outsiders in order to be able to offer jobs to members' kin. The work was strenuous and uncomfortable, but it paid better than anything else readily available to a blue-collar worker who had not finished high school. In dockworker families, taking a sixteen-year-old son to shape-up and calling in a favor to get him hired on was a rite of passage. Among Portland longshoremen, the most common paternal occupation was longshoreman. In Antwerp, 58 percent of dockworkers were the sons of dockworkers. The ratio in Manchester was three-quarters, and many of the rest had entered the docks with the help of their in-laws after marrying a dockworker's daughter. In Edinburgh in the mid-1950s, recalled longshoreman Eddie Trotter, "There wis nobody at all, other than a son, grandson, or a nephew or a brother o' a docker got a job as a docker." British prime minister Harold Macmillan, confronted with yet another strike threat, opined in 1962, "[T]he dockers are such difficult people, just the fathers and the sons, the uncles and nephews. So like the House of Lords, hereditary and no intelligence required."[16]

Harsh working conditions, economic uncertainty, and the insularity of docker life gave rise to unique mores. Dockworkers saw themselves as tough, independent men doing a very tough job. William Pilcher, studying longshoremen while working as one, found that his colleagues cherished and cultivated reputations as drinkers and brawlers. "They like to see themselves as rough-and-ready individuals, and that is the image that they present to outsiders and to one another," Pilcher observed. That self-image was also the public's image. A British survey published in 1950 placed dockers twenty-ninth among thirty professions in status, above only road-sweepers, at a time when dockers earned more than the average national wage. That judgment was the same among both men and women and among people of all social classes. Being a longshoreman meant belonging to a global fraternity of men with a common outlook on life and a common sense of exclusion from the mainstream.[17]

Labor militancy was a natural outgrowth of the dockworkers' situation. Longshoremen around the world fully understood that their well-being depended on collective action, because otherwise the large supply of men desperate to do manual labor would force wages to near-starvation levels. Their employers, in most cases, were not ship lines and terminal operators with assets and reputations to protect, but contractors hired to service a particular dock or a particular ship. This system allowed shipowners to evade responsibility for working conditions by claiming that not they but their contractors were in charge of dock labor. The lack of central authority on the management side was frequently mirrored on the union side. With no routine methods of resolving employment disputes, and with competing unions trying to prove their aggressiveness but often unable to impose settlements on their own members, strikes were frequent. A single grievance could bring an entire port to a standstill.

An eleven-nation study found that dockworkers, along with miners and seafarers, lost more workdays to labor disputes than any other professions. In Britain alone, dock strikes resulted in the loss of nearly 1 million man-days of labor from 1948 through 1951 and another 1.3 million in 1954. Dockworkers proudly represented the leading edge of labor radicalism.[18]

Solidarity was strengthened by the lessons of history. Longshore unions' power had waxed and waned in industrialized countries since the middle of the nineteenth century, and periods of union weakness inevitably brought heavier workloads and lower wages. After defeating a tumultuous strike in 1928, Australian dock operators slashed weekend pay and began hiring for half-day shifts, eliminating the single shift that had been a key union achievement. Across the United States, where the right to collective bargaining was not secured in law, shipping and stevedoring companies set out to break dock unions in the years after World War I and largely succeeded. Longshore wages in New Orleans fell from eighty cents an hour to forty cents after employers defeated the unions in 1923. West Coast employers rousted longshore unions in every port from Seattle to San Diego between 1919 and 1924 and then imposed lower wages and higher workloads. Demands for double shifts were common, and some ports tried to speed up loading by putting workers on piecework rather than hourly pay. After employers crushed the unions in Marseilles in 1950, "[i]t was a job with no rules," remembered French docker Alfred Pacini. Nothing speaks more eloquently to the traditional status of longshoremen than Edinburgh dockers' recollection of the greatest improvement after creation of the National Dock Labor Board in 1947: construction of an "amenity block" with individual lockers and showers, neither of which private employers had ever seen fit to provide.[19]

This history of antagonistic labor-management relations gave rise to two problems that plagued the shipping industry around the world. One was theft. Theft had always been a problem on the waterfront, and the growth of trade in higher-value products after World War II caused it to reach epidemic proportions. Some longshoremen justified thievery as a response to deteriorating economic conditions, but it remained a problem even where union contracts or government intervention had led to better wages: a British joke from the 1960s concerned a docker who was caught stealing a bar of gold and punished by having its value deducted from his next pay. "It wis the pilferin' that upset me," recalled a Scottish longshoreman of the 1950s. "It was terrible, terrible, terrible." Longshoremen prided themselves on such arcane skills as the ability to tap whiskey from a sealed cask supposedly stowed safely in a ship's hold. In Portland, small objects such as transistor radios and bottled liquor were usually stolen for personal use by family and friends, but not for sale. No such limits were observed in New York, where crime was rampant. Grace Line discovered that even sixty-kilo burlap bags of coffee beans were not immune from theft; the company purchased a sealed scale, protected against tampering by checkers who were aiding theft rings, to confirm the number of bags aboard trucks leaving the pier.[20]

The second problem arising from dockworkers' intense suspicion of employers was resistance to anything that might eliminate jobs. Wherever they secured a foothold, dock unions insisted on contract language to protect against a long history of employer abuses. The number of men needed to work a hatch, the placement of those men in the hold or on the dock, the maximum weight of a sling, the equipment they would use, and countless other details related to manning filled page after page in collective bargaining agreements. Shipping interests in Liverpool tried repeatedly to eliminate a practice known as the welt, under which half of each longshore gang left the docks, often for a nearby pub, while the other half worked; after an hour or two, the absentees would return and those who had been working would take a prolonged break. Ports the world over had seen strikes over employer efforts to alter work practices. In Los Angeles, labor productivity dropped 75 percent between 1928 and 1954 as union and management struggled over mechanization; West Coast ports handled 9 percent less cargo per work-hour in 1954 than in 1952. The Port of New York needed 1.9 man-hours to handle a ton of cargo in 1950, but 2.5 by 1956. In Britain, tonnage per man-year was nearly flat from 1948 to 1952, leaped by one-third thanks to a surge of cargo in 1953, and then sank again under the weight of stringent work rules.[21]

The solution to the high cost of freight handling was obvious: instead of loading, unloading, shifting, and reloading thousands of loose items, why not put the freight into big boxes and just move the boxes?

The concept of shipping freight in large boxes had been around for decades. The British and French railways tried wooden containers to move household furniture in the late nineteenth century, using cranes to transfer the boxes from rail flatcars to horse carts. At the end of World War I, almost as soon as motorized trucks came into wide civilian use, the Cincinnati Motor Terminals Company hit upon the idea of interchangeable truck bodies that were lifted onto and off of wheels with a crane. Farsighted thinkers were already proposing "a standardized unit container in the form of a demountable closed auto-truck body, which can be readily transferred by cranes between railroad flat cars, auto chassis, warehouse floors and vessels." The first American railroad to adopt the idea was the New York Central, which around 1920 introduced steel containers that fit side by side, six abreast, on shallow-bottomed railcars with drop-down sides.[22]

The mighty Pennsylvania Railroad, the nation's largest transportation company, became a powerful proponent of this new idea. The Pennsylvania's problem was that many of its customers did not generate a large amount of freight for a single destination. A small factory, for instance, might hold a boxcar on its siding for a week while filling it with goods for many different buyers. The railroad would have to attach this car to a freight train and haul it to its nearest interchange point, where the contents would be removed from the car, sorted into hand trucks, and reloaded into other boxcars headed toward different destinations. The Pennsy's alternative was a steel container just over nine feet wide, perhaps one-sixth the size of the average boxcar. The shipper might fill one of these containers with freight for Detroit, another for Chicago, another for St. Louis. The containers could be placed on a railcar by forklift, and at the interchange point, a forklift would simply move the containers to the proper connecting trains. Sorting loose freight at the transfer station cost 85 cents per ton, by the railroad's reckoning; transferring a five-ton container cost only 4 cents per ton, and also reduced damage claims and the need for boxcars.[23]

Some railroads sought to take advantage of the container not simply by lowering rates, but by changing the way they charged shippers. Since the onset of federal regulation in the 1880s, the Interstate Commerce Commission (ICC) had held firm to the principle that each commodity required its own rate, which of course was subject to ICC approval. With containers, though, the railroads were not handling commodities; the size and loaded weight of the container mattered far more than the contents. For the first time, they offered purely weight-based rates: the North Shore Line, running between Chicago and Milwaukee, charged 40 cents per 100 pounds to carry a 3-ton container, but only 20 cents per 100 pounds to carry a 10-ton container, with no adjustment at all for what might be inside. After four months of hearings in 1931, the commission ruled weight-based rates illegal. Although it found the container to be "a commendable piece of equipment," the commission said that the railroads could not charge less to carry a container than to carry the equivalent weight of the most expensive commodity inside the container. With that ruling, containers no longer made economic sense on the rails.[24]

Different container systems came into use on railways in other countries during the 1920s in response to a new competitive threat—the truck. Although long-distance truck transportation over primitive and often unpaved roads was impractical, trucks had obvious advantages for shorter hauls, and the railroads sought ways to reduce the truckers' cost advantage. In Australia, the Sunshine Biscuit Company used containers plastered with its advertising to ship its treats on open railcars with sides of wooden slats. The London, Midland, and Scottish Railway carried three thousand containers in 1927, and the French national railway promoted them as an efficient way for farmers to ship meat and cheese to the city. In 1933, it joined other railroads to form the International Container Bureau, an organization dedicated to making international container freight practical in Europe. Several U.S. and Canadian coastal ship lines tried carrying containers and truck trailers in the early 1930s, and Grace Line built wooden vans with metal reinforcing to cut pilferage of shipments between New York and Venezuela. The Central of Georgia Railroad formed Ocean Shipping Company to move loaded railroad cars between Savannah and New York—an idea that allowed the Central of Georgia to keep control of its freight, rather than handing it off to another railroad.[25]

Experimentation began anew after the war. Amphibious landing ships were recycled as "roll on–roll off" vessels to transport trucks along the coast, improving upon techniques originally developed to land troops and tanks in over-the-beach invasions. The International Container Bureau was reestablished in 1948, and the U.S. military began using small steel containers, called Conex boxes, for soldiers' personal belongings. The first ships designed for containers arrived in 1951 when Denmark's United Shipping Company opened a container service to move beer and foodstuffs among Danish ports. Dravo Corporation of Pittsburgh created the Transportainer, a steel box seven feet nine inches long, and more than three thousand were in use around the world by 1954. The Missouri Pacific Railway promoted its "speedboxes," aluminum containers on wheels, in 1951, and the Alaska Steamship Company began carrying both wood and steel containers from Seattle to Alaskan ports in 1953. Seatrain Lines, a ship line, approached containerization a different way, hoisting entire railcars on board ships and sailing them from U.S. ports to Cuba. All of these undertakings were modest in scope, but all had the same aim: to cut the cost of moving cargo through slow and inefficient ports.[26]

Yet these efforts were far from successful. "Contrary to what had been thought at first, the handling of containers led to hardly any cost savings," an influential European maritime expert admitted. A 1955 census found 154,907 shipping containers in use in non-Communist Europe. The number is large, but the containers were not: fully 52 percent of them were smaller than 106 cubic feet, less than the volume of a box 5 feet on a side. Almost all European containers were made of wood, and many had no tops; the user piled the goods inside and covered the load with canvas—hardly an efficient system for moving freight. The containers promoted by the Belgian national railway were meant to be slid up a ramp to fit *inside* truck bodies, requiring an extra stage of handling. American containers were typically made of steel, providing better protection but at enormous cost; one-quarter or more of the weight of a loaded container was the container itself.[27]

All over the world, the main methods for handling containers in the years after World War II offered few advantages over loose freight. "Cargo containers have been more of a hindrance than a help," a leading steamship executive complained in 1955. Many containers had metal eyes on top of each corner, requiring longshoremen to climb atop them to attach hooks before they could be lifted. The lack of weight limits meant that lifting could prove dangerous. Moving them with forklifts instead of winches, though, often damaged the containers. Large, expensive longshore gangs were still required to stow containers alongside loose freight in the holds of ships, where the boxes had to be maneuvered past built-in posts and ladders. "[I]t is certain that the goods would occupy far less space if they were stowed individually instead of in containers," the head of the French stevedores' association acknowledged in 1954. "This wasted space is quite considerable—probably over 10%." Ten percent of the ship's volume sailing empty amounted to a huge penalty for carrying cargo in containers.

For international shipments, customs authorities often charged duties on the container as well as the contents. And then there was the cost of sending emptied boxes back where they had come from, which "has always been a heavy handicap to container transport," Jean Levy, director of the French National Railway, admitted in 1948. Shipping food from a depot in Pennsylvania to an air base in Labrador cost 10 percent more using containers than with conventional methods, a 1956 study found—if the container was left in Labrador. When the cost of returning it empty to Pennsylvania was figured in, container shipping was 75 percent more expensive than loose freight.[28]

TABLE 2
Cargo Aboard the *Warrior*

| | *Number of Pieces* | *Percent of Weight* |
|---|---|---|
| Case | 74,903 | 27.9% |
| Carton | 71,726 | 27.6% |
| Bag | 24,036 | 12.9% |
| Box | 10,671 | 12.8% |
| Bundle | 2,880 | 1.0% |
| Package | 2,877 | 1.9% |
| Piece | 2,634 | 1.8% |
| Drum | 1,538 | 3.5% |
| Can | 888 | 0.3% |
| Barrel | 815 | 0.3% |
| Wheeled vehicles | 53 | 6.7% |
| Crate | 21 | 0.3% |
| Transporter | 10 | 0.5% |
| Reel | 5 | 0.1% |
| Undetermined | 1,525 | 0.8% |
| Total | 194,582 | 98.4% |

*Source: The SS Warrior*, p. 8.

By the early 1950s, there was little dispute that freight terminals were a transportation choke point. An unusual government-sponsored study, conducted in 1954, laid bare just how backward cargo handling was. The subject was the *Warrior*, a fairly typical C-2-type cargo ship, owned by Waterman Steamship Corp. The ship was chartered to the U.S. military, but on its run from Brooklyn to Bremerhaven, Germany, in March 1954, it carried a mix of cargo typical of merchant vessels, and was loaded and unloaded by civilian longshoremen. With government consent, the researchers had access to unusually detailed information about the cargo and the voyage.

The *Warrior* was loaded with 5,015 long tons of cargo, mainly food, merchandise for sale in post exchanges, household goods, mail, and parts for machines and vehicles. It also carried 53 vehicles. The cargo comprised an astonishing 194,582 individual items of every size and description.

These goods arrived in Brooklyn in 1,156 separate shipments from 151 different U.S. cities, with the first shipment arriving at the dock more than a month before the vessel sailed. Each item was placed on a pallet prior to storage in the transit shed. Longshoremen loaded the ship by lowering the pallets into the hold, where the men physically removed each item from its pallet and stowed it, using $5,031.69 worth of lumber and rope to hold everything in place. The longshoremen worked one eight-hour shift per day, excluding Sunday, and required 6 calendar days (including a day lost to a strike) to load the ship. Steaming across the Atlantic took 10½ days, and unloading at the German end, where longshoremen worked around the clock, took 4 days. In sum, the ship spent half the total duration of the voyage docked in port. The last of its cargo arrived at its ultimate destination 33 days after the *Warrior* docked at Bremerhaven, 44 days after it departed New York, and 95 days after the first Europe-bound cargo was dispatched from its U.S. point of origin.

The total cost of moving the goods carried by the *Warrior* came to $237,577, not counting the cost of the vessel's return to New York or interest on the inventory while in transit. Of that amount, the sea voyage itself accounted for only 11.5 percent. Cargo handling at both ends of the voyage accounted for 36.8 percent of the outlay. This was less than the 50 percent or more often cited by shipping executives—but only because Germany's "economic miracle" had yet to drive up longshore wages; the authors noted that port costs would have been much higher were it not for the fact that German longshoremen earned less than one-fifth the wages of U.S. longshoremen. Their conclusion was that reducing the costs of receiving, storing, and loading the outbound cargo in the U.S. port offered the best method of reducing the total cost of shipping. The authors went beyond the normal admonitions to improve longshoremen's productivity and eliminate inefficient work rules, and urged a fundamental rethinking of the entire process. "[P]erhaps the remedy lies in discovering ways of packaging, moving and stowing cargo in such a manner that breakbulk is avoided," they wrote.[29]

Interest in such a remedy was widespread. Shippers wanted cheaper transport, less pilferage, less damage, and lower insurance rates. Shipowners wanted to build bigger vessels, but only if they could spend more time at sea, earning revenue, and less time in port. Truckers wanted to be able to deliver to and pick up from the docks without hour upon hour of waiting. Business interests in port cities were praying for almost anything that would boost traffic through their harbors. Yet despite all the demands for change, and despite much experimentation, most of the industry's efforts to improve productivity centered on such timeworn ideas as making drafts heavier so that longshoremen would have to work harder. No one had found a better way to ease the gridlock on the docks. The solution came from an outsider who had no experience with ships.[30]

Chapter 3

# The Trucker

**The U.S. economy boomed** in the years just after World War II. The maritime industry did not. The entire merchant fleet had been commandeered by the government when the United States entered the war, and many ships did not revert to private control until July 1947, almost two years after the war ended. Coastal shipping had been all but closed down after German submarines sank several ships, and after 1945 coastwise traffic remained well below prewar levels. Trucks grabbed market share in domestic transportation, but the need to spend days painstakingly handling cargo each time a ship steamed into port kept the maritime industry from reducing costs enough to compete. "Until cargo handling costs can be reduced, there is little hope for coastwise revival," warned a California State Senate committee in 1951.[1]

Yet while the larger American ship lines were not particularly profitable, they were relatively sheltered. Foreign lines were barred from coastal service and routes to island territories, and a new American-owned competitor could not enter a domestic route without proving to the ICC that its entry would not harm other ship lines. Competition was also limited on international routes, where almost all ship lines belonged to cartels, known as conferences, that set uniform rates for each commodity. The U.S.-flag international lines received government subsidies to cover the higher wages of American crews, and both domestic and international lines—for regulatory reasons, international services were run by separate companies—had access to war-surplus ships. Inefficient though it was, the maritime industry thus felt little immediate pressure for change. Reshaping the business of shipping was left to an outsider with no maritime experience whatsoever, a self-made trucking magnate named Malcom Purcell McLean.

McLean was born in 1913 near the tiny town of Maxton, deep in the swamp country of southeastern North Carolina. Maxton, once called Shoe Heel, had been populated by Scottish Highlanders in the late eighteenth century. The local newspaper was the *Scottish Chief*, and local lore had it that Shoe Heel was renamed Maxton when a rail passenger shouted, "Hello, Mac!" from a train window and ten men responded. At the time of McLean's birth, Maxton Township, with about thirty-five hundred residents, was very rural and very poor. Electric lighting had arrived in Robeson County only in 1901. The town of Maxton, with about thirteen hundred inhabitants, had telephone service, but the surrounding area did not; as late as 1907, residents of Lumberton, the county's largest town, had to ride the train to Maxton to make long-distance calls.[2]

In later years, McLean took to portraying his life as a Horatio Alger story, in which his mother taught him business by giving him eggs to sell, on commission, from a crate at the side of the road. The reality was not quite so harsh. Although the family was far from wealthy, it was not without resources. McLean's father, also Malcolm P. McLean,* was "a member of a prominent and widely connected family," according to an obituary published in 1942. An 1884 county map shows half a dozen McLeans farming near Shoe Heel, and several other McLeans farmed or practiced law in Lumberton. Angus Wilton McLean, probably a cousin—his mother, like Malcom's, was a Purcell—started a bank and a railroad in Lumberton, served as assistant secretary of the United States Treasury in 1920–

* The younger McLean was named Malcolm at birth and continued to spell his name that way until after 1950, when he changed the spelling to Malcom. To avoid confusion, he is referred to as "Malcom" throughout this account.

21, and was governor of North Carolina from 1925 to 1929. Family ties may have helped the senior McLean obtain a job as a rural mail carrier in 1904 to supplement his income from farming. Upon young Malcom's graduation from high school in 1931, in the depths of the Great Depression, family ties got him work stocking shelves at a local grocery. Those local connections helped once more when an oil company needed a gas station manager in the nearby town of Red Springs, as a family friend lent McLean the money to buy his first load of gasoline.[3]

As recounted by McLean in the *American Magazine* in 1950, his rise began when he learned that a trucker earned five dollars for bringing the station's oil from Fayetteville, twenty-eight miles away. McLean proposed to do it himself. The station owner let him use an old trailer that had been rusting in the yard. McLean Trucking Company opened for business in March 1934, with McLean, still running the service station, as the sole driver. Soon after, family ties helped once more when a local man agreed to sell McLean a used dump truck on installments of three dollars a week. With the truck, McLean won a contract to haul dirt for the Works Progress Administration, a federal public-works program that at one point employed more than eleven hundred people in Robeson County. Even after hiring a driver, McLean earned enough to buy a new truck to haul vegetables from local farms. According to a much repeated tale, one trip found McLean so poor that he couldn't afford to pay the toll at a bridge along the way; he left a wrench with the toll collector as a deposit, redeeming it after selling his load in New York.[4]

This rags-to-riches tale fails to do justice to McLean's immense ambition. By 1935, at twenty-two years of age and with just one year of experience as a trucker, McLean owned 2 trucks and 1 tractor trailer, employed nine drivers who owned their own rigs, and had already hauled steel drums from North Carolina to New Jersey and cotton yarn to mills in New England. By 1940, as preparations for war revived the economy, six-year-old McLean Trucking owned 30 trucks and grossed $230,000. McLean built his operations during the war, gaining additional routes. A massive merger among seven of his competitors, which he opposed unsuccessfully all the way to the U.S. Supreme Court, barely affected the truck line. At the war's end in 1945, Malcom McLean controlled a thriving business with 162 trucks, mainly hauling textiles and cigarettes from North Carolina to Philadelphia, New York, and southern New England. Revenues in 1946 were $2.2 million, nearly ten times the level of 1940. McLean, already wealthy at age thirty-four, viewed this as just a beginning. As he wrote a few years later, "I saw that my only opportunity was to build and build and build, make a big trucking company out of a relatively small one."[5]

The economy of the late 1940s provided ample opportunity for a small trucking company to grow. As railway freight volumes languished, long-distance truck traffic more than doubled between 1946 and 1950. Getting a larger piece of the action, though, required the support of the Interstate Commerce Commission. The federal Motor Carrier Act of 1935 had brought interstate trucking under the authority of the ICC, which had regulated railroads since 1887. The ICC controlled almost every aspect of the business of common carriers—truckers whose services were on offer to the public. A common carrier could haul only commodities the ICC allowed it to haul, over ICC-approved routes, at ICC-approved rates. If a new firm wanted to begin service, or if an existing one wanted to serve a new route or carry a new commodity, it had to hire lawyers to plead its case at the commission. Any major change required hearings at which other truck lines and railroads had the opportunity to object. Regulation made trucking hugely inefficient; a trucker authorized to haul paper between Nashville and Philadelphia could not simply pick up a few tires or drums of chemicals to fill a half-empty truck, and might have to return home empty if authorized cargo were not available for the backhaul. The ICC's concern was not efficiency but order. Regulation protected the interests of established truck lines by limiting competition, and it protected the railroads by forcing truck lines to charge much more than railroad companies. More than anything else, the ICC wanted to keep the transportation industry stable.[6]

Regulation damped competitive spirit in the trucking industry. Showing the sort of ingenuity that would characterize his career,

McLean found ways around the regulators' obstacles. If winning new route authority was too arduous, why not buy a carrier that already had attractive routes? And if buying another truck line was too expensive, why not lease one? The labor unrest that followed the war left many truck lines struggling, and McLean repeatedly seized opportunities. Between 1946 and 1954, McLean Trucking bought or leased routes in at least ten different transactions, expanding its network from Atlanta to Boston. The company added six hundred trucks between 1947 and 1949, using the U.S. government as the unwitting financier: veterans were eligible for cheap government loans to set themselves up as independent truckers, so McLean encouraged veterans to become owner-operators, brought them together to purchase equipment in a single large order, and signed them up to haul freight for McLean Trucking.[7]

An obsessive focus on cutting costs was the key to McLean Trucking's success. The only way a truck line could attract much new business was by offering lower rates than competitors offered. A trucking company's salesman would call on a prospective client, learn how much cargo it generated for various destinations, and then study the rates that its current carrier had filed with the ICC. The truck line could then propose a lower rate to win the business, but only if it could prove to the ICC that the proposed rate would be profitable. In practice, this meant that a truck line could not underprice its competitors unless it had lower costs. Malcom McLean's sharp pencil was critical. In 1946, for example, McLean made a deal to lease the routes of Atlantic States Motor Lines, which had been closed down by a strike. Among its other attractions, Atlantic States was entitled to use highways that let McLean Trucking shave seventy miles off runs between North Carolina and the Northeast. The shorter trip meant fewer driver hours and therefore lower rates. By purchasing route authority from Garford Trucking in 1948, McLean Trucking gained southbound cargo from New England, so that trucks that hauled cigarettes north would not have to return empty—which meant that the company could charge less for the northbound trip.[8]

One hotly contested rate case illustrates McLean's handle on costs. In March 1947, McLean Trucking proposed to cut certain rates for cigarettes almost by half, charging $0.68 per hundred pounds to haul full truckloads from Durham, North Carolina, to Atlanta, and $1.10 to carry partial truckloads. At the time, other truck lines were charging $1.34 for full truckloads and $1.70 per hundred pounds for partial loads. McLean even wanted to underprice the much slower railroads, which protested the proposed rates as "unfair and destructive." Laying out its costs in great detail, McLean Trucking argued that tobacco products were cheaper to transport than other commodities because its administrative costs for cigarettes were 1.02 cents per vehicle-mile below its average for all freight, its sales and marketing costs 50 to 60 percent less, and its terminal costs 3 cents per vehicle-mile less than for full truckloads of other freight. After weighing such considerations as the density of tobacco products and McLean Trucking's insurance claims experience on cigarette shipments, the ICC rejected the proposed rate for partial truckloads but found the proposed full-truckload rate to be "just and reasonable," opening the way for McLean Trucking to vastly expand its business with the tobacco industry.[9]

Cost-saving innovations continually materialized as McLean Trucking grew. The company opened one of the earliest automated terminals in the industry, in Winston-Salem, North Carolina, using conveyors to transfer freight from one truck to another while saving labor. At a time when most trucks had gasoline engines, McLean Trucking was the first major company to install diesel engines in its tractors—and in an era when drivers typically bought fuel on their own, Malcom McLean arranged a corporate discount at service stations along the company's routes and told drivers to fuel only at those stations. The sides of the company's truck trailers were crenellated, not smooth; Malcom McLean claimed that experts at the University of North Carolina had told him crenellation would reduce wind drag and thus fuel costs. By the early 1950s, McLean Trucking was hiring young university graduates and putting them through one of the first formal management training programs in American business. Men just out of college would come to Winston-Salem, where their first task was learning to drive a truck. After six months of hauling freight, trainees were sent to a terminal and spent several months unloading trucks. Then came a stint in the office, learning the McLean Trucking method for making a proposal to a potential customer, which required careful analysis of the cost of serving the business. Only then were trainees dispatched to their first assignments, usually selling freight in Raleigh or Boston or Philadelphia.[10]

McLean Trucking quickly became known as a dynamic company in a very stodgy industry. By 1954, it had become one of the largest trucking companies in America, ranking eighth in revenue and third among all truck lines in after-tax profit. Assets, $728,197 in 1946, increased to $11.4 million in 1954 as the McLean fleet grew to 617 company-owned trucks. The only way to grow so fast was to borrow money. McLean Trucking's $200,000 of long-term debt in 1946 increased 31-fold to $6.2 million by 1951 as the company ordered more and more trucks. "He was a highly leveraged fellow," remembered Walter Wriston, who started lending to McLean on behalf of National City Bank in 1954 and later headed the company when, under the name Citibank, it became the largest bank in the world. "He understood cash flow. You'd go to a railroad in those days and talk about cash flow and they'd ask you what you meant."[11]

Heavy indebtedness, of course, was risky. Any slowdown in revenue growth could make it hard to service debt. Of necessity, a highly leveraged company had to focus on efficiency, for which Malcom McLean and his brother Jim, who ran day-to-day operations, had a passion. They knew every aspect of their business, and they knew how to squeeze out costs. Recalls one former employee, "When you reported for duty, you drove the truck through the gatehouse, you were weighed, and the truck was sealed. They started the tachometer, and you were given specific directions: 'You will proceed up Route 3A to Secondi's filling station, then you will proceed . . .' You didn't have any discretion at all." But after years behind the wheel themselves, the McLeans understood that the easiest way to control costs was to get employees involved. Holding down insurance and repair bills, for example, meant having safety-conscious drivers. Novices were trained by being paired with senior drivers on the run from Winston-Salem to Atlanta. The senior driver got a bonus of one month's pay if a man he had trained made it through his first year without an accident. The incentives were powerful: the veteran had a strong financial incentive to train the newcomer well, and the new driver understood that he had best drive very carefully if he wanted to stick around.[12]

Malcom McLean was not a man to sit and enjoy his success. The civic and philanthropic involvements common to successful businesspeople did not interest him. He was a restless soul, competitive, calculating, always thinking about business. "He wouldn't be able to sit still five minutes," a longtime colleague recalled before McLean's death. "You'd either have to play gin rummy with him or discuss business with him. Can't go quail hunting with Malcom without him betting on the first, the most, the biggest." His inventive brain churned out idea after idea for making money.[13]

One such brainstorm came in 1953, as McLean was fretting over increasing highway congestion and worrying that domestic ship lines, able to buy war-surplus cargo ships from the government for almost nothing, might undercut his trucking business. Rather than driving down crowded coastal highways, why not just put truck trailers on ships that could ferry them up and down the coast? By the end of that year, McLean was proposing to build waterfront terminals that would allow trucks to drive up ramps to deposit their trailers on board specially designed ships. The ships would move the trailers between North Carolina, New York City, and Rhode Island, circumventing the worsening traffic jams at a time when expressways were few and far between. At the port of arrival, other trucks would collect the trailers and haul them to their destinations.[14]

In the context of the 1950s, McLean's plan was revolutionary. Law and regulation ensured that trucks and ships had nothing in common: trucking companies ran trucks, and shipping companies ran ships. A few ship lines and barge companies carried trucks on their vessels, as McLean planned to do, but they were simply offering water transportation to any trucker who would pay. The idea that a truck line would use its own trucks to drive its own trailers on board its own ships, float the trailers down the coast, and then drive them to their destination at the other end violated the ICC's basic precepts. The truck-ship plan was startling as well because coastwise shipping was widely seen as a dying business. New York's docks were handling only half as much domestic cargo in the early 1950s as they had in the depressed 1930s. No one had invested significant money in coastal shipping in thirty years. McLean's interest was entirely a matter of cost. The ICC had regulatory authority over domestic shipping, and it allowed ship rates to be well below rail and truck rates to compensate for slower service. Sending his trucks by water would let McLean underprice other truckers running between North Carolina and the Northeast.

Late in 1953, a real estate firm working on McLean Trucking's behalf began looking for terminal sites. The timing could not have been better. The Port of New York Authority, an agency set up by the states of New York and New Jersey, was eager to expand its tenuous foothold in the port business. It had taken charge of the money-losing docks in Newark, New Jersey, in 1947, and it was eager to draw new business to what had been a sleepy lumber port. As it happened, the port at Newark, across the harbor from New York City, was uniquely positioned to serve McLean Trucking's needs. There was plenty of space to marshal trucks and easy access to the New Jersey Turnpike, which had opened in 1951. Even better, from McLean's point of view, the Port Authority had the power to issue revenue bonds; it could build the terminal and lease it to McLean Trucking, reducing the need for the company to raise funds. The Port Authority was so taken with McLean's concept that Austin Tobin, its executive director, and A. Lyle King, the director of marine terminals, became early and very public apostles of transporting truck trailers by train and ship.[15]

While the Port Authority prepared McLean Trucking's new waterfront terminal, Malcom McLean's own ideas were evolving. His 1953 proposal had involved buying S. C. Loveland, a small barge operator, in order to gain its coastal operating rights. Now, he was thinking bigger. In 1954, while still pursuing Loveland, he came upon the Waterman Steamship Corporation in the pages of a *Moody's* financial manual. Waterman, based in Mobile, Alabama, was a large, well-established operator sailing to Europe and Asia. Its tiny subsidiary, Pan-Atlantic Steamship Corporation, operated four ships along the coast between Boston and Houston. McLean immediately spotted the companies' attractions. Pan-Atlantic had been hurt badly by the 1954 longshore strike in New York, completing only 64 voyages the entire year, but it owned valuable operating rights to serve 16 ports. Its parent, Waterman Steamship, was debt-free, with assets including 37 ships and $20 million in cash. McLean made some preliminary soundings and found that Waterman could be bought for $42 million.[16]

What followed was an unprecedented piece of financial and legal engineering. First, to circumvent rules requiring ICC approval for a truck line to own a ship line, McLean created an entirely new company, McLean Industries, in January 1955. Although McLean Industries had publicly traded stock, it was clearly a family-controlled business; Malcom McLean was president, his brother James McLean was vice president, and his sister Clara McLean was secretary and assistant treasurer. Malcom, James, and Clara then put control of the trucking company in a trust, of which they were the beneficiaries. Malcom McLean kept $5 million of stock, but the trustees were authorized to sell the rest. As soon as the trust documents were signed, the McLeans resigned as directors of McLean Trucking, and within an hour McLean Industries took control of Pan-Atlantic. The country's best-known trucking magnate walked away from the business he had built in order to build a new one, based on some untested ideas about shipping.[17]

Several railroads protested the transaction, claiming that the McLeans were effectively controlling both McLean Trucking and the ship line in violation of the law. The ICC eventually agreed but noted that "the procedure followed was on the advice of counsel, and was not a deliberate violation of the act." In any case, in September 1955 the trustees sold off McLean Trucking's stock, making the legal issues moot. Malcom McLean did not fare badly in the process. He cleared $14 million on the sale of McLean Trucking. His net worth in 1955 was $25 million—the equivalent of $180 million in 2004 dollars. Asked later whether he had considered ways to shelter some of his wealth from the risks of entering the maritime business, his answer was an unequivocal "No." McLean explained: "You've got to be totally committed."[18]

Pan-Atlantic was just an appetizer. In May 1955, McLean Industries bid for Waterman itself. McLean and his bankers concocted a hugely complex financial transaction. McLean Industries would pay Waterman $75,000 to cease all domestic service and surrender its ICC operating certificate in an attempt to eliminate the ICC's jurisdiction over the purchase. Then, McLean Industries would borrow $42 million from National City Bank, an amount approaching the bank's legal limit for a single loan. McLean Industries would raise additional money through a $7 million issue of preferred stock. When the deal closed, Waterman's $20 million of cash and various other assets would be used to repay half the loan. Wriston's superiors at National City went apoplectic at the thought that $22 million of National City's money would still be at risk. Who knew whether anyone would use McLean's truck-ship service? Who would finance all the equipment? Would the trailers even survive a storm at sea? At the last minute, the bankers ordered the deal called off. Wriston phoned McLean at the Essex House, McLean's New York hotel, and told him, "You'd better get down here. Things are falling apart." When McLean arrived at National City's headquarters on Wall Street, Wriston advised that McLean himself would have to convince the bank's top loan executives to approve the loan. The bankers told McLean that the loan was too risky and Wriston too inexperienced. "He's just a trainee," one of them said. "He may just be a trainee, but he's going to be the boss of both of you pretty soon," McLean shot back. As McLean remembered later, "They said, 'Maybe we'll take another look.' " The loan was approved.[19]

But the deal was still not done, and a competing buyer, also financed by National City, had grown interested in Waterman. To avoid any chance of a slipup, the lawyers decided that the entire transaction needed to be completed simultaneously. On May 6, Waterman's board and McLean's bankers and lawyers convened in a Mobile boardroom only to realize that the board lacked a quorum. One of the Wall Street lawyers quickly took the elevator downstairs, stopped a passerby, and asked whether he wanted to earn a quick fifty dollars. The man was promptly elected a Waterman director, making a quorum. The Waterman board members then resigned one at a time, with each being replaced by a McLean nominee. The new board immediately voted to pay a $25 million dividend to McLean Industries, and with a phone call the money was wired to National City. As the meeting broke up, lawyers for the opposing bidder served the board with legal papers to prevent the transfer of the dividend, but the bank already had its money and McLean had Waterman. Typical of McLean's financial acumen, he laid out only $10,000 of his own cash to gain control of one of the country's largest ship lines through what later became known as a leveraged buyout. "In a sense, Waterman was the first LBO," Wriston recalled.[20]

McLean's prize was a formerly debt-free company whose bank loans and ship mortgages soared to $22.6 million at the end of 1955, nearly ten times its $2.3 million of after-tax income. In a step that set the norm for future leveraged buyouts, McLean disposed of unwanted Waterman assets to pay down debt; the sale of a hotel, a dry dock, and various other businesses raised almost $4 million within two months of the takeover. Now heavily in debt, McLean began maneuvering for a government handout. The federal government had become interested in ships carrying truck trailers, and Pan-Atlantic obtained $63 million of government loan guarantees for seven new roll on–roll off ships, designed to have trailers driven on board. The new ships would carry 288 truck trailers each and would reduce cargo-handling costs by more than 75 percent.[21]

The money was never spent, because McLean reconsidered his plan. He had realized that carrying trailers on ships was inefficient: the wheels beneath each trailer would waste a lot of precious shipboard space. Pondering that problem, McLean came up with a still more radical idea. A government maritime-promotion program made leftover World War II tankers available to ship lines very cheaply. Pan-Atlantic would buy two and convert them to haul truck trailer *bodies*—trailers detached from their steel beds, axles, and wheels. Subtracting the frames and wheels would reduce the space occupied by each trailer by one-third. Even better, the trailer bodies could be stacked, whereas trailers with wheels could not be. As

McLean envisioned it, a truck would pull the trailer alongside the ship, where the trailer body, filled with twenty tons of freight, would be detached from its steel chassis and lifted aboard ship. At the other end of the voyage, the trailer body would be lowered onto an empty chassis and hauled to its destination.[22]

The concept was costed out on Ballentine Beer, which McLean Trucking hauled from Newark. Analysts for the Port of New York Authority calculated that sending the beer to Miami on board a traditional coastal ship, including a truck trip to the port, unloading, stacking in a transit shed, removal from the transit shed, wrapping in netting, hoisting aboard ship, and stowage, would cost four dollars a ton, with unloading at the Miami end costing as much again. The container alternative—loading the beer into a container at the brewery and lifting the container aboard a specially designed ship—was estimated to cost twenty-five cents a ton. Container shipping would be 94 percent cheaper than breakbulk shipping of the same product, even allowing for the cost of the container.[23]

Tankers, of course, were not the ideal vessels for such a mission, but they reduced the financial risk. If no one wanted to ship containers on the return trip from Houston to Newark, the vessels could still make money carrying oil. McLean portrayed these "lift on–lift off" ships as "forerunners" of the roll on–roll off ships he intended to build with the government guarantees, but plans for the trailerships were pushed to the side and finally abandoned.[24]

The concept that became container shipping was Malcom McLean's. But in early 1955, when McLean jettisoned his plan to put entire truck trailers on Pan-Atlantic's ships and decided instead to carry just trailer bodies, he could not simply buy the equipment off the shelf. Small steel boxes were readily available, but it was obvious that lowering them into the hold and stowing them amid assorted bags and bales, the way other ship lines occasionally did, would bring little by way of cost savings. Truck trailer bodies could be purchased as well, but moving trailers weighing tens of thousands of pounds apart from their chassis and wheels was by no means a routine operation. McLean, impatient to build a business, demanded that his staff find a way to turn his concept into reality. In March, a Pan-Atlantic executive named George Kempton placed a call to Keith Tantlinger.

Tantlinger, then thirty-five, was chief engineer at Brown Industries in Spokane, Washington, and had already built a reputation as a container expert. Brown had been building truck trailers since 1932, and Tantlinger's job, along with designing trailers for trucking companies, involved speaking at industry meetings to promote Brown's products. In 1949, he had designed what was probably the first modern shipping container, a 30-foot aluminum box that could be stacked two high on barges operating between Seattle and Alaska or placed on a chassis and pulled by a truck. The order involved only two hundred containers, and despite much curiosity, no other orders followed. "Everybody was interested, but nobody wanted to reach for his pocketbook," Tantlinger remembered.[25]

McLean the trucker had never done business with Brown Industries. Now that he was in the shipping business, though, McLean wanted Tantlinger's expertise—immediately. The next morning, Tantlinger flew to Mobile, where Pan-Atlantic was based. "I understand you know everything there is to know about containers," was McLean's gruff greeting. McLean explained his plan. He proposed to use containers thirty-three feet long, a length chosen because the available deck space aboard the T-2 tankers was divisible by thirty-three. These boxes were at least seven times the size of any containers then in common use. Rather than having longshoremen stow them with other cargo in a ship's hold, he proposed to install metal frames, called flying decks or spardecks, above the tangle of pipes that covered the decks of his two tankers. The spardecks would hold the containers eight abreast. The idea was to attach six steel pieces, each a foot long with a small hole at the bottom, to the sides of each container. When the container was loaded on board ship, the steel pieces would slide vertically through slots in the frame of the spardeck, and a rod would be inserted through the holes, underneath the frame, to lock the container in place. Most important, the containers Pan-Atlantic planned to use would be designed to be shifted easily among ships, trucks, and trains.[26]

McLean's trucking superintendent, Cecil Egger, had begun experiments with two old Fruehauf truck trailers that had been strengthened with A-shaped steel brackets welded to each side. Tantlinger quickly saw that the system was unworkable: the containers were meant to be locked in placed with steel pieces protruding beneath them, making them impossible to stack, and the A-shaped brackets made the trailers too wide and too tall for the highways. Tantlinger told McLean that standard Brown containers, which used the aluminum sides and roof to bear most of the load, would do the job. McLean ordered two 33-foot containers, to be delivered in two weeks to the Bethlehem Steel shipyard in Baltimore, which was altering the tankers. On the appointed day, Tantlinger was to meet Pan-Atlantic executives for breakfast at the Lord Baltimore Hotel. When they failed to arrive, he called the shipyard and learned that the men were already there. Tantlinger rushed to the shipyard, where Malcom and Jim McLean, Kempton, and Egger were jumping up and down on the roof of a container. Tantlinger had told Malcom McLean that the wafer-thin aluminum roof was strong enough to keep the container rigid, and the McLean group was trying, unsuccessfully, to disprove his claim. Sold on the merits of Brown's containers, McLean ordered two hundred boxes and demanded that the reluctant Tantlinger move to Mobile to be his chief engineer.

Part of Tantlinger's job was to convince the American Bureau of Shipping, which sets standards for maritime insurers, that the *Ideal-X* would be seaworthy when loaded with containers, while the U.S. Coast Guard wanted assurance that the containers would not endanger the ship's crew. After negotiation, the Coast Guard agreed to a test. Pan-Atlantic asked trucking company workers to load two containers with cardboard boxes filled with coke briquets, a cargo of average density and negligible cost. The boxes were lashed to the spardeck of one of the converted T-2s. The ship then sailed back and forth between Newark and Houston, the Coast Guard checking the load after each voyage, until a trip though heavy seas persuaded that maritime agency that loaded containers were safe. Photos of the test, showing the stacks of cardboard boxes dry and firmly in place after each voyage, got the Bureau of Shipping's approval.

And then there was the matter of loading. Most cargo ships in the 1950s had winches that allowed them to load and unload in any port, but a standard shipboard winch could not shift a twenty-ton container without destabilizing the ship. The solution took the form of two huge revolving cranes at a disused shipyard in Chester, Pennsylvania. The cranes, with booms seventy-two feet above the dock, were designed to move on tracks installed on the dock, parallel to the ship. Pan-Atlantic dismantled them, cut twenty feet out of their structures, and shipped them off to Newark and Houston, while port workers at both locations reinforced the piers to accommodate the added weight and installed the rails and large power supplies the cranes required. Hanging from the cranes was another money-saving piece of equipment newly invented by Tantlinger, a spreader bar stretching the entire length and width of a container. The spreader eliminated the need for longshoremen to climb ladders to the roof of each container and attach hooks dangling from the crane. Instead, the crane operator, sitting in a cab sixty feet above the dock, could lower the spreader over a container and engage the hooks at each corner with the flip of a switch. Once the box had been lifted and moved, another flip of the switch would disengage the hooks, without a worker on the ground touching the container.[27]

McLean wanted to start Pan-Atlantic's new service in 1955. The government did not move quite so fast. Not until late 1955, after months of hearings, did the ICC overrule objections from the railroads and authorize Pan-Atlantic to carry containers between Newark and Houston. Delays in gaining Coast Guard approval pushed the start date back further. On April 26, 1956, one hundred dignitaries enjoyed lunch at Port Newark and watched the crane place a container on the *Ideal-X* every seven minutes. The ship was loaded in less than eight hours and set sail the same day. McLean and his executives flew to Houston to watch the arrival. "They were all waiting on Wharf II for the ship to arrive, and as she came up the channel, all the longshoremen and everybody else came over to look," one witness recalled. "They were amazed to see a tanker with all these boxes on deck. We had seen thousands of tankers in Houston, but never one like this. So everybody looked at this monstrosity and they couldn't believe their eyes." For McLean, though, the real triumph came only when the costs were tallied. Loading loose cargo on a medium-size cargo ship cost $5.83 per ton in 1956. McLean's experts pegged the cost of loading the *Ideal-X* at 15.8 cents per ton. With numbers like that, the container seemed to have a future.[28]

Pan-Atlantic's Sea-Land Service was in business, meeting a schedule of one weekly sailing in each direction between Newark and Houston. Pan-Atlantic itself was barred from owning trucks, but it contracted with trucking companies to pick up shipments at customers' loading docks and to carry arriving containers to their final destinations. Between April and December, Pan-Atlantic completed forty-four container voyages along the East and Gulf coasts. In a typical McLean touch, his engineers figured out that through the addition of small deck extensions, the tankers' capacity could be increased from 58 containers to 60 and then 62. If there was a way to squeeze a dollop of extra revenue from the aging tankers, McLean would find it.[29]

Again, the railroad and trucking industries did their best to close down the show. They protested vehemently that McLean's takeover of Waterman without ICC approval was a blatant violation of the Interstate Commerce Act. Although Waterman had renounced its domestic operating rights to escape ICC jurisdiction, that renunciation had not been accepted by the ICC—and Pan-Atlantic's request for "temporary" authority to take over the Waterman rights, keeping them in the corporate family, made the entire deal look suspicious. In November 1956, an ICC examiner agreed. Although Malcom McLean was a "man of vision, determination and considerable executive talent," the examiner said, his purchase of Waterman without commission approval broke the law. As punishment, he recommended that McLean Industries be forced to divest Waterman. The ICC rejected the examiner's recommendation in 1957, leaving McLean in control of both Pan-Atlantic and Waterman, and, more important, of Waterman's large fleet.[30]

Malcom McLean was by no means the "inventor" of the shipping container. Metal cargo boxes of various shapes and sizes had been in use for decades, and numerous reports and studies supported the idea of containerized freight before the *Ideal-X* set sail. An American steamship operator, Seatrain Lines, had operated specially built ships holding railway boxcars in metal cells as early as 1929, lifting the boxcars on and off with large dockside cranes. These ample precedents have led historians to downplay the nature of Malcom McLean's achievements. His container was just a "new adaptation of a long-used transportation formula whose birth dates to the early years of the twentieth century," French historian René Borruey asserts. American historian Donald Fitzgerald concurs: "Rather than a revolution, containerization of the 1950s was a chapter in the history of development of maritime cargo transportation."[31]

In a narrow sense, of course, these critics are correct. The high cost of freight handling was widely recognized as a critical problem in the early 1950s, and containers were much discussed as a potential solution. Malcom McLean was not writing on a blank slate. Yet the historians' debate about precedence misses the transformational nature of McLean's accomplishment. While many companies had tried putting freight into containers, those early containers did not fundamentally alter the economics of shipping and had no wider consequences.

Malcom McLean's fundamental insight, commonplace today but quite radical in the 1950s, was that the shipping industry's business was moving cargo, not sailing ships. That insight led him to a concept of containerization quite different from anything that had come before. McLean understood that reducing the cost of shipping goods required not just a metal box but an entire new way of handling freight. Every part of the system—ports, ships, cranes, storage facilities, trucks, trains, and the operations of the shippers themselves—would have to change. In that understanding, he was years ahead of almost everyone else in the transportation industry. His insights ushered in change so dramatic that even the experts at the International Container Bureau, people who had been pushing containers for decades, were astonished at what he had wrought. As one of that organization's leaders confessed later, "we did not understand that at that time a revolution was taking place in the U.S.A."[32]

## Chapter 4

# The System

**A dock strike loomed** over East Coast ports in the autumn of 1956. Facing the prospect that the Pan-Atlantic and Waterman fleets would sit idle, Malcom McLean decided to use the time to advantage. Six of Waterman's C-2 freighters were transferred to Pan-Atlantic's control. They were sent to Waterman's shipyard in Mobile, which had been closed after World War II but was reopened to convert them into pure containerships. The idea was to build a honeycomb of metal cells in the holds so that 35-foot containers, two feet longer than those carried on the *Ideal-X*, could be lowered in and stacked five or six high. The ships were to be rebuilt and back at sea by 1957. Of course, there was no model of a pure containership, the metal cells did not exist, and no one had ever stacked containers five or six high. How tightly should the containers fit into the cells? How would a stack of six containers behave when a ship rolled in heavy seas? And how could the vessels be unloaded at ports where there were no land-based cranes? As was his way, McLean did not preoccupy himself with such details. He simply told his staff to get the job done.[1]

The C-2s, unlike Pan-Atlantic's T-2 tankers, had been designed to carry large amounts of mixed cargo in their five holds, and altering them posed no great problem. The decks were widened from 63 feet to 72 feet, and the hatches were expanded so that the entire container storage area would be accessible from above. The cells to hold the containers inside the ship were a tougher challenge. At the Alabama State Docks in Mobile, Keith Tantlinger built a mock-up 20 feet high. The cell guides, vertical strips of steel with a 90-degree angle to hold the corners of a container, were mounted on hydraulic jacks, which could be raised and lowered to simulate a heeling ship. A crane tried to deposit and remove a container from the cell while it was at various angles, and instruments measured the stresses and strains on the container and the cell as it tilted this way and that. After hundreds of tests, Tantlinger concluded that each cell should be 1¼ inches longer than the container it was meant to hold and ¾ of an inch wider; smaller dimensions made it too hard for the crane operator to ease the box into the cell guides, but larger ones allowed the container to shift too much. The cells were built and installed in the holds, giving the C-2s the ability to carry 226 containers, almost four times the load of the *Ideal-X*.[2]

Bigger ships with bigger loads would make loading and unloading vastly more complicated. The methods used for the smaller T-2s were no longer good enough: with 226 containers, a loading rate of one container every seven minutes would require a vessel to spend more than twenty-four hours in port to take on a full load. Every aspect of the operation needed to be redesigned for faster handling. Tantlinger invented a new trailer chassis, with edges sloped so that a container being lowered by a crane would be guided into place automatically. A new locking system allowed a longshoreman to secure or release the container by raising or lowering a handle at each corner of the chassis, doing away with the labor-intensive routine of using iron chains to prevent the box from slipping off the truck. These changes meant that a truck could deliver or take on a container and quickly drive away without occupying precious space at dockside. The containers themselves were redesigned with heavy steel corner posts to support the weight of more containers above them, and a new refrigerated version had the cooling unit set within the profile of the container, so that it could be stacked along with nonrefrigerated boxes. New doors were designed with the hinges recessed within the rear corner posts rather than protruding from the sides.

All of these new containers had a special steel casting built into each of their eight corners. The casting contained an oblong hole designed to accommodate the most critical invention of all, the twist lock. This device, with one conical section pointing down and another up, could be inserted into the corner castings of containers as they were stacked. When one was lowered upon the other, a longshoreman could quickly turn the handle and lock the two boxes tightly together. By pulling the handle the other way, a worker could release the two boxes in seconds when it was time to discharge the ship.[3]

Not until the cells and containers had been designed could Pan-Atlantic focus on the other critical component of its new operation, the cranes. The big dockside cranes in New York and Houston were inadequate to meet the new demands, and the other ports McLean wanted to serve lacked large cranes altogether. Shipboard cranes were the obvious answer, but existing shipboard cranes were not big enough to lift a 35-foot container weighing 40,000 pounds. No established maritime crane manufacturer could design and deliver a test model within the 90 days left in McLean's ambitious schedule. In desperation, Tantlinger, who knew of the logging industry from his years in Washington State, proposed calling companies that manufactured diesel-powered logging cranes. Robert "Booze" Campbell, whose engineering firm helped redesign the ships and terminals, came upon the Skagit Steel & Iron Works in Sedro-Woolley, Washington.

Skagit Steel's owner, Sidney McIntyre, had never worked on ships and was unfamiliar with electric cranes, but he agreed to build one. He was, in Campbell's description, "a mechanical genius." Within ninety days, Skagit Steel produced an enormous crane, which rode on a huge gantry that bridged an entire ship. The C-2s had their wheelhouses amidships, so each vessel required two cranes, one fore and one aft. The cranes moved backward and forward on rails placed along the ship's sides and could travel across the width of the vessel, stopping immediately above any container and hoisting it vertically.

Long, folding arms allowed the cranes to travel out over the dock to pick up and lower containers.[4]

The combination of cells and gantry cranes allowed the containers to be handled with unprecedented speed. Once the first column of cells had been unloaded, the ship could be loaded and unloaded simultaneously, in assembly-line fashion: each time the crane traveled to the dock to deposit an incoming trailer on an empty chassis, it would pick up an outgoing trailer and place it into an empty cell. With two cranes, each loading and unloading fifteen boxes an hour, the *Gateway City*, the first of the converted C-2s, could be emptied and reloaded in just eight hours. The new ships were "[t]he greatest advance made by the United States merchant marine in our time," said Congressman Herbert Bonner, chairman of the Merchant Marine Committee. Tantlinger was not so certain. Before the *Gateway City*'s first voyage, on October 4, 1957, he dropped by the F. W. Woolworth store in Newark and purchased all of the store's modeling clay. He cut the clay into small pieces with his pocket knife and wedged several pieces in the narrow spaces between the corners of the top containers and the metal frames of the cells. When the *Gateway City* docked in Miami three days later, he retrieved the clay to see how much the containers had shifted. The indentations on the clay revealed that they had moved by only 5/16 of one inch—proof, at last, that a stack of containers in the hold would not sway dangerously as a containership rolled at sea.[5]

Pan-Atlantic had four of its six pure containerships in service by the end of 1957, with a ship sailing south from New York or east from Houston every four and a half days. The last two converted C-2s joined the fleet early in 1958. The *Ideal-X* and its sister tankers were sold off, along with 490 of the original 33-foot containers and 300 matching chassis. Pan-Atlantic's Sea-Land Service, its capacity five times larger than it had been a year earlier, seemed poised for explosive growth.[6]

Instead, it sailed into trouble. McLean planned to use two of the all-container ships to open service to Puerto Rico in March 1958. Puerto Rico was a potentially lucrative market. As an island, it relied on ships to provide almost all of its consumer goods. As a U.S. commonwealth, it was subject to the Jones Act, a law requiring that cargo moving between U.S. ports use American-built ships with American crews. Limited competition allowed the few carriers serving Puerto Rico to charge very high rates, and McLean figured that Pan-Atlantic's containers could easily grab market share. He figured without the longshoremen. When the first containership arrived from Newark, longshoremen in San Juan refused to unload the containers. Four costly months of negotiation ensued, with two ships sitting idle. Pan-Atlantic finally bent to union demands to use large, twenty-four-man gangs to handle containerships, and regular service opened in August. The delay, plus the cost of getting rid of the now obsolete tankers, drove McLean Industries dangerously into the red. A net loss of $4.2 million for 1958 nearly wiped out the earnings retained during the company's first three years.[7]

McLean was not deterred. Pan-Atlantic's problems, he determined, were rooted in the maritime industry's passive, slow-moving culture. Domestic ship lines, such as Pan-Atlantic, operated in a highly regulated environment that left little room for entrepreneurial spirit. American-owned lines operating internationally, such as Waterman, were allowed to join international rate-making cartels. U.S.-flag ships, using American crews, had the exclusive right to carry the huge flow of U.S. government shipments, including military cargo, and many lines received government operating subsidies as well. This sheltered culture led to excesses like Waterman's headquarters building in Mobile, with its revolving globe in the lobby and the lavish executive suite on the sixteenth floor. It did not breed the sorts of creative, aggressive, hungry employees that suited Malcom McLean. McLean decided it was time for a culture change. In June 1958, Pan-Atlantic, which now ran only containerships, moved to a new headquarters in a converted pineapple warehouse near the Newark docks, while Waterman, the traditional breakbulk ship line, was deliberately left behind in Mobile.

The new Pan-Atlantic office had a very different atmosphere. Malcom McLean had a simply furnished glass-fronted office facing a large, open floor on which desks were lined up side by side. Every morning, McLean wandered the floor to check on the latest cash flow statement or the status of shipbuilding plans, disregarding hierarchy to get the information he wanted. The company's tone, though, was set by his sister Clara. Her desk was in the middle of the floor, where she could keep an eye on everything and everyone. She knew who had come in late. She decorated the office; managers who were promoted into glass-fronted offices of their own found that she had selected their furnishings for them, right down to the art. "If you put a picture or a calendar on the wall, you got a note from Clara the next morning," one recalled. She set the rules: coffee nowhere but the coffee room, no personal phone calls, desks cleared every night. She personally reviewed every single time card and approved every hire.[8]

Malcom McLean was not the only shipping magnate with an interest in containerization. In 1954, as McLean was leasing terminals for his proposed roll on–roll off service on the East Coast, the Matson Navigation Company began to sponsor academic research on cargo handling. Matson, based in San Francisco, was thinking about containers as well, but its approach was the polar opposite of McLean's.

Matson, established in 1882, had been a loosely managed, family-dominated company that grew from a single ship in Hawaii into a transportation conglomerate. It owned California oil wells, oil tankers, and tanks in the Hawaiian Islands to store the oil. It owned passenger ships and built hotels on Waikiki Beach to attract passengers. It owned Hawaiian sugar plantations and the ships to carry sugar to the mainland. For a few years after World War II, it even owned an airline. None of this made much money, and the company's underlying problem was that many of its big shareholders didn't want it to make much money. The board of directors included representatives of major Hawaiian sugar and pineapple growers whose main interest was a cheap way to get their products to market. Whether the shipping service made a profit was almost incidental.[9]

Things began to change in 1947, when the Matson family convinced veteran steamship executive John E. Cushing to postpone his planned retirement and serve for three years as president. Cushing put the company on a budget for the first time and took a serious interest in addressing dismally low productivity. In 1948, Matson installed a revolutionary mechanized system to ship sugar to the mainland in bulk rather than in hundred-pound bags. Bulk sugar had required large investments—huge bins at the Hawaii end to hold the raw sugar, a special fleet of trucks to carry the sugar from mills to the pier, conveyors to move the sugar from the trucks to the top of the bin, and more conveyors to recirculate it within the bins, so the sticky substance would not solidify in place. These outlays had brought vastly lower costs. Sugar had given Matson a feel for what automation could achieve. Shortly after Cushing's departure, the company decided to look into mechanizing the handling of the general cargo it carried between the West Coast and Hawaii.[10]

Matson moved deliberately. Pan-Atlantic, under McLean's control, was a scrappy upstart building a brand-new business, and it risked little by acting quickly. Matson had no such haste; it had a large existing business to protect, and its directors were tight with the purse strings. After commissioning outside studies for two years—the same two years it took Malcom McLean to move from a concept to a functioning business—Matson created an in-house research department in 1956. The man recruited to run it was Foster Weldon, a geophysicist most recently involved in developing the Polaris nuclear submarine.

The contrast with Pan-Atlantic could not have been more stark. McLean's engineers, people like Keith Tantlinger and Robert Campbell, were no intellectual slouches, but they had worked in industry, not academia, and they were well advised not to flaunt their pedigrees in public. Weldon was a professor at the prestigious Johns Hopkins University in Baltimore and a well-known figure in the new science of operations research, the study of efficient ways to manage complex systems. Pan-Atlantic's initial technology had been designed on the fly, using obsolete tanker ships, shipbuilding cranes, and containers whose length was determined by the size of the tankers, on the assumption that it could all be improved once the business was up and running. Weldon found this catch-as-catch-can strategy bewildering. "All transportation companies have their own pet theories on the detailed equipment requirements comprising a 'best' container system, but there are no quantitative data relating even such gross characteristics as container size to the economics of a total transportation operation," he wrote pointedly. His goal, as he defined it, was to develop good data and use them to find the *optimal* way for Matson to embark upon container shipping.[11]

Weldon quickly came upon the issues that would shape Matson's approach. About half of the company's general cargo was suitable for shipment in containers, but the flow was out of balance: for every ton the company shipped from Hawaii to the mainland, it shipped three tons from the mainland to Hawaii. Revenues from the west-bound run would need to cover the cost of returning large numbers of empty containers west to east. Even worse, much of Matson's business came from food processors in California sending small loads to mom-and-pop grocery stores in the islands. Matson would need to consolidate these small shipments to fill whole containers in California, and would then have to open the containers in Honolulu and parcel out the loads for various destinations. This would make container shipping expensive. On the other side of the equation, though, Weldon found that by eliminating the need to transfer individual pieces of cargo from trucks to ships and back again, containers would eliminate almost half the cost of Matson's existing business. "[T]his cost has increased steadily in the past and will continue to do so indefinitely as long as the operation remains a manual one," he concluded. "There is certainly no indication of a change in the current trend of spiraling longshore wages with no corresponding increase in labor productivity." Given the urgent need to automate, Weldon conceived of a way to make the container work: if Matson could load those small shipments into containers in route-sequence order, delivery trucks could collect the containers in Honolulu and proceed immediately on their routes. The goods for each store would be handled only when the truck arrived there, making containerization on the Hawaii run economically viable.[12]

Given that containers made sense, how big should they be? Weldon's analysis pointed out that the smaller the size, the greater the number of loads that would fill entire containers going directly from shipper to recipient, with no reloading. On the other hand, two 10-

foot containers would take twice as long to load on a ship as one 20-foot box, making poor use of the company's investment in cranes and ships. After analyzing thousands of Matson shipments by computer—a task that in 1956 required feeding in thousands of punch cards—Weldon's researchers concluded that vans of 20 to 25 feet would be most efficient in the Hawaii trade: larger containers would travel with too much empty space, while containers shorter than 20 feet would require too much loading time. They recommended that Matson start out by carrying containers on deck, as Pan-Atlantic had, with conventional breakbulk cargo in the holds. By converting six of its fifteen C-3 cargo ships to carry containers on deck, Matson would be able to offer weekly container service between Honolulu and both Los Angeles and San Francisco. Weldon found that this arrangement would be profitable even if the container business stayed small. If the business grew, the company could convert additional ships to carry only containers. Containerization, he concluded, "would appear to present the fortunate circumstance of a promising initial course of action offering the option of going as far as desired and stopping at any point that prudent planning dictates."[13]

Matson management accepted Weldon's recommendations in early 1957. Leslie Harlander, a newly minted naval architect, was put in charge of the engineering. Harlander was told to hire a staff and begin detailed planning for every aspect of a container operation. He was given clear guidance to be careful about money. Every choice had to be justified based on whether it offered a higher return on investment than the alternatives.[14]

Harlander and his brother Don, an engineer who specialized in cranes, began to lay out their requirements for cranes in July 1957. In October, they went to Houston to observe the first arrival of Pan-Atlantic's newly rebuilt *Gateway City*. The *Gateway City* was a C-2 ship, slightly smaller and slower than the World War II–vintage C-3s in Matson's fleet, and it was equipped with Sea-Land's two novel shipboard cranes. With both cranes working, the *Gateway City*'s turnaround time was no longer than that of the much smaller *Ideal-X*. As the Harlanders saw firsthand, though, the shipboard cranes had shortcomings. Pan-Atlantic's two crane drivers each sat high above the deck facing two colored lights. A green light told one driver that he could move the crane trolley over the side of the ship to deposit a container on the dock, while a red light told the other driver to wait. If both cranes accidentally dangled forty-thousand-pound containers over the side at the same time, the unbalanced weight could capsize the vessel. Matson, with plans to serve only a small number of large ports rather than many small ones, had no need to put up with this risk. The first big decision was an easy one: land-based cranes were the way to go.[15]

These would not be leftover cranes adapted from some other use like the cut-down shipyard cranes Pan-Atlantic had pressed into service in 1956. The original Pan-Atlantic cranes were revolvers, known in a shipping trade as "whirleys." They did well enough at picking up a container from the deck of a ship and swinging it in an arc toward the dock, but their design made it difficult to lower the container precisely atop a trailer chassis, which slowed down the entire operation. Matson's cranes were designed from scratch, with a requirement that they be able to unload an incoming container and load an outgoing box within five minutes—a cycle two minutes shorter than that of Pan-Atlantic's first cranes. The Matson cranes were to have booms that stretched ninety-five feet from the dock, more than enough to span the entire width of the ships in Matson's fleet. The operator would control a trolley to move the lifting beam out over the ship, lower the lifting beam to pick up a container, hoist the container, and then travel toward the dock at speeds up to 410 feet per minute. At high speed, these movements would have left each container swinging from the long hoist cables, far above the deck. Les Harlander designed a special lifting spreader to solve the swing problem, testing its feasibility by building a model with his son's Erector Set over Christmas of 1957.[16]

Weldon's work had concluded by recommending a container 20 to 25 feet long. Harlander had the job of getting a design developed. In late 1957, Matson engaged Trailmobile, a manufacturer of truck trailers, to build two prototype containers and two chassis. Another contractor constructed two lifting spreaders and a steel frame that would simulate a container cell within a ship. Months of testing followed. Gauges to measure strain were attached to the equipment, and the stresses were established as containers of various weights and densities were lowered into the cell, lifted out again, and placed on the chassis. The test cell was set at various angles to determine just how much clearance was needed between the containers and the vertical angle bars that formed the corners of the cell. Loaded boxes were stacked to measure the pressures on the bottom container, and lift trucks were run inside the containers to measure the strain on the floors.

When the results were in, Harlander's team decided that the most economical size for Matson was 8½ feet high and 24 feet long, 11 feet shorter than Pan-Atlantic's containers. The specifications took into account Weldon's finding that each pound of weight saved was worth 20 cents, each additional cubic foot inside the container worth $20. To improve structural integrity, the roof would be a single sheet riveted in place rather than several panels attached with sheet metal screws, the design Trailmobile used for highway trailers. Steel corner posts would have to be able to support 120,000 pounds—the weight of several stacked containers, and much more than the posts in Pan-Atlantic's first containers could support. The doors, two layers of aluminum with stiffeners between, were designed to dovetail rather than to meet in a straight line, to withstand twisting pressure due to a ship's rolling in a heavy sea. The floor would be Douglas fir with tongue-and-groove joints. Special attachments to make the containers compatible with specific cranes and forklifts were ruled out on grounds of cost. "It takes very little in the way of extra features to add, say, $200 to the cost of a container," Harlander commented. "There would be a marked change in the total profit picture if the equipment costs were, say, 10 percent higher than they need to be to do the job satisfactorily."[17]

Early in 1958, as McLean was preparing to open Pan-Atlantic's new route to Puerto Rico, the Pacific Coast Engineering Company (PACECO), the lowest of eleven bidders, won the contract to build Matson's first crane. PACECO was not comfortable with the unusual design, and it declared that it would not be responsible for swinging containers, problems with the trolley, or difficulties working as fast as Matson specified. Harlander agreed that Matson would take responsibility for the design, and PACECO began work on an A-shaped monstrosity rising 113 feet from the dock, with legs 34 feet apart so that two trucks or two railcars could pass beneath the crane. Trailmobile built 600 containers and 400 chassis to Matson's specifications. Matson developed a lashing system so that containers could be stacked up to five-high on deck, depending on their weight, without risk of damage at sea.[18]

Meanwhile, Weldon's research department pursued its quest for optimality by investigating the most efficient way to use Matson's fleet. Renting time on an IBM 704 computer at several hundred dollars a minute, the researchers built a fully fledged simulation model of the business, incorporating data on volume and costs for more than three hundred commodities at every port the company served at every time of year. Then they added in data on port labor costs, the current utilization of docks and cranes, and the load aboard each ship, to provide real-time answers to practical questions: Should a big Hawaii-bound ship call at Hilo and Lanai, or should it transfer its cargo to a feeder ship at Honolulu? What time of day should a vessel depart Honolulu so as to minimize total costs of delivering a load of pineapple to Oakland? Such simulations were new in the 1950s and had never been used in the shipping industry.[19]

Matson entered the container era on August 31, 1958, when the *Hawaiian Merchant* sailed from San Francisco with 20 containers on its deck and general cargo in its hold. The *Hawaiian Merchant* and five other C-3s were soon carrying 75 containers at a time, painstakingly loaded by old revolving cranes while the first of Matson's new cranes was being erected in Alameda, on the east side of San Francisco Bay. On January 9, 1959, the world's first purpose-built container crane went into operation, loading one 40,000-pound box every three minutes. At that rate, the Alameda terminal could handle 400 tons per hour, more than 40 times the average productivity of a longshore gang using shipboard winches. Similar cranes were installed in Los Angeles and Honolulu in 1960.[20]

By then, Matson had moved to phase two of the plan that Weldon had laid out at the start of 1957. The *Hawaiian Citizen*, another

C-3 freighter, was modified to carry containers stacked six high and six abreast in its holds as well as on its deck. Four vertical steel angle bars, attached to the ship's structure, were installed to constrain each stack of containers within the holds. At the top of each angle bar, a large steel angle helped guide the containers as the crane lowered them into place. The hatches were expanded so that every stack of containers was accessible to the crane, making the hatch covers so large, 52 feet by 54 feet, that the crane would have to lift them out of the way before starting work on the containers beneath. One of the five holds was outfitted with a cooling system and electrical hookups for refrigerated containers, and lights in the engine room gave warning if the temperature within any of the 72 refrigerated containers was too high or too low. After the hold was loaded and the hatch covers put into place, additional containers could be stacked two-high atop the covers, giving the ship a capacity of 408 25-ton containers. Maintaining stability was a constant problem, especially on heavily loaded runs to Hawaii; when necessary, Matson solved this by organizing the containers before loading so that the heaviest would go at the bottom of each stack, lowering the vessel's center of gravity.

The $3.8 million conversion was completed in six months, and in May 1960 the *Hawaiian Citizen* began sailing a triangular route between Los Angeles, Oakland, and Honolulu. When the vessel arrived in port, the longshoremen first removed the lashings from the deck containers. The crane lifted the deck containers onto chassis pulled by transporters, which took them to the marshaling yard for onward shipment. Once the deck was clear, the crane lifted the hatch covers over one row and unloaded the first cell, occupied by a stack of six containers. The crane then switched to two-way operation. A transporter pulling an outbound container would pull beneath the crane, alongside one with an empty chassis. Every three minutes, the hoist would dive into the ship, lift an arriving container, move it to the waiting chassis, then pick up the outbound container from the other chassis and return to the ship. As it finished each row, the crane would move along the dock to position the boom directly over the next row. Instead of spending half its time in port, like other ships, the *Hawaiian Citizen* was able to spend twelve and a half days of each fifteen-day voyage at sea, making money. Matson's cautious directors were so pleased that they agreed to spend $30 million for containerships by 1964.[21]

By now, everyone in the close-knit maritime industry was talking containers. The talk, however, far outstripped the action. Aside from Matson in the Pacific and Pan-Atlantic, now renamed Sea-Land Service, on the Atlantic coast, very few ship lines were putting containers to routine use. Carriers needed to replace their war-era fleets, but they were afraid to do so at a moment when the shipping industry seemed to be on the cusp of technological change.

It was easy enough to conclude that containers would change the business, but it was not obvious that they would revolutionize it. Containers, said Jerome L. Goldman, a leading naval architect, were "an expedient" that would do little to reduce costs. Many experts considered the container a niche technology, useful along the coast and on routes to U.S. island possessions, but impractical for international trade. The risk of placing multimillion-dollar bets on what might prove to be the wrong technology was high. Sea-Land's shipboard cranes were indeed radically new, but they soon developed a reputation for maintenance problems that caused ships to be delayed. American President Lines, which sailed across the Pacific, created a container that attached to a single pair of wheels so that a truck could pull it without a chassis, but had to abandon the idea once the company added up the cost of giving every container extra structural elements to replace the chassis. The experience of Grace Line offered a graphic warning. Grace had won a $7 million government subsidy to convert two vessels into containerships and spent another $3 million on chassis, forklifts, and 1,500 aluminum containers, only to have longshoremen in Venezuela refuse to handle its highly publicized ships. Having badly misjudged the politics and the economics of container shipping, it would eventually sell the ships to Sea-Land at a loss. As a Grace executive noted ruefully, "The concept was valid, but the timing was wrong."[22]

Sea-Land itself was finding the container business difficult. Its Puerto Rico service was struggling against Bull Line, which controlled half the southbound trade and 90 percent of shipments from Puerto Rico to New York. Bull opened a trailership service in April 1960 and added containerships in May 1961, skimming some of the shippers McLean had hoped to convert to containers. Business on the mainland was not much better. A few food and drug companies, such as Nabisco and Bristol Myers, signed up right away to ship from New York–area factories to Houston, and Houston's chemical plants used containers to send fertilizers and insecticides to the Northeast. Most big industrial companies, though, were not desperate for container shipping. Ideas such as a combined sea-air service, with Sea-Land carrying cargo from New York to New Orleans and an airline taking it onward to Central America, found few takers. The cargo flow through Pan-Atlantic's home terminal at Newark jumped from 228,000 tons in 1957 to 1.1 million tons in 1959, as the Puerto Rico service began—and then abruptly stopped growing. Another longshore strike in 1959 did serious damage. Revenues fell. From 1957 through 1960, Sea-Land's container shipping business lost a total of $8 million. McLean Industries was forced to suspend its dividend.[23]

In desperation, McLean tried in 1959 to buy Seatrain Lines, the only other coastal ship line in the East and an opponent of Waterman's efforts to secure operating subsidies on international routes. Seatrain's management turned him down. Competitors traded rumors that McLean Industries was near bankruptcy. Waterman, unprofitable without subsidies, was put up for sale, minus the cash and many of the ships that had made it so attractive to McLean in 1955.[24]

The problem, McLean decided, was the maritime mind-set: Pan-Atlantic's staff, experienced in the slow-moving ways of the maritime industry, did not know how to sell to an industrial traffic manager who cared not about ships, but about getting freight to the customer on schedule at low cost. McLean brought in a team of aggressive young trucking executives to turn the business around. He had agreed not to poach McLean Trucking employees when he gave up the trucking company in 1955. Now, former McLean

Trucking employees, many of them still in their twenties or early thirties, began moving into key positions at Pan-Atlantic, alongside young talent head-hunted from other big truck lines.

"They were just recruiting," one of those hires remembered. "It was like a football draft. You recruit the best quarterback." Many were invited to Newark without being told what job McLean had in mind for them. When they arrived, they were given intelligence and personality tests—a rare practice in the 1950s. McLean wanted people who were smart, aggressive, and entrepreneurial; the wrong test scores meant no job offer. Education did not matter; although Malcom McLean had a box at the Metropolitan Opera, intellectual airs were frowned upon, and new hires were advised to fracture their grammar to fit in with a crowd of truckers. "When we had nothing else to do, we would stand and pitch pennies," remembered naval architect Charles Cushing, an MIT graduate who joined the company in 1960. "They don't teach you to pitch pennies at the Wharton School."[25]

Those who made the grade were given large responsibilities. Bernard Czachowski was hired from McLean Trucking to oversee Pan-Atlantic's vital relations with the independent truck lines that picked up and delivered its freight. Kenneth Younger, from Roadway Freight, came to manage the Puerto Rican business. Paul Richardson, who had entered McLean Trucking's management training program out of college in 1952 and had stayed with the truck line when McLean spun it off, signed on as New England sales manager in 1960 and within eight months was in charge of sales nationwide. Richardson's secret weapon was a simple form with the pompous title "Total Transportation Cost Analysis." The form provided a side-by-side comparison of the costs of shipping a product by truck, rail, and containership, including not just transportation rates, but also local pickup and delivery, warehousing, and insurance costs. Salesman were instructed to add up each column to show the saving containers would bring, and then multiply by the number of loads the company shipped over the course of a year. The bottom line was the total annual saving, a number much more likely to be large, and memorable, than the traditional measure of a few dollars per ton.[26]

The Pan-Atlantic name was dropped in early 1960, and the ship line was rechristened Sea-Land Service to emphasize that it was a new venture on the leading edge of the freight industry. The work was seven days a week, an exciting, demanding environment. Memos were not wanted. Conflict among executives was a given; managers were expected to meet, thrash out their differences, and act. Performance was measured constantly, and rewarded not with cash but with stock in the fast-growing company. Decades later, those early Sea-Land employees remembered the years when they were creating the container shipping industry as the best time of their lives. "It was a hard-charging, fast-charging company. Malcom would give us assignments and we didn't ask questions, we just went out and did 'em," one said. Malcom McLean—universally called Malcom behind his back, but addressed by every single employee as Mr. McLean—presided over it all, constantly checking the numbers, making sure that the cash flowed.[27]

After a stinging $1.5 million loss in 1960, McLean sought to cope with adversity in his usual way: by plunging deeper into debt. In 1961 Sea-Land bought four World War II tankers and lengthened them by inserting large sections, known as midbodies, built in a German shipyard. These "jumboized" vessels could carry 476 containers—twice as many as Sea-Land's existing containerships, eight times as many as the *Ideal-X*. Competitors complained that the German reconstruction made Sea-Land's vessels ineligible to sail domestic routes as "American" ships, but to no avail. The government approved McLean's application to put the ships into service between Newark and California in 1962, making Sea-Land the only intracoastal ship line. The unbalanced trade made the economics of the intracoastal route treacherous: the eastbound service, heavy with canned fruits and vegetables from California's Central Valley, handled ten thousand tons a month, but California-bound ships carried only seven thousand tons and lots of empty containers. Those same economics, though, assured that there would be no serious competition on the intracoastal route. There simply was not enough freight.[28]

Even as Sea-Land expanded to the West Coast, McLean kept a close eye on Puerto Rico. Puerto Rico was an attractive market for U.S. ship lines. The economy was growing by leaps and bounds under the commonwealth government's economic development program, Operation Bootstrap. The program, featuring generous tax incentives, lured hundreds of U.S.-based manufacturers to what in the 1950s had been an impoverished and heavily agricultural island. They would import their raw materials from the U.S. mainland, use cheap Puerto Rican labor for assembly, and ship their products back to the United States. Private fixed investment in Puerto Rico had more than doubled between 1953 and 1958, and the island's economic output was growing 8 to 10 percent per year. This boom meant rapidly rising demand for shipping—and, thanks to complex U.S. laws governing the maritime sector, only U.S. domestic ship lines could handle the trade. Foreign-owned companies and the U.S. companies subsidized to sail international routes were ineligible.[29]

Sea-Land had been sailing to San Juan since 1958, but its service was less than exemplary. It owned no terminal. Incoming containers with freight for multiple customers were unstuffed in old aluminum warehouses near the dock, where the contents often sat for months because there was no system for notifying customers that their freight had arrived. Containers trucked elsewhere on the island tended to disappear, to be converted into shops, storage sheds, or homes. "It was chaotic," recalled an executive in the Puerto Rico operation. Sea-Land's efforts to gain market share in Puerto Rico had made little headway. Bull Insular Line, the dominant carrier, controlled more than half the shipments from the mainland to Puerto Rico and 90 percent of the freight headed north.[30]

In March 1961, McLean Industries made a surprise offer to buy Bull Line. The $10 million bid was an enormous stretch for a company that was at the limit of its resources. McLean Industries' huge loss in 1960 had wiped out all of its retained earnings. Sea-Land had negative net worth of $1.1 million, although McLean's accounting made the company's situation look worse than it really was. Bull Line was heavily indebted as well, having lost money in the two prior years trying to compete with Sea-Land. Its owners were eager to sell out. The attraction for McLean was that the deal would give Sea-Land a near-monopoly in the Puerto Rico trade—which is exactly why federal antitrust authorities opposed it. Bull's directors received government telegrams advising them not to proceed with the sale to McLean, and they quickly found another buyer. McLean was left to seek revenge by trying to block Bull's efforts to acquire two used ships from the navy.[31]

Then came a remarkable stroke of good fortune: the company that purchased Bull Line, a privately owned maritime conglomerate, had expanded its way into financial trouble. It first stopped reconstruction on the two ships it had acquired for Bull Line, and then, in June 1962, stopped sailing altogether. As Bull Line collapsed into bankruptcy, McLean was able to grab the two ships. Overnight, Sea-Land became the dominant carrier to an island that was almost totally dependent upon U.S. shipping. Before new competitors could move in, it quickly consolidated its position, scheduling containerships from Newark to San Juan every two days and adding sailings from the West Coast and Baltimore. Sea-Land spent more than $2 million on two new terminals in San Juan in 1962 and 1963. In a politically deft move, it also opened routes to the Puerto Rican ports of Ponce and Mayaguez. Neither city had much besides canned tuna to ship out in containers, but providing container service earned McLean the goodwill of Teodoro Moscoso, the creator of Operation Bootstrap and a powerful figure in Puerto Rico's economic development.[32]

The expansion of Sea-Land's Puerto Rico service coincided with a remarkable flourishing of the island's economy. In the 1950s, Operation Bootstrap had attracted mainly small, labor-intensive factories to Puerto Rico. Many workers gained regular wage employment for the first time, and the resulting rise in personal income drove a surge in consumer spending. Retail sales rose 91 percent between 1954 and 1963 after adjustment for inflation. A large share of that merchandise came from the mainland, filling the southbound ships in the Puerto Rico trade. As the island's rising wages began to make it less attractive for labor-intensive factories, Operation Bootstrap undertook a concerted drive to bring in large, capital-intensive manufacturers. Manufacturing, only 18 percent of Puerto Rico's economic output in 1955, reached 21 percent in 1960 and 25 percent by 1970, with most of the growth coming in nontraditional sectors such as pharmaceuticals and metal products. Total trade between Puerto Rico and the mainland nearly trebled during the 1960s, and almost all of it went by ship.[33]

Sea-Land benefited from this boom—but it also helped cause it. Puerto Rico's shipping-dependent economy had been a prisoner of high transport costs. Between 1947 and 1957, as U.S. prices overall were rising 31 percent, rates per ton for shipping between the mainland and Puerto Rico increased about 50 percent. Federal regulators approved five general rate increases over that decade, effectively taxing Puerto Rican consumers to cover the inefficiencies of U.S.-flag ship lines. McLean's push into the Puerto Rico trade in 1958 began to shake up this rate structure, which benefited mainly Bull Line. Over the ensuing decade, by Sea-Land's estimates, the cost of shipping consumer goods from New York to San Juan fell 19 percent, and the average rate per ton for freight shipped in full truckloads fell by a third. Lower southbound rates for industrial components and northbound rates for finished products magnified the advantages of opening factories in Puerto Rico, and McLean Industries established a new subsidiary to help manufacturers locate there. By 1967, Sea-Land was carrying 1,800 containers each week between the island and the U.S. mainland, half of them to or from Puerto Rican factories.[34]

Its unassailable position in Puerto Rico provided a firm base for Sea-Land's growth. Sea-Land owned 7,848 containers, 4,876 chassis, and 386 tractors at the end of 1962. By the end of 1965, it had expanded to 13,535 containers and controlled 15 containerships calling at 15 ports, using Puerto Rico as a hub to serve the Virgin Islands. At the center of this expanding empire was a new office building at Port Elizabeth, New Jersey, where the berths at the new Sea-Land terminal, the first purpose-built container terminal anywhere, were visible out the window. The building, like the rest of the Port Elizabeth complex, was built by the Port of New York Authority, without a nickel of Sea-Land's money. "A lot of people thought Malcom was building a big pagoda," recalled Gerald Toomey, who was recruited to Sea-Land in 1962. "He knew what he was doing. You put a pencil to what that building cost and what it's saved the company, it turned out to be a very good deal."[35]

Sea-Land was a large company by 1963, with nearly three thousand employees, and an increasingly difficult one to manage. Computers had arrived in 1962, but only for administrative purposes such as payroll; at Port Elizabeth, Sea-Land kept track of its incoming and outgoing boxes on magnetic boards high on the walls of its octagonal control room, with an employee reaching a long pole to move the corresponding metal piece on the board each time a container was moved in the yard. At the end of each day, photos were taken of the board to provide a permanent record. Containers had a way of vanishing, especially in Puerto Rico, where a lack of warehouse space led many recipients to store goods in the containers they arrived in; headquarters produced an "aging report" listing containers that had not been seen for a week, and local supervisors frantically worked the phones to try to locate missing boxes before a manager called. Loading required teams of vessel planners to pore over sheets listing the weight and destination of each container as they figured out the best way to load each ship. Computers would not begin to take on that job until 1965.[36]

Malcom McLean could no longer be involved in every decision. Yet his basic approach to management remained unchanged. McLean was still a daily presence at headquarters. "It wasn't unusual that when you came to work, he'd say, 'Good morning, how are you doing this morning?' " recalled a long-time Sea-Land accountant. "Malcom was a good salesman. He'd give the impression that he knew you." When a building for consolidating container loads was needed in Baltimore or Jacksonville, McLean would go and choose the site. When refrigerated containers were needed, managers would spend two days debating how many to buy, only to hear McLean say, "I appreciate the exercise, but I've already signed a contract for five hundred." When the chance came to buy Alaska Freight Lines in 1963, McLean hardly bothered to investigate the company's finances, much less such operational issues as access to Anchorage harbor in the winter; McLean was in a hurry, and the chance to break into the Alaska trade quickly was too good to pass up.[37]

Above all, he kept his eye on the money. Teletypes clattered constantly as outlying terminals sent booking information to headquarters. Clerks updated records showing how many days each container had carried revenue traffic, how many tons it had hauled, how many dollars it had grossed. Geographic analysis documented the land transportation patterns of Sea-Land cargo. Monthly financial reports revealed how much revenue Sea-Land received from each commodity shipped from Newark to Texas, reminding all that an eighteen-ton container of liquor was twice as profitable as a four-ton container of toys. Weekly reports documented cash flow. And there was an endless stream of demands for better cost control. Shaving 1.6 cents off the cost of handling 100 pounds in Ponce could save $14,300 a year. One more container lift per gang hour would save $180,000. Limiting long-distance telephone calls to three minutes would save $65,000. "Probably more attention was paid to financial results there than you find in any company around today," remembered Earl Hall, later Sea-Land's chief financial officer. In 1961, its sixth year, Sea-Land's container business had finally moved into the black. So long as McLean was involved in running it, Sea-Land never lost money again.[38]

Chapter 5

# The Battle for New York's Port

**For the Port of New York Authority,** which was providing Pan-Atlantic a home, the arrival of containerization was a godsend. For New York City, it proved to be a disaster. City officials wasted enormous sums in a futile attempt to keep the city at the center of a shipping industry whose changes New York could not possibly accommodate. Despite their costly efforts, the local economy was left devastated as new technology made the nation's largest port obsolete.

In the early 1950s, before container shipping was even a concept, New York handled about one-third of America's seaborne trade in manufactured goods. New York's role was even larger when measured in dollars, because the port had increasingly come to specialize in high-value freight. This success was not easily earned, for the city had some important disadvantages as a port. The city's piers—283 of them at midcentury, with 98 able to handle oceangoing vessels—were strung out along the Manhattan and Brooklyn waterfronts. The main railroad connections, however, were across the harbor and across the Hudson River, in New Jersey. Freight trains arriving from points north, south, and west were sent to the railroads' large yards located inland, where the cars were sorted by destination and moved by switch engine to one of the railroad terminals that lined the New Jersey side of the harbor. Each railroad owned a fleet of lighters, barges pushed by tugboats, to carry its freight cars across the harbor, either to or from its own New York terminal or to the dock being used by an oceangoing ship. Getting tires from Akron to a Europe-bound vessel thus entailed repeated shunting and shifting. It was economically viable only because the Interstate Commerce Commission, the federal regulator, required railroads to charge the same rates to Brooklyn and Manhattan as to New Jersey; in effect, they were forced to throw in the lighter trip across the harbor for free to keep New York competitive with other East Coast ports.[1]

The growth of the trucking industry starting in the 1920s made the inadequacy of New York's piers even more apparent. By midcentury, about half the cargo headed to or from the docks traveled by truck rather than by train. After coming through the Lincoln or Holland Tunnel, truckers had to navigate dockside streets so congested that, in 1952, the city barred all but pier-bound vehicles from Twelfth Avenue, the waterfront street in midtown Manhattan. If they were headed to the Brooklyn docks, truckers coming from the west had to fight their way through Manhattan to cross one of the East River bridges. Trucks normally waited in line an hour or two just to enter a pier to pick up or deliver at a transit shed, a warehouse adjacent to the dock. The transit sheds were usually designed with truck (or, in some cases, rail) loading docks on one side of the structure and access to ships on the opposite side. Outbound cargo would be taken off the truck with a forklift or by hand, stored in the transit shed until the ship arrived, and then handled again to get it onto the dock, with each operation adding yet more expense.[2]

Delivering by truck meant engaging a "public loader," a type of enterprise unique to New York. A public loader was a gang that claimed the sole right to load and unload trucks on a particular pier, backed by the muscle of the International Longshoremen's Association, the dockworkers' union. Shipping interests, mayors, governors, and the Teamsters union, which wanted its members to handle the work, had tried for decades to get rid of public loaders. The men who did the loading were members of a thoroughly corrupt ILA branch, Local 1757, and were ostensibly owners of the "cooperative" for which they worked. In reality, however, the public loaders were secretly controlled by leaders of the ILA, which had joined forces with a trucking organization to create a "Truck Loading Authority" that published "official" rates for loading—5½ cents per 100-pound bag of almonds or marble chips; 6½ cents per 100 pounds of auto parts, tires, or fish guts; 8 cents per 100 pounds of canned beer—with all hours after 5 p.m. paid at time-and-a-half. Other firms that sought to handle unloading encountered vandalism and outright violence. Shippers that tried to circumvent the public loaders' illicit monopoly by using their own workers to unload were liable to find that the ship would sail with their cargo sitting on the pier. Even after the newly established Waterfront Commission banned public loaders in December 1953, thugs continued to control access to the docks.[3]

The port was a vastly important source of jobs in New York City. In 1951, as operations were returning to normal after the war, more than 100,000 New Yorkers were employed in water transportation, trucking, and warehousing, not counting railroad employees and workers in the municipal ferry system. Another 14,000 New Yorkers worked in "transportation services," such as brokerage and freight forwarding, handling the complexities of international trade in an age when each leg of a complicated journey had to be arranged, and paid for, separately. More than one-third of all "transportation services" workers nationally were located in New York. About three-fourths of the nation's wholesale trade in the early 1950s was transacted through New York, even if the goods did not always pass through the city. Across the country, about 1 in 25 private-sector workers (excluding railroad employees) worked in merchant wholesaling in 1951, but the ratio in New York was 1 in 15.[4]

Then there were the factories situated on the waterfront for ease of shipping. Food-processing plants had located along the Hudson River and the Brooklyn waterfront during the first quarter of the twentieth century, and dozens of factories making dyes, paints, pharmaceuticals, and specialty chemicals dotted the shore from Long Island City in Queens to Bay Ridge in Brooklyn. At midcentury,

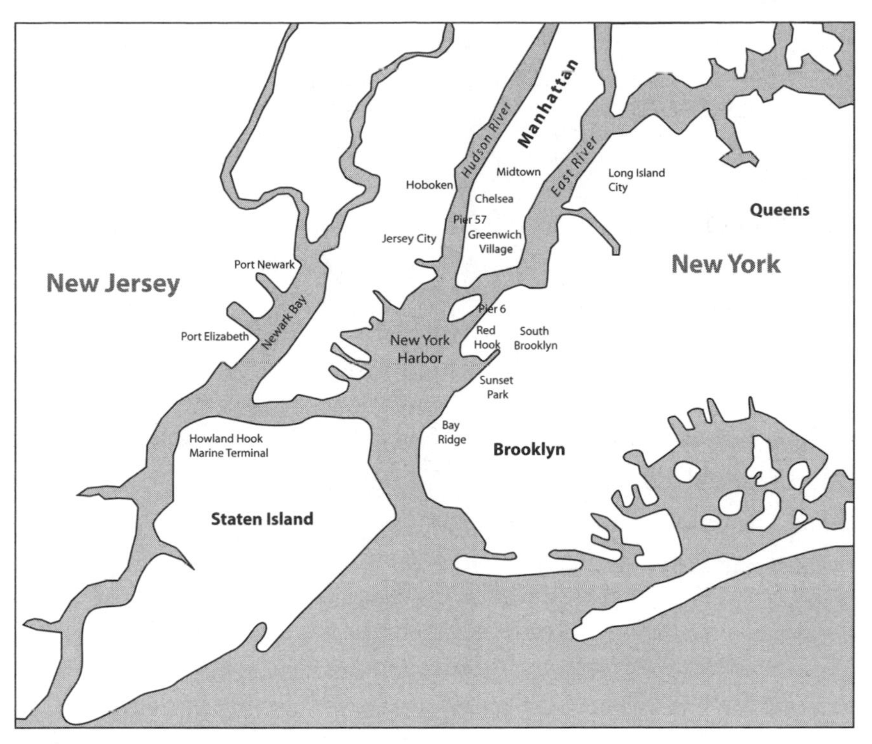


**The port of New York.**

New York's expanding manufacturing sector occupied more than 33,000 chemical workers, 78,000 workers in food processing, and thousands more in shipbuilding and electrical machinery, industries that needed inexpensive freight transportation. In 1956, according to a conservative estimate, 90,000 manufacturing jobs within New York City were "fairly directly" tied to imports arriving through the Port of New York.[5]

Marine construction and ship repair employed thousands more. Add in the lawyers, bankers, and insurance brokers who serviced the shipping business, and the livelihoods of half a million workers may have depended directly on the port. The area near Bowling Green, in lower Manhattan, was thick with shipping company offices, served by the insurers a few blocks away on John Street. Brooklyn,

Table 3
Port-Related Employment in New York City, 1951

| *Industry* | *Number of Workers* | *Number of Firms* |
|---|---|---|
| Merchant wholesaling | 206,315 | 22,135 |
| Water transportation | 67,453 | 637 |
| Trucking and warehousing | 36,164 | 3,494 |
| Chemicals and allied products | 33,472 | 1,129 |
| Transportation services | 13,968 | 1,030 |
| Pulp, paper, and boxes | 12,977 | 294 |
| Primary metal industries | 11,452 | 249 |
| Stone, clay, glass manufacturing | 9,880 | 590 |
| Ship repair | 9,469 | 84 |
| Meat products | 7,345 | 183 |
| Petroleum refining | 1,161 | 7 |
| Grain-mill products | 1,061 | 30 |
| Total | 410,717 | 29,862 |
| Memo: New York City employment | 3,008,364 | |

*Source*: U.S. Census Bureau, *County Business Patterns* (1951)

the most populous borough, had less shipping-related office work but more waterfront employment, with 13 percent of all jobs in the borough located directly on the docks.[6]

This powerful economic engine was already beginning to miss a few strokes in the years after World War II. Its location had helped the Port of New York gain market share during the war, as the refineries and military terminals in Brooklyn and along the New Jersey waterfront dispatched thousands of ships across the North Atlantic. In 1944, when it moved nearly one-third of all U.S. waterborne exports, New York handled twice as much cargo as in 1928 and five times as much as in the worst Depression year, 1933. Even during the war, though, experts were warning of the parlous state of the docks. Those warnings seemed to be confirmed after the war, as cargo traffic slumped owing to the lack of imports from a prostrate Europe. Although European recovery briefly boosted exports, the Korean War put the U.S. economy back on a war footing and devastated foreign trade. The total value of imports and exports at all U.S. ports sank from $18.5 billion in 1951 to $15.6 billion three years later, with exports hit particularly hard as factories switched production from consumer goods to war matériel.[7]

New York was losing the battle for that export traffic. World War II had stimulated economic growth in the West and the South, and factories in Dallas and Los Angeles were much less likely to ship through the Port of New York than were plants in Rochester and Cleveland. The impending opening of the St. Lawrence Seaway in 1956 would permit direct steamship traffic between Great Lakes ports and Europe, with one forecast predicting that it would divert 8 percent of New York's exports and 3 percent of its imports by 1965.[8]

High land freight rates were a further handicap. New York officials were prone to complain that the railroads unfairly favored Philadelphia, Baltimore, or Norfolk, but the truth was that railroads and truckers could serve those points at lower cost; railcars could reach the piers without being floated across the harbor, and truckers faced much less congestion. New York's rate disadvantage was even larger for truck freight than for rail freight, as sending a load by truck from Cleveland to the New York docks could cost four dollars more per ton than sending it to Baltimore. Truckers frequently sought to add the cost of New York port delays to customers' bills, charging sixty to eighty cents per ton more to deliver to the piers than to other Manhattan locations and generating a flood of complaints to the Federal Maritime Board.[9]

Many of the port's other problems, however, were of its own making. After three decades of labor peace from 1915 to 1945, labor turmoil became routine after the war. Some or all of the docks were closed by strikes in 1945, 1947, 1948, 1951, and 1954. Between 1945 and 1955 the International Longshoremen's Association, the legally recognized union throughout the port, battled with the Communist-backed National Maritime Union and with the American Federation of Labor, which ejected the ILA on corruption charges in 1953 and then set up a new American Federation of Longshoremen in an effort to supplant it. With the demise of public loaders, the Teamsters union sought to claim the right to load and unload trucks on the piers, precipitating violent clashes between Teamsters and Longshoremen in 1954. Wildcat strikes on individual piers were common until the ILA, abetted by shipping interests that preferred one corrupt but reliable bargaining partner to constant conflict among competing unions, won a series of elections and regained control late in the decade. Throughout the 1950s, the high risk of labor disruption encouraged shippers to use other ports.[10]

Crime drove shippers away from New York as well. Cargo theft was rampant; most goods were packaged in small boxes or crates, so stealing wristwatches, liquor, or almost anything else was not particularly difficult. The bistate Waterfront Commission, created in 1953 after urgings from New York governor Thomas E. Dewey, made inroads against racketeering by banning public loaders and taking control of pier hiring. It deliberately sought to reduce the workforce and thereby raise longshoremen's incomes in hopes that they would have less need to steal. Even after the Waterfront Commission barred 670 ex-convicts from longshore jobs, though, one in five longshoremen had a criminal record. Cargo theft remained a massive problem—so much so that both the Port Authority and the City of New York refused to cooperate with the filming of a comedy starring James Cagney lest the title, *Never Steal Anything Small*, give moviegoers the wrong impression.[11]

And if land-transport costs, labor concerns, and crime were not enough to deter businesses from shipping through New York, there were the port's decrepit facilities. The East River pier at Roosevelt Street dated to the 1870s, the Hudson pier at West Twenty-sixth Street to 1882. The city-owned pier at Christopher Street had been built in 1876. These piers, and dozens like them, were narrow fingers protruding into the harbor, designed for the days when ships would turn ninety degrees from the channel, point their bows toward the shore, and tie up to the dock for days on end. Some piers were not even wide enough for a large truck to turn around. For the privilege of leasing one of these obsolete facilities, ship lines paid between $0.96 and $2.00 per square foot per year, three to six times the going rate in other East Coast ports. The city had launched a program to renovate and fireproof its piers in 1947, but officials judged the cost of building new piers to be prohibitive. Many piers were literally collapsing into the water. Abandoned pilings and floating debris from fallen piers were obstacles to navigation as well as an eyesore. "By 1980, it will be hard to find space in a whaling museum for piers that met the requirements of 1870 and were condemned as obsolete as long ago as 1920," Port Authority executive director Austin E. Tobin commented in 1954.[12]

Despite its name, the Port of New York Authority was a latecomer to maritime affairs. The major activity of the bistate agency since its founding in 1921 had been building and operating bridges and tunnels; after its early efforts to rationalize the tangle of rail lines and terminals in the New York region were beaten back by the railroads, the Port Authority retreated from involvement in freight transportation.[13] But, as political scientists Wallace S. Sayre and Herbert Kaufman noted in 1960, the independence and broad political support enjoyed by New York's public authorities, including the Port Authority, encouraged them to "seek out new outlets for their energies." In the 1940s, the governors of both New York and New Jersey asked the agency to get involved with shipping, for entirely different reasons. New York governor Dewey thought that the Port Authority might be able to push organized crime off the docks. New Jersey governor Walter Edge wanted it to develop piers on the New Jersey side of the harbor. Tobin and Port Authority chairman Howard Cullman jumped at the opportunity, calculating that taking on some port projects could build support for the Port Authority's expansion into the business they most wanted it to enter: airports.[14]

In 1947, the New York World Trade Corporation, a new state authority backed by key business leaders, proposed to take over all of the city's docks and later to acquire all private docks and waterfront warehouses as well. New York mayor William O'Dwyer rejected the plan and asked the Port Authority to look at the city's piers. After a three-month study, the Port Authority offered to sell $114 million of revenue bonds and build thirteen new steamship berths, four railroad carfloat terminals, and a 1.5-million-square-foot produce terminal, while paying the city $5 million per year in rent. This would have been no small undertaking: the amount involved, the equivalent of almost $900 million in 2004 consumer prices, was more than the city had spent on its docks over decades. The proposal quickly encountered heavy fire. The ILA was opposed. So was the city's Department of Marine and Aviation, which ran the docks; it had waged a bitter and unsuccessful battle to keep the Port Authority from taking over the city's two main airports in 1947, and it did not want to give up another of its functions. Most of all, city politicians did not want the Port Authority on their turf. City officialdom was convinced that the piers were a potential gold mine, not a badly outdated piece of infrastructure. As Robert F. Wagner, then Manhattan borough president and a member of the city's governing Board of Estimate, asked later, "The piers were making money; why didn't they take over the sanitation department instead?" The Board of Estimate rejected the Port Authority's offer in 1948 and turned down a revised proposal in 1949.[15]

While New York officials thought they could modernize the city's piers without the Port Authority's involvement, the financially troubled city of Newark, New Jersey, had no such illusions. Its money-losing municipal docks were in a state of physical collapse. Newark agreed to lease its docks (and its airport) to the Port Authority late in 1947. Between 1948 and 1952 the agency spent $11 million to dredge channels and rebuild wharves. It then announced construction of the biggest terminal yet on the New Jersey side, designed for the Waterman Steamship Company, which would be moving across the harbor from Brooklyn. The Waterman terminal would have a fifteen-hundred-foot wharf running parallel to the shore for faster docking and easier loading—a feature no New York City pier could match. Watching the construction in Newark and the defection of a major steamship operator, New York's city controller suggested that perhaps the city should give up its docks after all. "For some time the Port Authority dock control plans have looked good to us," editorialized the *New York World-Telegram*. "Continued rejection can mean only that the city wants to hang on to waterfront control for political purposes." A Port Authority spokesman declared that the agency was not inclined to begin negotiations with New York City again.[16]

Late in 1953, as the Waterman terminal neared completion, the Port Authority first heard of McLean Trucking's desire to build a terminal on New York Harbor. A trucking company was an odd candidate to lease prime waterfront land, and its plan to drive trucks aboard ships was even odder. The timing, however, could not have been better. Port Authority officials were eager to draw additional business to build upon Port Newark's success, and the agency was uniquely positioned to serve McLean Trucking's needs. On the Newark waterfront it could offer space to marshal trucks, nearby rail lines, and easy connections to the newly built New Jersey Turnpike. Thanks to its ability to issue revenue bonds, the Port Authority had the means to finance any new facilities that would be required. All of these were advantages that New York City could not match. Malcom McLean and A. Lyle King, the agency's director of marine terminals, quickly struck a deal.[17]

The Port Authority proceeded to exercise its new waterfront muscle. After signing McLean, it proposed to build a terminal for rubber importers at Port Newark—a terminal whose prospective tenants would relocate from cramped quarters in Brooklyn. In mid-1955, it finally got a toehold on the New York side of the harbor by purchasing two privately owned miles of Brooklyn waterfront, wharves that it had declined to acquire twice previously but now found politically opportune to buy. Proclaiming its interest in Brooklyn provided cover for another investment in New Jersey: a $9.3 million, four-berth terminal at Newark for Norton Lilly & Co. in November 1955, which led to that ship line's move from Brooklyn to the New Jersey side of the harbor.[18]

Then came the most aggressive move of all. On December 2, 1955, New Jersey governor Robert Meyner announced that the Port Authority would develop a 450-acre tract of privately owned tidal marsh just south of Port Newark. The new Port Elizabeth, the largest port project ever undertaken in the United States, was planned eventually to accommodate twenty-five oceangoing vessels at once, enabling New Jersey to handle more than one-fourth of all general cargo in the Port of New York. Previously, the Port Authority had shown little interest in Elizabeth's marshlands. McLean's idea of putting truck trailers on ships changed that view entirely. Now, port planners foresaw a resurgence of coastal shipping, and the new Port Elizabeth would have ample wharf and upland available for "the proposed use of large shipping containers on specially adapted vessels." There might not even be a transit shed, the most expensive part of pier construction. The first containership had yet to set sail, but the Port Authority was making clear that the future of container shipping would be in New Jersey, not in New York.[19]

The frenzy of activity on the New Jersey side of the harbor caused alarm in New York City. In the past, the New Jersey docks had been notable for their lack of activity; the modest traffic through Port Newark, mainly lumber, accounted for only a couple of percent of the port's nonoil cargo through the 1940s. As ship operators relocated from New York, however, its share would surely grow. With the amount of general cargo flat, every ton handled in New Jersey meant one ton less handled in New York, draining jobs from the city.[20]

This simple calculus was a problem for New York politicians. Robert F. Wagner, familiar with the docks from years as Manhattan borough president, had been elected mayor in 1953 after assembling an unusually broad coalition of labor unions and ethnic groups. The one major bloc he failed to capture was the Italians, who voted overwhelmingly for incumbent mayor Vincent Impellitteri. Gaining support from the group that supplied most of New York's dockworkers may have been part of Wagner's motivation in boosting Department of Marine and Aviation outlays to $13.2 million, more than double the previous level, in his first capital budget, announced in late 1954. Verbal weapons were soon unsheathed. In the summer of 1955, city marine and aviation commissioner Vincent O'Connor charged the Port Authority with trying to "sabotage" city efforts, in the face of "a growing City determination to meet the challenge of its waterfront without yielding its precious waterfront properties to Port Authority control." O'Connor, a lawyer, was close to the ILA, and he shared its concern about the loss of jobs. That September, Mayor Wagner made pier reconstruction one of his top four capital-spending priorities, along with education, transit, and pollution control.[21]

Concern about the docks reached to Albany as well. New York governor Averell Harriman was sensitive to city objections that the Port Authority was promoting New Jersey at New York's expense, but he also knew that the city lacked the money to rebuild its piers. A week after the plans for Port Elizabeth were announced, Jonathan Bingham, a top Harriman aide (and former campaign speechwriter for Wagner) called Matthias Lukens, Tobin's deputy, and Howard Cullman, the agency's chairman, to report that the governor was "disturbed" about Norton Lilly's move from Brooklyn to New Jersey. "He also expressed the opinion he was not sure that we should be spending such money to take business away from New York City," Lukens reported in a confidential memo for his files. According to Cullman, "[Bingham] said he understood completely that the New York piers were in shocking condition, but he did not think the Governor should come out publicly and say the Port of New York Authority should run them."[22]

The container was not yet reality in 1955, and given Malcom McLean's status as a shipping-industry outsider, his plans had so far drawn little attention. With Mayor Wagner committed to keeping shipping in New York, O'Connor came forth with a six-year plan to build new piers and transit sheds, and the city began to pump large amounts of money into the docks. The 1956 capital budget included $14.8 million for waterfront construction as the initial installment on a port program that was estimated to cost $130 million. The plans were state-of-the-art for the mid-1950s, with piers parallel to the shoreline, separate terminal levels for passengers and freight, and paved patios that allowed trucks to back up to high loading docks on the land side of the transit sheds. There would be five new warehouses to handle rail freight lightered across the harbor and a big new terminal for Cunard's transatlantic passenger liners. The crown jewel, clearly intended as a slap in the Port Authority's face, was a $17 million pier with a cargo and passenger terminal for Holland-America line. After sixty-six years in New Jersey, that company would buck the trend that the Port Authority had unleashed and would move to Manhattan.[23]

After decades of inflation, the raw numbers are inadequate to convey the scope of the city's plans. Mayor Wagner's proposed six-year port reconstruction scheme was to cost $130 million in 1956 dollars—the equivalent of $800 million in 2004 dollars. Across the country, the growing Port of Los Angeles had spent $25 million on construction over the ten-year period from 1945 to 1954; Wagner proposed to spend two-thirds of that amount on the Holland-America terminal alone.

None of these proposals, of course, could do much about the underlying problems of the city's docks. Costs were simply uncompetitive with those at other ports. The fundamental geographic disadvantages remained. The new lighter terminals might make it easier to handle rail freight destined for New York, but rail freight intended for an outbound ship would still have to be lightered across the harbor, off-loaded onto a pier, and then reloaded onto an ocean-going vessel. Trucks headed for the docks would still have to fight traffic in the Holland and Lincoln tunnels and along the waterfront. And, of course, rebuilt wharves would do nothing about the port's labor problems, problems so severe that the reopening of one of the first piers rebuilt by the city was delayed by a dispute over which ILA members would receive priority in hiring. O'Connor told ILA leaders directly in the summer of 1955 that the union's practices were "a stumbling block for the city in its efforts to rent good piers in certain areas."[24]

Wagner's own City Planning Commission was skeptical of O'Connor's port projects and recommended that the city restart negotiations to transfer its docks to the Port Authority; it "felt the Port Authority could assure greater development and utilization of the port with benefits to the City's economy." The mayor was unresponsive. Large-scale building was a hallmark of Wagner's tenure, and he had no intention of ceding waterfront reconstruction to an agency over which he had no control. Wagner was close to organized labor, and the city's labor leaders rightly feared that a Port Authority takeover would mean abandonment of some of the docks. Wagner's lack of an ethnic base in New York politics—"There weren't too many German-Americans who voted in New York," recalled Thomas Russell Jones, an influential black politician of the era—made it essential for him to seek support in black, Irish, and Italian neighborhoods reliant on waterfront jobs. In this he succeeded: in his first reelection campaign, in 1957, Wagner captured about half the Italian vote, a big improvement over 1953. Business backed the port renovation effort as well. The Downtown–Lower Manhattan Association, a new civic group started by David Rockefeller of Chase National Bank, urged that all piers in lower Manhattan, save four on the East River, be retained for commercial shipping. "We support the present program of the Department of Marine and Aviation to continue to seek suitable piers in this area and for their modernization and rental on a self-sustaining basis," the association said in its initial plan, released in 1958.[25]

Port spending took on unprecedented proportions. In September 1957, Mitsui Steamship Company agreed to move to a new $10.6 million city-owned terminal in Brooklyn, and Holland-America signed a twenty-year lease for the new terminal in Manhattan. By 1957, O'Connor was envisioning $200 million of waterfront investment by 1962—the equivalent of $1.4 billion in 2004 dollars. Talk of selling the piers to the Port Authority subsided. For their part, Tobin and King were now convinced that the container was the future, and the Port Authority lost interest in taking over city piers that would never have the acreage or transport connections for containers. Although the Port Authority was proceeding with plans to turn twenty-seven outmoded piers in Brooklyn into twelve modern ones, the agency understood that it was in a race to recover its investment before container shipping made the reconstructed piers obsolete. "We already knew that we were building something [in Brooklyn] that would pay itself back, but it wasn't the future," recalled Guy F. Tozzoli, then the Port Authority's head of port planning. The agency's greater concern was that the city was unleashing a subsidy war that could depress pier rents. Tobin attacked the "utter inadequacy" of the city's lease with Holland-America, contending that it involved a city subsidy of $458,000 a year, creating a "new policy of undercutting established pier rental levels by subsidizing private shippers." O'Connor fired back that the "port octopus" was exerting "all its propaganda efforts to thwart the City in the desire of New York to keep its waterfront under the control of its citizens rather than yield it to a bi-state group which thrives on its lack of direct responsibility to the public."[26]

The City Planning Commission, meanwhile, was promoting the view that the port might not be the city's economic future after all. It wanted new office and residential buildings along the East River in lower Manhattan, and suggested in 1959 that rebuilding derelict piers for shipping was not the best use of precious waterfront land. O'Connor responded by enlisting the support of Robert Moses, the city's powerful parks commissioner and a member of the Planning Commission, and then by attacking the Planning Commission itself. Wrote O'Connor: "The assertion, elaborately made by the Commission, that the potential of the Port of New York must be judged by its recent past, rather than by an affirmative anticipation of its future, is an example of negative, rather than constructive planning. It would appear to be inconsistent with the dynamism of New York."[27]

Left unsaid was that much of the city's investment was already going to waste. In 1955, when O'Connor first proposed building five new terminals to handle cross-harbor lighter traffic, lighters moved 9.5 million short tons of cargo between New Jersey and the New York City docks. By 1960, after the city had spent $10 million on new lighter terminals, one-third of that traffic had vanished, and the trend was inexorably downward. The rebuilt Pier 57 on the Hudson River, custom-designed for Grace Line's combined passenger and freight service, was modern enough, but the rapid expansion of air travel had made it obsolete almost before it opened. New piers alone clearly would not be enough to preserve the pattern of port commerce in New York City. The container, hardly noticed by New York officials, was about to become the final nail in the coffin.[28]

Within six months of its start, Pan-Atlantic's container service was carrying 120 containers a week between Newark and Houston. The Pan-Atlantic terminal in Newark had become a busy transshipment hub where longshoremen consolidated smaller shipments into full containers. In early 1957, after barely nine months of operation, Pan-Atlantic leased six additional acres in Newark, twelve times its original space, to store containers and chassis. After a government-sponsored study found that container shipping cost 39 percent to 74 percent less per ton than conventional shipping, the Propeller Club, the association of top shipping company executives, devoted a full day of its 1958 convention to containers. No one could doubt that conventional shipping would soon be in trouble.[29]

As container traffic surged, so did Port Newark's fortunes. Newark's cargo tonnage doubled between 1956 and 1960 while tonnage on the New York side declined slightly, taking New Jersey's share of total port traffic from 9 percent to 18 percent in just four years. Pan-Atlantic, renamed Sea-Land Service in 1960, accounted for more than one-third of Newark's general cargo and 6 percent of all general cargo in the Port of New York. All of this was achieved in the once moribund domestic trade, which had now shifted almost entirely out of Manhattan.[30]

A stone's throw to the south of Sea-Land's Newark terminal, dredges and bulldozers were beginning to shape Port Elizabeth. After two years of planning and after overcoming protests from wary local officials, the Port Authority had embarked in 1958 on a massive construction project: a 9,000-foot channel, 800 feet wide and 35 feet deep, directly opposite Port Newark; thousands of feet of wharf frontage; rail lines; and roadways up to 100 feet wide. Port Authority planners projected that Elizabeth would handle 2.5 million tons of container traffic each year, four times the level then being handled in Port Newark. The differences from New York's dock reconstruction were plain. In a 1961 speech discussing New York City's port redevelopment, marine and aviation commissioner O'Connor did not utter the word "container," and the piers he was building were meant to serve vessels carrying mixed freight, passengers, and baggage. Port Elizabeth, by contrast, was designed from the start as a port for containers. Fortuitously, building on marshland required the Port Authority to start by dredging a channel, filling the wharf area with dredged spoil, and then allowing the fill to settle. Wharves and roadways did not get under construction until 1961, by which time Malcom McLean's container concepts were even further developed. As eventually built, Port Elizabeth's first berths each had about eighteen acres of paved area alongside, to cut down on the cost of moving containers from storage to ship. The design, the Port Authority's magazine explained, "permits a continuous flow of trailers to shipside in 'assembly line' fashion."[31]

The new Sea-Land terminal at Port Elizabeth, opened in 1962, operated on a scale that was inconceivable in New York City. McLean won government permission to sail from Newark to the West Coast through the Panama Canal, and Sea-Land's traffic soared: the Port of New York handled more domestic general cargo in 1962 than in any year since 1941. Almost all of this cargo moved across the Sea-Land pier in New Jersey. Almost none of it moved through New York City. The leisurely port calls of the early 1950s were becoming a memory. A mixed load of containers and breakbulk freight—the kind of load New York City's new piers were built to handle—was an economic drain, because the cost of extra port time to handle noncontainerized cargo ate up the savings from containerization. With no room to store thousands of containers and chassis and no way to handle the hundreds of trucks and railcars coming to meet every ship, New York City's docks were in no position to compete.

For the port as a whole, containerization still remained a sideshow in 1962. Containers accounted for only 8 percent of the Port of New York's general cargo, entirely in domestic trades. None of the port's international traffic, which remained in Manhattan and Brooklyn, was in containers. Yet the trend was ominous. As Sea-Land expanded in the Caribbean, the island traffic that had once flowed through Bull Line's pier in Brooklyn moved to Sea-Land's complex in Elizabeth. New Jersey's share of the port's general cargo reached 12 percent in 1964.

Despite yet more huge investments by the city, including a $25 million pier to handle high-speed ships on order by United States Lines, prospects for the city's piers grew dimmer by the day. The Department of Marine and Aviation requested another $40 million for pier construction in 1964–65. The ILA, desperate to fend off competing claims to use of the urban shoreline, proposed that new waterfront developments in Manhattan should combine piers with apartments. But the combative O'Connor was gone, and the City Planning Commission was not afraid to take on his successor, Leo Brown, in the Wagner administration's waning days. "We believe it is neither necessary, desirable, nor indeed feasible to 'turn back the clock' and attempt to rebuild two more miles of Manhattan waterfront for cargo piers," the commission warned in 1964. In any case, fundamental problems had not been solved. Shipping executives continued to complain about petty corruption on the docks and about the "chaotic conditions that exist in the transfer of cargo between land and water carriers along the waterfront." New concrete was not enough to make ship lines want to dock in New York.[32]

The Port of New York Authority was expanding without cease as container shipping became an international business. By 1965, half a dozen ship lines announced plans to launch container services to Europe in 1966, and dozens of new ships were on order. There was no longer any question of handling this business in Manhattan, or even in Brooklyn. Only Port Elizabeth had the space to accommodate the surging demand for container facilities.

The Port Authority rushed expansion at Port Elizabeth late in 1965, with five new piers and sixty-five acres of paved storage areas. At the time, no fewer than seven steamship lines expressed interest in moving across the harbor from the outmoded docks in New York City. Just ten months later, the agency moved ahead with yet another expansion, which would enable Port Elizabeth to handle twenty containerships at a time. The container tide was running so strong that the Port Authority no longer needed to pretend that Manhattan and Brooklyn would recover their places in the maritime universe. "[A]s we go through the next ten years in the Port of New York there's no question in our minds but that a lot of the cargo will have to go from the center part of the harbor where the big city buildings are over to the Newark-Elizabeth location," Port Authority maritime director Lyle King told a television audience. "As a matter of fact, they are talking that way now, with the plans for new container ships." When New York officials demanded that the Port Authority build container terminals in Brooklyn and Staten Island in return for permission to erect the World Trade Center, they won only promises that the Port Authority would take a closer look. So far as New York City's opinion-makers were concerned, it had become perfectly acceptable for the Port of New York to be located in New Jersey. "The Port Authority, a bistate body, must view New York harbor as an entity and locate its facilities on the basis of geography and economics, not politics," intoned the *New York Times*.[33]

The numbers tell the tale of New York's port. In 1960, with only Sea-Land allowed to ship containers under the ILA contract, containerized freight accounted for less than 8 percent of the port's general cargo tonnage. More than three-quarters of all general cargo still came into Brooklyn and Manhattan. In 1966, with Port Elizabeth's first phase up and running, nearly one-third of the port's general cargo was crossing the docks in New Jersey, and 13 percent was being shipped in containers. "The Port of New York—America's container capital" became the Port Authority's advertising slogan around the world. Financial interests began to speak openly of other "worthwhile activities" that could be located on Manhattan's waterfront, such as apartment complexes and marinas. Manhattan's docks had fallen so silent that one ILA official accused city marine and aviation commissioner Leo Brown of doing "a fine job as a parking-lot operator."[34]

New York City dockers and politicians fought back by seeking to block the World Trade Center and picketing city hall. "If [the Port Authority] can put money into Elizabeth and Newark, why can't they spend some in New York, to help create some permanent jobs to replace those lost by the moving of the Brooklyn Navy Yard?" asked Robert Price, deputy mayor under John V. Lindsay, in 1966. The problem, he said, was simple unfairness: "New York City handles two thirds of the deep-sea cargo and has gotten only one third of the Port Authority's investment." All the Port Authority could offer in response was the promise that the relatively modern Brooklyn docks would continue to handle breakbulk cargo, although, "[w]ith breakbulk operations diminishing, it is unlikely that new conventional piers will be built in the near or distant future."[35]

The Lindsay administration's public bluster notwithstanding, officials now recognized that the Manhattan docks had no future. In 1966, parks commissioner Thomas Hoving requested permission to convert Pier 42 in Greenwich Village to recreational use; over its protestations, the Department of Marine and Aviation was forced to cede the pier's upper story. By the following year, a dozen carriers had placed their first orders for new vessels meant to carry nothing but containers, both in their holds and on deck. No fewer than sixty-four of these gigantic ships were under construction, and the Port Authority touted a study showing that 75 percent of the Port of New York's general cargo could move in containers. When the ILA's Manhattan locals sought a meeting with Lindsay to demand that the city build new piers to save their jobs, even the new marine and aviation commissioner, Herbert Halberg, advised that "to build marine terminals in Manhattan, in the quantity requested, is not at present good economic planning based on the needs of the marine industry, nor good city planning."[36]

The union made a last-ditch effort to preserve the old port by hiring Vincent O'Connor, Wagner's marine and aviation commissioner, to lobby for pier construction. O'Connor delivered a plan for a combined ship/rail/truck terminal in lower Manhattan with an airplane landing strip on the roof. Another scheme called for a "vertical pier" over the East River, using technology developed for automated parking garages to lift containers from shipboard to storage places high in the sky. Such fantasies were of no use. "With few exceptions, all of the major ocean carriers operating containerships at the Port of New York are berthing at Elizabeth," the Port Authority reported in 1969. When proposals for a new passenger ship terminal reached the front burner in 1970, Lindsay decided to get the city out of the port business at long last. "Dear Austin," he wrote Port Authority chief Tobin in language unthinkable a few years earlier, "After considering the alternatives available to us, I am convinced that the entity best able to construct and operate the terminal is the Port Authority." The passenger terminal would eventually be built in Manhattan—but the agency, soon to be renamed the Port Authority of New York and New Jersey, had no further opposition from city government as it developed a vast new port well away from its geographic roots.[37]

As containers supplanted conventional ships, New Jersey's share of the port's general cargo reached 63 percent in 1970. Two years later, 549,731 containers crossed the New Jersey docks. In New York City, though, only destruction was visible. Tonnage at the Port Authority's Brooklyn docks fell 18 percent between 1965 and 1970. "The container is digging our graves and we cannot live off containers," ILA president Thomas Gleason complained, and he was not far wrong. In 1963–64, Manhattan employers used 1.4 million days of longshore labor. Hirings slid below a million in 1967–68, breached 350,000 in 1970–71, and dropped to 127,041 in 1975–76—a 91 percent decline in longshore employment in twelve years. Total employment at marine cargo-handling businesses in Manhattan, including office work, fell from 19,007 in 1964 to just 7,934 in 1976. The situation in Brooklyn was better thanks to the Port Authority's investments, but not for long. Two years after Manhattan's longshore employment began its protracted decline, Brooklyn's followed, dropping from 2.3 million hirings in 1965–66 to 1.6 million in 1970–71 and to just 930,000 in 1975–76. By the time the Waterfront Commission closed its hiring hall at the Bush Docks in 1971, employment there and at the adjoining Brooklyn Army Terminal had fallen 78 percent in a decade. Brooklyn's once mighty cargo-handling industry was just a shadow of its former self.[38]

On the New Jersey side, meanwhile, growth exceeded all forecasts. Stevedores and ship lines were complaining of a labor shortage. Forty ship lines were operating from Port Newark and Port Elizabeth in 1973. The new port's relentless expansion led to a 30 percent increase in hirings between 1963 and 1970, despite the efficiencies of containerization.

By the middle of the 1970s, the New York docks were mostly a memory. Lighters carried a grand total of 129,000 tons of freight to waiting ships in 1974—less than one-tenth of the load moved in 1970, one-fiftieth as much as in 1960. Some shipping remained in Brooklyn, but Piers 6, 7, and 8, fully rebuilt in the late 1950s and known as "Little Japan" for their tenants, emptied out as five Japanese carriers moved to New Jersey. Bull Line, whose Puerto Rico business was a mainstay of the Brooklyn docks, shrank drastically before closing altogether in 1977. The four-pier complex on the Hudson River north of Fourteenth Street, reconstructed as a state-of-the-art terminal for United States Lines in 1963, stood vacant and unrentable, a monument to the city's costly unwillingness to accept that its time as a port was over. When new tenants finally appeared, years later, the Chelsea Piers reopened for an entirely different use: recreation.[39]

The decline of the docks reverberated through New York City's economy, most strongly in the poorest neighborhoods of Brooklyn. In 1960, there were only 23 census tracts, of the 836 in the borough, in which at least 10 percent of the labor force worked in the trucking and maritime industries. On a map, these tracts form a belt parallel to the waterfront, from Atlantic Avenue in the north to Sunset Park in the south. They had much in common: large numbers of immigrants, mainly Italian; low incomes; and very low education levels. In Tract 67 in South Brooklyn, 57 percent of adults had fewer than eight years of schooling. In Tract 49 in what is now Cobble Hill, 64 percent of adults had not gone beyond the eighth grade. Tract 63 in South Brooklyn was home to 1,071 employed workers—including only four with college degrees. By 1970, transportation-industry employment had fallen sharply throughout this entire district, and the population had plummeted. The depths of the decline could be seen in a housing study conducted a few years later: in Sunset Park and Windsor Terrace, an area adjoining the docks with more than 100,000 residents, not a single privately owned housing unit was completed in 1975.[40]

The revolutionary changes in cargo handling had far more dire implications for off-dock workers in transportation and distribution. Between 1964 and 1976, the number of trucking and warehousing workers rose nationally, but the number in New York fell sharply after 1970. With fewer vessels calling at New York City, fewer trucks were needed to deliver and collect cargo at the piers. Transit warehouses were abandoned or put to far less labor-intensive uses, such as parking. An entirely different pattern of goods distribution took hold. Sealed containers filled with export freight were delivered to Newark and Elizabeth, where they were stacked in the open until the vessel arrived; only small loads to be consolidated in single containers now required sorting in a warehouse. Imported containers were hauled straight from the pier to new single-story warehouses built on large plots in central New Jersey and eastern Pennsylvania. There, businesses could enjoy lower labor costs, while benefiting from the growing network of superhighways giving easy access to the port. Trucking and warehousing employment in these areas tracked national trends much more closely than in New York.

Employment in wholesaling, traditionally one of New York's leading industries, was hurt as well, even as it was growing strongly across the country. If employment change in Manhattan and Brooklyn had mirrored national trends in these sectors from 1964 to 1976, the two boroughs would have added 200,000 jobs, most of them suitable for manual or clerical workers. Instead, New York lost more than 70,000 jobs in these port-related industries, while similar employment nationally rose 32 percent.

The changes in transport costs induced by containerization hit manufacturing, too, eliminating not only factory-floor jobs but also related trucking and distribution work as plants moved out of New York. Factory employment in New York City had begun to fall in the mid-1950s, a decade before the container came into widespread use, yet the city retained a surprisingly robust factory sector into the 1960s. In 1964, New York's five boroughs were home to just over 30,000 manufacturing establishments employing nearly 900,000 workers. Almost two-thirds of the city's manufacturers were located in Manhattan, where the apparel and printing industries dominated.

The factory sector held steady through 1967, then abruptly collapsed. Between 1967 and 1976, New York lost a fourth of its factories and one-third of its manufacturing jobs. The scope of this deindustrialization was shockingly widespread, with forty-five of forty-seven important manufacturing industries experiencing double-digit declines in employment.[41]

How much of the loss of industry can be blamed on the container? There can be no definitive answer, as containerization was only one of many forces affecting manufacturers during the late 1960s and the first half of the 1970s. This period saw the completion of expressways that opened up suburban acreage to industrial development. New York's high electricity costs pushed out some factories. The general shift of population to the South and West accelerated, leaving New York factories poorly situated to serve expanding markets. The economic downturn of the early 1970s contributed to a fall in manufacturing employment nationwide, and New York's outmoded factories, often housed in antiquated buildings with little land on which to expand or rebuild, bore the brunt of this shrinkage.

There can be no doubt, however, that containerization eliminated one of the key reasons for operating a factory in New York City: ease of shipment. A New York City location had long offered transport-cost advantages for factories serving foreign or distant domestic markets, as local plants could get their goods loaded on ships with much less handling than could factories inland. The container turned the economics of location on its head. Now, a company could replace its crowded multistory plant in Brooklyn or Manhattan with a modern, single-story factory in New Jersey or Pennsylvania, could enjoy lower taxes and electricity costs at its new home, and could send a container of goods to Port Elizabeth for a fraction of the cost of a plant in Manhattan or Brooklyn. This is exactly what occurred: while industry fled the city, 83 percent of the manufacturing jobs that left New York between 1961 and 1976 ended up no further away than Pennsylvania, upstate New York, or Connecticut.[42]

In 1962, the Brooklyn waterfront was still lined with piers crowded with ships, vast transit sheds, and large, multistory factory buildings literally a stone's throw from the docks. The shift of shipping to New Jersey through the 1960s, combined with the closing of the Brooklyn Navy Yard in 1966, helped destroy the industrial base of one of the largest manufacturing centers in the country. Long known for having a disproportionately large share of the New York region's manufacturing, Brooklyn was remarkable for its disproportionately small share of manufacturing activity by 1980. Economic conditions were so bad that Booklynites abandoned the borough in droves. The population fell 14 percent between 1971 and 1980. Inflation-adjusted personal income fell for eight consecutive years. Not until 1986 did Brooklyn workers regain the income level they collectively enjoyed in 1972.[43]

The container was not the sole cause of the surprising and painful economic changes of the 1960s and 1970s, but it was an important cause. Container technology developed far more quickly and affected transportation industries far more significantly than even its most ardent proponents had imagined. New York was only the first established shipping center whose economy would be transformed in ways that were unimaginable before the container arrived on the scene.

Chapter 6

# Union Disunion

**The dislike between** Teddy Gleason and Harry Bridges was almost visceral. Gleason, a voluble Irishman born hard by the New York docks, held together the International Longshoremen's Association, the union representing dockworkers from Maine to Texas, with a combination of personal charm, warm humor, endless patience, and more than occasional tolerance of corruption. Bridges was an Australian-born ascetic, a man whose role in the brutal battles that brought the International Longshoremen's and Warehousemen's Union (ILWU) to power in Pacific ports made him a legend among his members. The two men disagreed about almost everything, including how their unions should respond to the threat that automation posed to longshoremen's jobs. For a decade beginning in 1956, they struggled with very similar issues in very different ways. Both leaders understood from the outset that automation could put tens of thousands of jobs at risk and transform shoreside labor—their members' labor—into almost an incidental expense. They ended up finding different ways to win extraordinary benefits for their members—in return for allowing the container to reshape the long-established pattern of life on and around the docks.

It was in New York, where Gleason had run a local ILA union and then become the chief deputy to "Captain" William Bradley, the international president, that automation first arose as a major labor-relations issue. The New York Shipping Association, the group of stevedoring companies and ship lines that negotiated the local contract with the ILA, offered an unusual proposal in 1954. Shippers were starting to send their export cargo to the dock already tied on wooden pallets, intending that the entire pallet be transported as a single unit. Since pallets were easy to move with forklifts on the pier and to stow with forklifts inside the ship, the Shipping Association asked the union to handle them with gangs of only 16 men at each hatch, rather than the normal complement of 21 or 22. The ILA quickly did the math: the companies' proposal could mean the loss of up to 30 jobs on a vessel with five hatches. The union objected, and the Shipping Association retreated.[1]

Pan-Atlantic's venture at Port Newark two years later initially attracted little attention from the union. Port Newark, like all parts of New York harbor, operated under ILA contracts. Gleason knew Malcom McLean—the ILA had organized some McLean Trucking warehouse workers in 1939—and the union agreed to handle Pan-Atlantic's containers when the *Ideal-X* first set sail in 1956. Some union leaders made clear their distaste for the container, but the ILA had a host of more pressing concerns: it was in internal turmoil; it faced yet another attempt to oust it as the sole union on New York's docks; its portwide contract was set to expire on September 30, 1956; and its top bargaining demand, a single contract covering the entire Atlantic and Gulf coasts, was meeting strong management resistance. Members at Port Newark were worried about preserving their system for assigning work, which sought to equalize their earnings. In the grand scheme of things, two small ships carrying a few containers were not a priority at union headquarters on West Fourteenth Street in Manhattan. Besides, as an ILA official later told Congress, Pan-Atlantic's was a new operation that added longshore jobs rather than removing existing jobs. The servicing of Pan-Atlantic's containerships, along with the work of consolidating small shipments into full containers at Pan-Atlantic's terminal, was divided between two ILA locals, one black and one white, on the understanding that most of the twenty-one men in each gang were not needed for loading and unloading and would stay out of the way.[2]

Automation became a serious issue when the ILA negotiated new contracts in the fall of 1956. After observing McLean's container operation, the New York Shipping Association sought freedom for employers to hire only as many longshoremen as they wanted "for any type of operation not practiced at this time." An even more ominous issue arose in New Orleans, where the employers wanted to define longshore work as moving cargo from a point of rest on the wharf to the ship—language that would have shut the ILA out of work loading or emptying containers or moving them within the terminal area. Both proposals were eventually dropped, and after a ten-day strike the union achieved much of its main goal, winning a single contract covering wages and pensions from Maine to Virginia. Gleason, the union's chief negotiator in New York, gave absolutely no ground on the automation-related demands, but the battle lines had been drawn. As a presidential board of inquiry noted dryly after the strike, proposals for smaller gang sizes "were a bone of contention with the union."[3]

By 1958, the ILA had seen off competing unions in New York and was free to turn its full attention to automation. The first returns on container shipping were in, and they were alarming. "A containership can be loaded and unloaded in almost one-sixth of the time required for a conventional cargo ship and with about one-third of the labor," McLean Industries told shareholders after two years of operation. Other ship lines were examining containers, and, unlike Pan-Atlantic, they wanted to move consolidation of small shipments away from the docks to inland sites, where it would be out of the ILA's jurisdiction. The match was lit by Grace Line, a company specializing in the South America trade. In November 1958, Grace docked its new *Santa Rosa* on the Hudson. The vessel was designed such that workers could load containers and mixed cargo by rolling them through doors in the side rather than hoisting them through hatches in the deck. Citing the ease of loading, Grace asked to hire only five or six men for each hatch. ILA Local 791 promptly refused to work the ship. When the company held firm, the ILA announced a boycott of all containers, except Pan-Atlantic's, unless they had been filled by ILA members. Fred Field, head of the council of New York ILA locals, angrily accused the ship lines of "soliciting freight in prepackaged containers."[4]

With tensions mounting, the ILA stopped work on November 18 and convened twenty-one thousand longshoremen at Madison Square Garden to hear about the threat of mechanization. Union leaders demanded that employers "share the benefits" and insisted that they would not accept smaller gang sizes. The issue was critical to the union: if Grace Line had its way, far fewer men would be needed on the docks. Intense negotiations led to a temporary compromise in December: the ILA agreed that companies using containers before November 12, 1958—including Grace—could continue to use them, so long as they hired twenty-one men for each hatch. On December 17, the port's labor arbitrator declared, "The problem of containers is well on the road to an amicable solution by both parties."[5]

It was not. Negotiations over container use resumed in January 1959 but got nowhere. The issue festered until August, the start of bargaining on new contracts for all East Coast and Gulf Coast ports. At the most important talks, covering New York Harbor, the ILA demanded that ship lines "spread the fruits of automation." It offered to eliminate one or two longshoremen from each gang. In return, it sought a six-hour workday and a requirement that every container, whatever its origin, be "stripped and stuffed"—that is, emptied and then reloaded—by ILA members on the pier. Stripping and stuffing, of course, were entirely make-work, and would have eliminated any cost savings from containerization. A few days later, the New York Shipping Association countered with a general concept: employers would protect the jobs of regular longshoremen in return for unlimited freedom to automate.[6]

In a conventional labor-management relationship, a management proposal to guarantee jobs in return for the right to automate would have led to intense negotiations. Negotiating with the ILA, however, meant an endless series of distractions. With almost no warning, the union scheduled a membership vote for September 14—two weeks ahead of contract expiration—on whether the ILA should affiliate with the national AFL-CIO labor federation, six years after the old American Federation of Labor had expelled it for corruption. All other business was suspended while union leaders tried to convince members to vote yes. Only after the referendum's narrow passage did contract negotiations resume, just a few days before the September 30, 1959, expiration date. The talks had a positive tone, and, with details unsettled a few hours before the contract expired, Gleason and New York Shipping Association president Alexander Chopin agreed on a fifteen-day extension. Field protested that the extension violated the ILA's long-standing credo, "no contract, no work," and his Manhattan local promptly went on strike. A few hours later, after separate negotiations covering southern ports failed, longshoremen from North Carolina to Texas walked out. Faced with two uprisings, Gleason canceled the contract extension and endorsed a strike—only to run afoul of powerful Brooklyn ILA boss Anthony Anastasia. Anastasia, a volatile Italian immigrant with no love for Gleason and the other Irishmen who dominated the ILA's leadership, directed his own members to work and accused Gleason of backing the strike only to benefit Field. A court temporarily ended the chaos on October 6 by ordering the ports reopened for at least eighty days.[7]

The employer side was no more united than the union. Every ship line had its own plans for automation, and the only company that truly understood the container business, Pan-Atlantic, was not at the bargaining table. When serious negotiations resumed at the start of November, the Shipping Association rejected the union's proposal to strip and stuff all containers at the pier, but agreed to a container tax to compensate longshoremen hurt by containerization. The details proved sticky. The employers offered severance pay for displaced dockers. The union wanted a guarantee of dockworkers' incomes instead; it dismissed severance pay as impractical because, in an industry where workers were hired by the day, automation was likely to mean less work for everyone rather than total unemployment for some.

The outcome, in December 1959, was a three-year contract stating that New York employers would have the right to automate in return for protecting longshoremen's incomes. Beyond that broad principle, all details were left to arbitration. "What the shippers did was give us a piece of the pie," one ILA leader crowed. "Their savings with containers will be tremendous and they just passed on some of the cash to us." The Shipping Association told a similar story. "The steamship industry and shippers are in a position for the first time to fully test and evaluate the economies that might result from these new developments," Shipping Association president Vincent Barnett wrote his members. Civic boosters, long preoccupied by the decline of New York's docks, touted the pact as the instrument of the port's salvation. The new contract "may give New York a jump on competing ports in developing the use of huge containers in international shipping," the *Herald Tribune* explained. The *New York Times* thought that it "should open the door to a flood of containers."[8]

The flood did not come, because nothing had been agreed but generalities. Three arbitrators, including Gleason, a management representative, and a neutral third member, spent months pondering the details, trying to navigate between the ship lines' concern that a royalty on each container would "become another long-term mortgage on the industry" and the ILA's worry that the carriers would find ways to avoid paying royalties. Finally, in the autumn of 1960, the arbitrators ruled by a two-to-one vote that employers in the Port of New York could use container-handling equipment without restriction—in return for paying $1.00 per ton for every container moving on a containership, $0.70 per ton for each container on ships designed both for containers and mixed freight, and $0.35 per ton for containers being carried on conventional breakbulk ships. As a sop to the union, the arbitrators dictated that when ship lines or stevedore companies stuffed or unstuffed containers at their own terminals, they would have to employ ILA labor.[9]

With the 1960 arbitration award, the Port of New York was officially open to any ship line wishing to carry freight in containers. The reality was otherwise. The arbitrators had ordered that the container royalties be paid into a fund, but they had refused to say anything about how the fund should be spent. In addition, the arbitrators had neglected to define the term "container." Gleason, the union's representative on the arbitration panel, predicted that these omissions would cause further union-management conflict. He was right.

The ILA's Pacific coast counterpart, the International Longshoremen's and Warehousemen's Union, took an entirely different tack in addressing waterfront automation.

The ILWU had a long history of difficult and at times violent relations with employers in the Pacific ports. The union, then the Pacific division of the ILA, gained recognition only after a bloody coastwide strike in 1934, and staged 1,399 legal and illegal stoppages over the next fourteen years in order to assert its rights. The net result of this constant conflict was a large body of rules, both written and unwritten, governing port operations in great detail. One formal rule provided that, once assigned to a job at a particular hatch of a particular ship, a worker would do only that specific job until the ship sailed; if loading was complete at one hatch but not at another, an idle worker from the first hatch could not be shifted to help out at the second. An important "hip pocket" rule, codified nowhere but enforced by the gang foreman as required, provided that a trucker delivering palletized cargo to a pier would have to remove each item from the pallet and place it on the dock. Longshoremen would then replace the items on the pallet for lowering into the hold, where other longshoremen would break down the pallet once more and stow each individual item—all at a cost so high that shippers knew not to send pallets to begin with.[10]

Developing such rules, a top ILWU official admitted later, "took no end of imagination and invention." The union regarded them as indispensable to preserve jobs and maintain uniform costs among competitors. The stevedoring firms with which the ILWU negotiated were willing to accept the rules to avoid the alternative of endless wildcat strikes. Louis Goldblatt, the union's longtime secretary-treasurer, claimed that the stevedores actually liked many of the rules, because the ship lines paid them a premium of 30 percent for each man-hour worked. Perversely, the more man-hours required to discharge and load a ship, the more profit the stevedore could make.[11]

The other reason strict work rules were accepted is that there was little choice. The stevedores' association had attempted to loosen many of the rules in contract negotiations in 1948. Unwisely, it did so by mounting a personal attack on Harry Bridges. The union president, a political radical from his Australian youth, made no secret of his socialist sympathies, and the employers labeled him a Communist and declared that they would not deal with Communists. By so doing, they merely enhanced his reputation on the docks. The ILWU walked out when the contract expired, and the union's leadership was so successful in promoting solidarity that members stayed out through a ninety-five-day strike. Finally, the major ship lines brought the conflict to an end by pushing aside the stevedores' association and their own rabidly anti-Communist counsel and taking charge of negotiations. The union achieved its greatest desire: it was finally able to negotiate face-to-face with the ship operators that ultimately paid for its services, rather than with the financially tenuous middlemen at the stevedoring companies.[12]

The largest of the Pacific ship lines, Matson, was facing a financial squeeze, and it persuaded the others that it was time for "a new look" in labor-management relations. The companies agreed to leave the work rules alone, in return for a contract clause allowing stevedores to use new devices and methods so long as individual workers did not face speedups. Innovation would no longer automatically trigger a strike. If a gang thought it was being asked to perform dangerous or excessive tasks, a union representative and a supervisor would try to work things out while unloading or loading continued; if no settlement could be reached at the job site, the dispute would move quickly to higher levels and, if needed, to binding arbitration. These provisions created a new openness, with the union frequently bending the rules to permit new equipment and smaller gangs so long as workers received a portion of the savings. Faced with cargo volumes one-quarter smaller than before the war, the ILWU, in the words of two California labor experts, accepted that "[r]adical measures were necessary to halt the decline in maritime commerce."[13]

The amount of cargo handled per man-hour through the early 1950s, however, remained dismally low. A congressional investigation of the ports of Los Angeles and Long Beach in 1955 uncovered such informal practices as "four-on, four-off," a custom that had begun as a brief rest break for half of the eight holdmen in each gang and had expanded to the extent that workers often worked for only half of their shifts. The investigation left the ILWU cornered and friendless. It had long been plagued by allegations that it was a Communist front, and the government had sought repeatedly to deport Bridges, notwithstanding his status as a naturalized U.S. citizen. The Congress of Industrial Organizations, the leftist side of the labor movement, had expelled it for alleged Communist ties in 1951, and after the AFL and the CIO merged in 1955, Bridges was fearful that the Teamsters and other AFL-CIO unions would seek to challenge its jurisdiction over the docks. Even its former parent union, the ILA, wanted nothing to do with the ILWU, despite its own isolation from the rest of the labor movement; when Bridges wrote ILA president William Bradley to offer support during the 1956 East Coast dock strike, Bradley fired back that Bridges's support was undesired. Bridges, a sophisticated tactician, was painfully aware of his union's vulnerability to government pressure, and he knew that ending contract abuses and improving productivity were essential to keep the government out of union affairs. "You have got our promise and the employers have got our promise that we will go down there [to the rank and file] and persuade and push and do our best," Bridges told the congressional committee.[14]

The impending launch of Pan-Atlantic's container service in the East, and the ongoing study of container usage by Matson, the largest West Coast ship line, made it clear that shipowners were intent on automating cargo handling. Although many members of his union opposed concessions of any sort, Bridges, protected by his credentials as a militant uncorrupted by his dealings with the bosses, began to argue publicly that the union needed to think ahead. "Those guys who think we can go on holding back mechanization are still back in the thirties, fighting the fight we won way back then," he said.[15]

Its origins in the West's turn-of-the-century labor radicalism, its remarkable victories in the strikes of 1934 and 1948, and the ideology of its leaders gave the rank and file unusual power within the ILWU. A change in the union's position on work rules and automation could not be imposed from the top; it would have to be endorsed by the coastwide caucus of representatives elected by their local unions, and then approved by a vote of the entire longshore membership. The task of selling the need for a new approach fell to Bridges. He first presented the issue to the union's negotiating committee, of which he was a member. In March 1956, as the ILWU caucus debated priorities for the upcoming contract negotiations, the negotiating committee urged that the union accept automation in return for higher wages and shorter working hours:

> [M]uch of our past effort has gone into a somewhat unsuccessful attempt to retard the wheels of industrial mechanization progress. In many cases, these efforts have only resulted in our eventual acceptance of the new device, accompanied by our loss of jurisdiction over the new work involved. . . . We believe that it is possible to encourage mechanization in the industry and at the same time establish and reaffirm our work jurisdiction, along with practical minimal manning scales, so that the ILWU will have all of the work from the railroad tracks outside the piers into the holds of the ships.[16]

This point of view was highly controversial within the union. The 1948 contract had left the ILWU firmly in control of the docks in almost every Pacific port. All longshoremen were either full ILWU "A-men" or else "B-men" who were hired as extras when all the A-men were employed, and hoped to get enough experience to be admitted to the union as A-men themselves. Most A-men belonged to regular gangs, with the same group dispatched together from the hiring hall under their elected gang boss, who was also a union member. The handling of each ship was supervised by a walking boss, an ILWU member as well. The stevedoring companies' superintendents, nominally in charge, understood that it was usually wiser to get along with the union than to insist on strict enforcement of the contract. This cozy arrangement, which gave the longshoremen unusual control over their workplace, seemed to be threatened by the joint "Statement of Principles" put forth by the ILWU and the Pacific Maritime Association at the start of their 1956 negotiations. The key provision read simply: "There shall be no requirement for employment of unnecessary men."[17]

Bridges put the Statement of Principles to a membership vote and received only a tepid endorsement, with 40 percent of ILWU members voting no. He clearly had no mandate to agree to changes in manning requirements. Instead, the union and the Maritime Association signed a contract dealing with normal economic matters, and arranged to address mechanization and work rules in separate talks.

Those talks began early in 1957 but quickly foundered on employers' complaints that union members were ignoring the existing contract. J. Paul St. Sure, head of the Pacific Maritime Association, made clear that the shipowners were unwilling to trade money for elimination of work rules until they were certain that Bridges could make ILWU locals live up to whatever bargain he struck. That hard-nosed stance led to a year of intense debate among centralizers and decentralizers within the ILWU. In a surprising speech to the union's caucus in April 1957, Bridges demanded that locals abide by contract language and improve productivity. Opposition, though, was still too strong to overcome; automation issues were referred to the Coast Labor Committee for further study. The committee, consisting of Bridges, one member from the Northwest, and one from California, reported to another union caucus in Portland in October that shippers were increasingly demanding to pack their own cargo into pallets, vans, or other loads that would be handled as single units on the docks. It estimated that up to 11 percent of longshore work hours could be lost as these practices spread. "There is nothing, except our willingness to handle them, that prevents a very considerable increase in unit loads, made up by the shipper," the committee wrote. It painted a stark choice: "Do we want to stick with our present policy of guerilla resistance or do we want to adopt a more flexible policy in order to buy more specific benefits in return?"[18]

The report opened the way to a remarkable debate among rank-and-file members at the caucus. For the first time, men from up and down the coast had a chance to learn in detail about the changes under way in the shipping industry. "Every longshoreman started talking about what can be done under mechanization and still maintain jobs and income, benefits, pensions, and so forth," recalled a labor journalist who was on the scene. Delegates from Los Angeles and Long Beach, where practices such as needlessly unloading and reloading pallets were most entrenched, opposed any compromise. "Perhaps we have the most to lose of any local on the coast," one Los Angeles delegate complained. Delegates from Bridges's home local in San Francisco led the support for negotiations over automation, arguing that the union should make sure that members shared the benefits of new methods of work rather than trying to stop them. After two days of debate, a voice vote backed Bridges's proposal to begin informal negotiations about automation. On November 19, the union wrote the Pacific Maritime Association offering to discuss new methods and elimination of work rules, with the desire "to preserve the present registered force of longshoremen as the basic work force in the industry, and to share with that force a portion of the net labor cost saving to be effected."[19]

Again, the employers were less than enthusiastic. "Many of them felt that this was a form of bribing the men on the job to do the job they were hired to do in the first place," St. Sure explained. Bridges and St. Sure, who had developed a close working relationship, decided that the automation issues were too complex to resolve before the contract expired in June 1958, so they put their immediate focus on one very fundamental change in the contract. The union had won a six-hour workday in 1934, but an unwritten rule prohibited an employer from calling a halt after six hours if loading or unloading was not finished; although the contract guaranteed a minimum of only four hours' pay when a longshoreman was hired, a "normal" shift was nine hours—six at straight time and three at time-and-a-half. The contract Bridges and St. Sure negotiated in 1958 turned longshoring into regular, full-time work. Longshoremen were guaranteed a full eight hours' pay each day—at straight time. This benefited some men but hurt others, because the loss of overtime hours paid at 150 percent of base wage meant less income for many workers. Only 56 percent of ILA members voted to ratify the contract.[20]

The start of Matson Navigation's West Coast–Hawaii service in 1959 made negotiations about automation urgent. "There were specialized cranes that were built specifically for Matson's operation," a former Los Angeles longshoreman recalled. "Well, after the worker saw that, or read about it in the *Dispatcher* [the ILWU newspaper], it didn't take long to sink in that this was the coming way the cargo was going to be moved." The leadership warned the union caucus in April of "such rapid changes in shipping that within even a few years the industry might take on a completely new appearance." The Pacific Maritime Association, though, downplayed the risk of job loss. "We feel it will be years before the present work force will be affected at all by automation," St. Sure told ILWU negotiators in May 1959. Bridges apparently shared that view. "Harry didn't seem to believe that containerization was going to be that important," said the *Dispatcher*'s former editor.[21]

Against that background, the employers made a concrete offer in 1959: in return for the elimination of work rules, they would guarantee that all A-men who had been on the roster in 1958 would at least equal their 1958 earnings in future years, and that employment would shrink only as longshoremen quit or retired. The union produced a counteroffer in November. In return for each man-hour saved by more efficient methods of cargo handling, it asked the employers to pay one hour's average wage into a compensation fund. Trouble was, no one knew how much money might be involved. St. Sure finally grabbed a number out of thin air and offered the union $1 million in compensation for all work that might be lost owing to automation prior to June 1960. Bridges naturally asked for more, making a counteroffer of $1.5 million, and a temporary deal was struck. In return for $1.5 million and a guarantee that no A-men would be laid off, the union agreed that the employers had the right to change methods of work over the coming months. Negotiations on a permanent arrangement would continue.[22]

Months of serious study and dialogue ensued, involving the ILWU, the Pacific Maritime Association, and a variety of dissident factions within both groups. When formal negotiations reopened on May 17, 1960, St. Sure announced that the employers would not sign another interim agreement on automation; they wanted a complete contract. The union again proposed that employers contribute to a worker compensation fund based on man-hours saved. The ship lines had supported just such a concept in 1959. Now, however, they changed their tune, offering flat annual payments to buy out the old work rules for a fixed price, with no obligation to share future cost savings. Three days later, Bridges accepted the idea in principle. The union threw a figure on the table: $5 million per year over four years, an amount equivalent to about twenty cents each year for each man-hour worked in 1959.[23]

Dozens of bargaining sessions followed before the landmark Mechanization and Modernization Agreement was finally signed on October 18, 1960. On the management side, small coastal carriers, Japanese steamship lines, and stevedoring companies all demanded exemptions from contributing to the guarantee fund, and St. Sure had to threaten resignation to obtain unanimous support from the Pacific Maritime Association's executive committee. The political problems on the union side were even worse. The ILWU local in San Francisco had agreed to terms for handling Matson's new containership, *Hawaiian Citizen*, but when the vessel called at Los Angeles in August 1960, just as the mechanization talks were reaching a critical stage, ILWU Local 13 refused to work the ship. The Maritime Association promptly shut the entire port, and several ship lines threatened to move next door to Long Beach, where a different union local held sway. The Los Angeles Board of Supervisors responded with a proposed ordinance making port employees civil servants with no right to strike, an idea that was anathema to the ILWU. Bridges was forced to crack down hard on Local 13. Port, union, and Maritime Association officials signed an unusual agreement setting out penalties for men who refused to work as directed. St. Sure and Bridges made a joint appearance before the Board of Supervisors promising to install a full-time arbitrator to deal quickly with any labor disputes in the port. The Los Angeles docks reopened within a couple of weeks, but bad feelings lingered between the ILWU's local officials and its international leaders.[24]

Two months later, when the draft Mechanization and Modernization Agreement was presented to the ILWU's October caucus, delegates knew that it meant the end of an era. "It is the intent of this document that the contract, working and dispatching rules shall not be construed so as to require the hiring of unnecessary men," read the key clause. The word "container" did not appear, but the language gave management the right to change working methods for all types of cargo so long as this did not result in unsafe conditions or "onerous" workloads; the union could file a grievance if it believed conditions were onerous. The ILWU retained control of cargo sorting on the dock, but containers and pallets arriving at the dock fully loaded would no longer be emptied and repacked by longshoremen.

In return for near-total flexibility, the employers agreed to pay $5 million per year. Part of the money would support retirement: longshoremen with 25 years of service would receive $7,920, or approximately 70 weeks' base pay, upon retirement at age 65, plus the $100 monthly ILWU pension. Workers aged 62 to 65 would be paid $220 a month until age 65 if they would retire early. The rest of the money guaranteed all A-men average weekly earnings equivalent to 35 hours of work, whether or not their services were needed on the docks. Anyone hired as a longshoreman after the agreement was signed would never be eligible for the guarantee because, as a union spokesman explained, "they will not have given up anything."[25]

The caucus demanded numerous changes before sending the draft for a membership vote. More than one-third of the ILWU's members voted no. Some opponents, such as San Francisco's famed longshoreman-philosopher Eric Hoffer, were outraged on ideological grounds. "This generation has no right to give away, or sell for money, conditions that were handed on to us by a previous generation," Hoffer stormed. Dockers in Los Angeles, still angry that Bridges had interfered in their local labor dispute and upset about the loss of work unstuffing and restuffing containers, rejected it by nearly two to one. The local in Seattle backed Bridges; so did his home local in San Francisco, where the unusually old workforce—nearly two-thirds of San Francisco longshoremen were 45 or older—liked the retirement provisions. Members in those two cities provided most of the votes to approve the contract.[26]

The Mechanization and Modernization Agreement brought surprises all around. The initial result, predictably, was a wave of retirements. With incentives encouraging older longshoremen to leave the workforce, the number over age 65 fell from 831 in 1960 to 321 in 1964, and the number between 60 and 65 dropped by one-fifth. Contrary to expectations on both sides, though, income guarantees for active dockers proved unnecessary. Rather than a labor surplus, the docks experienced a labor shortage thanks to an increasing flow of cargo. Large numbers of B-men were admitted as A-men for the first time in years.[27]

The agreement delivered everything the ship lines had hoped for in terms of productivity. Labor productivity had been flat for nearly two decades prior to 1960. The employers' new ability to change work methods for noncontainerized cargo drove up tonnage per man-hour 41 percent in five years, and overall productivity, adjusted for changes in the mix of cargo, doubled within eight years. Shippers could send their canned goods, bagged rice, flour, and similar products on pallets without having to pay longshoremen to unpack and repack the pallets. Iron and steel were handled by two men on the dock rather than four or six, and six cotton bales were now sent for export prepacked on a single pallet weighing 3,000 pounds—too heavy under the old rules, but permissible under the new. Tonnage per man-hour in sugar rose 74 percent between 1960 and 1963, in lumber 53 percent, in rice 130 percent. In the agreement's third year, West Coast ports used 2.5 million fewer man-hours of labor than the previous contract would have required, a figure equal to 8 percent of all the labor those ports had employed in 1960.[28]

Contrary to the union's expectations, these massive productivity gains came from sweat, not automation. "The evidence suggests that the employers, for the most part, devoted their effort to trying to squeeze more physical labor from the workforce, rather than innovating or undertaking new investment," wrote economist Paul Hartman after a careful analysis of the trends. Sacks grew larger, and sling loads, no longer bound by the former weight limit of 2,100 pounds, increased to as much as 4,000 pounds. The result was much harder physical work for the men in the hold, who had to shove these heavy loads into place. Extremely large sling loads, long prohibited by contract, were soon known on the docks as "Bridges loads."[29]

Bizarrely, the parties now switched sides. The union demanded that the employers mechanize *faster* to eliminate these physical burdens. "We intend to push to make the addition of machines compulsory," Harry Bridges told management negotiators in 1963. "The days of sweating on these jobs should be gone and that is our objective." The ship lines were hesitant to spend the money. The ILWU responded by filing grievances against the lack of machinery on docks and in holds. After one of the strangest arbitration proceedings ever to occur in any industry, the employers were ordered in June 1965 to provide longshoremen with more forklifts and winches.[30]

Through 1966, West Coast shipping interests paid $29 million into funds that provided early retirement, death, and disability benefits, along with wage supplements to displaced longshoremen. This proved to be a hugely profitable investment. In 1965 alone, by one estimate, the ship lines saved $59.4 million owing to the Mechanization and Modernization agreement, nearly twelve times their annual payment. The improved efficiency, which contributed to a surge in carriers' profits, came at a time when the container had barely begun to make itself felt in the Pacific ports. Container shipments accounted for a scant 1.5 percent of all general cargo tonnage at Pacific coast ports in 1960 and less than 5 percent in 1963. When the container finally arrived in force, leading to unimagined productivity increases, it brought yet more surprises. The port of Los Angeles, where longshoremen had been so certain that automation would destroy jobs, was to flourish beyond all expectations, while the port of San Francisco, whose longshoremen had been the strongest proponents of the Mechanization and Modernization Agreement, would wither. As they negotiated over automation in 1960, though, neither management nor labor was able to foretell what the container would do. The law of unanticipated consequences prevailed. As Bridges confessed later, "Frankly speaking, the ILWU was caught off guard, as were many shipping companies."[31]

The Mechanization and Modernization Agreement set the rules for the U.S. Pacific coast and was immediately extended to western Canada. In the East, the politics of the fractious International Longshoremen's Association would not allow such a sweeping settlement of automation issues. The ILA represented longshoremen from Maine to Texas, but it negotiated separately with different groups of employers up and down the coast. There was no coastwide contract such as there was in the West. Nor was there a Harry Bridges, a union leader trusted and respected enough to win member support for an otherwise suspect deal. The ILA's headquarters had minimal power over local union leaders, who could do pretty much as they pleased. "It is the only anarchist union," columnist Murray Kempton wrote aptly in the *New York Post*.[32]

William Bradley, the ILA's president from 1953 through 1963, was a genial man who took the title "captain" from his days as a tugboat operator. He was named to head the union after longtime president Joseph Ryan was forced out on corruption charges in 1953. Having never worked the docks, Bradley won little respect among longshoremen in Brooklyn and Manhattan, much less those in Houston and Savannah. In 1961, dissidents demanding strict contract enforcement and maintenance of twenty-one-man gangs won union posts at Anthony Anastasia's Local 1814 in Brooklyn. In the midst of contract negotiations the following year, the choleric Anastasia tried to withdraw his local from the ILA and bargain on independently. Bradley relied on Teddy Gleason, officially the union's chief organizer and executive vice president, to sort out such internal political complications as well as relations with employers. Gleason, who had once been a pier supervisor—a management job—had the waterfront in his blood; his father and grandfather had both been longshoremen, and he and his twelve siblings had been raised a few blocks from the docks in lower Manhattan. His longshore work, though, had been as a checker counting cargo, not as a holdman physically lifting coffee and cement. The Irishmen among his rank and file were not sure that he was tough enough, and the Italians and blacks, the other main ethnic groups working the docks, were not sure they wanted yet another Irish American leader. This was not an environment in which bargainers could make progress on a subject as emotional as automation.[33]

The union's internal political problems were rooted in unpleasant economic realities. Although the port as a whole was prospering, Manhattan's piers were not. The number of men hired at the five Manhattan hiring halls fell 20 percent from 1957–58 through 1961–62, while hiring in Brooklyn and New Jersey increased. Urban redevelopment projects, such as the proposed World Trade Center, threatened the removal of piers that would never be replaced, and the congestion along the entire Hudson River waterfront made it obviously unsuitable for new container operations. Brooklyn longshoremen, by contrast, saw no immediate threat to their jobs. Container operations had not even begun in many ports, including Philadelphia and Boston, and therefore were not a priority for local leaders. With these very different situations leading powerful local presidents to stake out differing views, the ILA had great difficulty coming up with a united approach to the container.[34]

The arbitrators' temporary compromise of 1960, protecting jobs but allowing the employers unlimited use of containers and machinery in return for payments into a royalty fund, barely mattered. A royalty board was set up to manage the expected flood of money, but a drop in port traffic during the economic slowdown of 1960–61 meant that there was little by way of royalties. Gleason alleged that other ship lines were trying to evade royalty payments by encouraging shippers to pack their cargo on pallets rather than in containers. "This is a clear-cut threat to our existing collective bargaining agreement and to the royalty program," he charged in late 1961. Longshore hours worked in December 1961 were down 4 percent from the previous year and down 20 percent from December 1956, but there was still no compensation being paid out to the men whose work had diminished.[35]

Job security in the face of automation thus became the overwhelming union concern as contract negotiations began in 1962. Job security, though, played differently in different places. New York leader Frank Field demanded that the ILA negotiate for a portwide seniority system: business at his own Local 858's docks, in lower Manhattan, was drying up, but the customary pier-specific seniority meant that displaced Manhattan longshoremen could not easily find work on other piers. The leaders of other locals had no desire to reduce their own members' job security by giving Field's members priority in hiring.[36]

Those internal divisions were all too apparent as the ILA entered negotiations for the 1962 contract. The union opened with demands that all New York longshoremen involved in handling prepacked cargo of any sort receive an extra two dollars per hour, and that all containers be assessed a penalty of two dollars per ton payable to the royalty fund. Anastasia, whose Brooklyn local had seen little impact from containers, publicly criticized his own union's proposals as ridiculous and again threatened to withdraw from the ILA. The Shipping Association simply ignored the union's demands, instead proposing that ship lines be allowed to handle containers with eight-man gangs and other cargo with gangs of sixteen, and that crane operators be removed from the union's jurisdiction. These changes, multiplied by the 560 gangs in the port, would have been cataclysmic for the ILA. Economic consultant Walter Eisenberg warned Gleason that the employers' proposal would lead to a sharp increase in container traffic and would save employers between $108 million and $144 million over the life of a three-year contract. The union thought that this money rightfully belonged to its members, but the employers considered it the unearned fruits of featherbedding, to which workers had no claim. The talks stalled, even with federal mediation, because Gleason lacked the political strength to agree to any contract that would eliminate work. The union, he pledged to his angry members, would "not sell out jobs like Bridges did."[37]

Neither side was prepared for anything like the Mechanization and Modernization Agreement on the West Coast. After high-level political soundings in Washington, the mediators proposed that the

ILA and the New York Shipping Association sign a one-year contract while undertaking a joint study of automation and job security. The Shipping Association agreed reluctantly. The union refused, and a strike closed the entire port at the end of September 1962. President Kennedy ordered the union back to work for an eighty-day "cooling-off period" and named three professors to investigate the dispute. Like the federal mediators a month earlier, the professors suggested a joint labor-management study. The employers refused unless the union agreed not to strike during the year. The ILA wanted no study that might cost jobs in the long run. The mediators' hint that the employers might eventually reduce the workforce by paying workers to retire made Gleason irate. "We don't want to sell jobs," he insisted in late October. "The West Coast sold their men out, but here on the East and Gulf Coast, we don't do that." The professors withdrew. Two days before Christmas 1962, the cooling-off period expired, and the union walked out once again.[38]

Kennedy named three men to try to mediate: Republican senator Wayne Morse, who had formerly worked as a labor mediator; Harvard Business School professor James Healy; and Theodore Kheel, a New York labor lawyer. On January 20, 1963, nearly a month into the strike, they announced a proposal: the union would get a one-year contract with large pay and benefit increases, and the secretary of labor would study the job security issues and make recommendations. The ILA and the Shipping Association would attempt to implement the recommendation, but if they failed, they would select a neutral board to do the job. Superficially, the plan seemed to favor the union; the employers faced a large increase in wage and benefit costs with no assurance of greater productivity. Gleason criticized the proposal, then accepted it, while the Shipping Association futilely objected. The union, seemingly victorious, returned to work.

The appearance of a union victory was misleading. A separate statement by the mediators could be read in no other way than as a warning to the ILA: "We wish . . . to emphasize our strong belief that the capacity of this industry to support wages and benefits to which the employees are entitled cannot continue without serious impairment in the absence of marked improvement in manpower utilization." The implication was that if the union remained unwilling to make a deal on containers, the government stood ready to impose one.[39]

As the Labor Department studied port automation through the rest of 1963, the ILA endured yet another internecine feud. Gleason, officially the executive vice president but clearly the union's most powerful figure, launched a campaign to replace Bradley as president. The hapless Bradley, nominally the boss, accused Gleason of having subjected the union to an "unnecessary strike" over automation. Gleason was not wildly popular, but even his critics acknowledged the need for a strong hand at the helm, and the union convention kicked Bradley into the new post of president emeritus. With Manhattan local leader John Bowers elected to Gleason's old job over the head of the Negro local in Mobile, George Dixon, the ILA remained under New York Irish domination as it faced the container issue head-on.

If Gleason's ascent altered the bargaining environment, so did the continuing decline of New York City's docks. With containers accounting in 1963 for more than 10 percent of the entire port's general cargo for the first time, and with Robert Wagner, the mayor who had made the docks a priority, preparing to retire, resolving the container situation became urgent for the New York contingent within the ILA. Fear that automation would destroy the union was the major issue at a conference of the ILA's southern locals in June 1964. When the Department of Labor released its study of the Port of New York at the start of July, Gleason offered an unexpected response: "The time may well be ripe to institute in this industry a guaranteed annual wage."[40]

The 1964 contract negotiations took on an unusually conciliatory tone. The New York Shipping Association proposed smaller gangs and more flexibility in work assignments, as the Labor Department report urged. In return, it offered increased pensions and early retirement, a guarantee that each man hired would get eight hours' work, severance pay for men permanently displaced, and an annual income guarantee for regular longshoremen. When the ILA rejected any concessions on gang sizes, federal mediators were requested once more. The mediators named by President Johnson in January 1964 urged that employers fund a guaranteed income for permanent longshoremen who showed their availability for work. In return, the mediators proposed that employers be permitted to transfer workers from hatch to hatch, and from one task to another, and that the size of general-cargo gangs be pared to seventeen men by 1967. Gleason was willing to concede smaller gangs, but the proposal to let workers do multiple jobs sparked outrage among the checkers, who feared that their less strenuous record-keeping jobs might vanish. Against his own desires, Gleason was forced to take the union out on strike again in September 1964.[41]

The Johnson administration, increasingly concerned about inflationary labor settlements, ordered the dockers back to work and imposed an eighty-day cooling-off period. This time, the ILA and the New York Shipping Association negotiated without the usual histrionics. In return for a massive wage and benefit increase all along the coast, including three additional paid holidays and a fourth week's vacation, the union agreed to reduce the gang size for all general cargo, including containers, to seventeen men by 1967. Starting in 1966, employers in New York would pay a royalty on every container passing through the port, with the funds to be used to guarantee qualified longshoremen the equivalent of sixteen hundred hours of work each year so long as they checked in at the hiring hall, even if they rarely got hired. This Guaranteed Annual Income would be paid until retirement age, creating a permanent subsidy for displaced dockers. A union flyer summed up the huge changes that the new contract would bring: "This agreement takes the industry from a completely casual workforce to a stable, secure livelihood."[42]

Where the ILA was concerned, though, nothing was ever simple. Just before Christmas, as the cooling-off period expired, wildcat strikes began in Baltimore, Galveston, and New York. Then, in a secret ballot on January 8, 1965, ILA members in New York shocked the union leadership by rejecting the new contract, income guarantees and all. Gleason scheduled a revote, but not before hiring a public relations firm and, in an extraordinary act for the head of a secretive union, going on the radio to explain the contract. The second time around, ILA members in New York voted yes, but the next day members in Baltimore voted no. A separate dispute broke out in Philadelphia, followed by an unrelated ILA walkout in most ports in the South. Not until March 1965 were new contracts establishing a guaranteed annual income in place in New York and Philadelphia. The way for containerization was clear—in two ports. In most other cities along the East and Gulf coasts, containerization had not even been addressed.[43]

The Mechanization and Modernization Agreement on the Pacific coast and the Guaranteed Annual Income in the North Atlantic were among the most unusual, and most controversial, labor arrangements in the history of American business. They were products of a time in which the permanent disappearance of work owing to automation was a matter for thoughtful discussion. The United States government, particularly the Department of Labor, was undertaking serious studies of automation's impact in hopes of better assisting affected workers, and organizations such as the American Foundation on Automation and Employment were holding much noticed conferences. President Kennedy addressed the issue himself in 1962: "I regard it as the major domestic challenge, really, of the 60s, to maintain full employment, at a time when automation, of course, is replacing men."[44]

For organized labor, automation was a front-burner issue. Two-thirds of the labor leaders responding to a survey identified it as unions' most serious concern. Automation is "rapidly becoming a curse to this society," AFL-CIO president George Meany told the labor federation's annual convention in 1963. The substitution of machinery for manpower was threatening to unions, blurring long-established jurisdictional lines and raising bargaining costs by reducing the number of workers in a plant, and displacement could be devastating to workers. Many workers in the 1960s lacked basic reading and mathematical skills, and education levels were low enough to make retraining problematic: half of U.S. factory production workers had no more than a tenth-grade education.[45]

Individual unions and employers tried to grapple with automation in their own ways. New York electricians bargained for a five-hour day in 1963 to spread the available work. The United Auto Workers proposed a flexible workweek, rising to forty-eight hours when the unemployment rate fell below a specified level and falling below forty hours, to save jobs, when unemployment was high; the automakers rejected that plan, but they eventually accepted the union's proposal for a fund to continue workers' incomes in the event of layoffs. The Airline Navigators' Association agreed to surrender jobs at Trans World Airlines in return for up-front cash payments, plus three years of severance pay and health insurance. The United Mine Workers, the International Typographers Union, the International Ladies' Garment Workers, and the American Federation of Musicians all tried to bargain for contracts that protected their members as employers sought to automate.[46]

The longshore agreements were seen as models for addressing these concerns. They by no means resolved all problems. "The union gave up more than it should have," former ILWU secretary-treasurer Louis Goldblatt insisted in 1978. "It did not get all it was fundamentally entitled to, such as recapturing all work on the waterfront." There were many struggles yet to come over union control as container terminals moved away from the docks. On both coasts, the Teamsters union challenged labor contracts that promised off-dock stuffing and unstuffing work to the longshore unions, disputes that the courts eventually settled in the Teamsters' favor. The use of ships that carried entire barges loaded with containers posed a novel set of labor-relations challenges, and union representation of the office workers whose computers increasingly controlled container operations would be a source of dispute for decades.[47]

More problematic for many longshoremen were the social changes stemming from the disappearance of the timeworn pattern of waterfront work. Traditional skills, such as knowing how to stow cargo aboard a breakbulk ship, lost value. Older men, whose seniority had enabled them to climb from the depths of the hold to less demanding work on the deck, found that smaller gang sizes made deckmen's jobs too stressful. Fathers could no longer bring their sons into high-paying, albeit dangerous and demanding, waterfront jobs, because the jobs themselves were vanishing. Longshore families, now receiving stable incomes, were free to move from tough waterfront neighborhoods to comfortable suburbs, dealing a blow to the class solidarity that came from isolation. The days of long-established gangs working together, of casual conditions that let men work when they wanted to work and fish when they wanted to fish, would never be regained, as work once marked by independence and freedom from control became a high-paying but highly structured job. "They're turning this job into a factory job," complained New York longshoreman Peter Bell. Agreed Sidney Roger, editor of the ILWU newspaper in San Francisco, "I've heard so many men say, this is exactly what they said: 'It's no fun any longer working on the waterfront. The fun is gone.' The fun has to do with the men working together, a sense of camaraderie."[48]

Despite these discontents, the longshore unions' tenacious resistance to automation appeared to establish the principle that long-term workers deserved to be treated humanely as businesses embraced innovations that would eliminate their jobs. That principle was ultimately accepted in very few parts of the American economy and was never codified in law. Years of bargaining by two very different union leaders made the longshore industry a rare exception, in which employers that profited from automation were forced to share the benefits with the individuals whose work was automated away.

Chapter 7

# Setting the Standard

**ontainers were the** talk of the transportation world by the late 1950s. Truckers were hauling them, railroads were carrying them, Pan-Atlantic's Sea-Land Service was putting them on ships, the U.S. Army was moving them to Europe. But "container" meant very different things to different people. In Europe, it was usually a wooden crate with steel reinforcements, 4 or 5 feet tall. For the army, it involved mainly "Conex boxes," steel boxes 8½ feet deep and 6 feet 10½ inches high used for military families' household goods. Some containers were designed to be shifted by cranes with hooks, and others had slots beneath the floor so they could be moved by forklifts. The Marine Steel Corporation, a New York manufacturer, advertised no fewer than 30 different models, from a 15-foot-long steel box with doors on the side to a steel-frame container with plywood sides, 4½ feet wide, made to ship "five-and-dime" merchandise to Central America. Of the 58,000 privately owned shipping containers in the United States, according to a 1959 survey, 43,000 were 8 feet square or less at the base, while a mere 15,000, mainly those owned by Sea-Land and Matson, were more than 8 feet long.[1]

This diversity threatened to nip containerization in the bud. If one transportation company's containers would not fit on another's ships or railcars, each company would need a vast fleet of containers exclusively for its own customers. An exporter would have to be cautious about putting its goods into a container, because the loaded box could go only on a single carrier's vessel, even if another line's ship was sailing sooner. A European railroad container could not cross the Atlantic, because U.S. trucks and railroads were not set up to handle European sizes, while the incompatible systems used by various American railroads meant that a container on the New York Central could not readily be transferred to the Missouri Pacific. As containers became more common, each ship line would need its own dock and cranes in every port, no matter how small its business or infrequent its ships' visits, because other companies' equipment would not be able to handle its boxes. So long as containers came in dozens of shapes and sizes, they would do little to reduce the total cost of moving freight.

The United States Maritime Administration decided in 1958 to put an end to this incipient anarchy. Marad, as it was known, was an obscure government agency, but it held enormous power over the maritime industry. Marad and a sister agency, the Federal Maritime Board, dispensed subsidies to build ships, administered laws dictating that government freight should travel in U.S.-flag vessels, gave operating subsidies to U.S. ships on international routes, and enforced the Jones Act, the venerable law dictating that only American-built ships, using American crews and owned by American companies, could carry cargo between U.S. ports. The wide variety among containers increased its financial risk: if a ship line took Marad's money, built a vessel to carry its unique containers, and then ran into financial problems, Marad could end up foreclosing on a ship that no one would want to buy. Marad's desire to set common standards was supported by the navy, which had the right to commandeer subsidized ships in the event of war and worried that a merchant fleet using incompatible container systems would complicate logistics. The situation was urgent: several ship lines were seeking subsidies to build vessels to carry containers, and if standards were not set quickly, each carrier might go off in its own direction. In June 1958, Marad named two committees of experts, one to recommend standards for container sizes and the other to study container construction.

The problems the committees faced were not entirely novel. The railway industry, for example, had gone through a standardization process. The gauge—the distance between the inside faces of a pair of rails—on North American railroads varied between 3 feet and 6 feet during the nineteenth century. Trains on Britain's Great Western Railway, with a gauge of 7 feet, could not travel on lines with the most common British gauge of 4 feet 8.5 inches. In Spain, gauges varied from 3 feet 3.3 inches to 5 feet 6 inches, and the multiplicity of gauges in Australia foreclosed long-distance rail transport well into the twentieth century. In some cases, the gauge had been chosen more or less randomly. In others, builders deliberately sought to prevent their line from interconnecting with others that might compete for traffic. Over time, these differences worked themselves out. The Pennsylvania Railroad took over lines in Ohio and New Jersey after the Civil War and converted them to its own gauge. When Prussia proposed a railway link to the Netherlands in the 1850s, the Dutch narrowed their lines so that trains could run through from Amsterdam to Berlin.[2]

The railway precedent suggested that ship lines might eventually make their container systems compatible without a government dictate. Yet the analogy is misleading. The gauge that became "standard" on railways had no particular technical superiority, and standardization had almost no economic implications; the width of the track did not determine the design of freight cars, nor the capacity of a car, nor the time required to assemble a train. In the shipping world, on the other hand, individual companies had strong reasons to prefer one container system to another. The first carrier with fully containerized ships, Pan-Atlantic, used containers that were 35 feet long, because that was the maximum allowed on the highways leading to its home base in New Jersey. A 35-foot container would have been inefficient for carrying canned pineapple, Matson Navigation's biggest single cargo, because a fully loaded container would have been too heavy for a crane to lift; Matson's careful studies showed that a 24-foot box was best for its particular mix of traffic.

Grace Line, which was planning service to Venezuela, worried about South America's mountain roads and opted for shorter, 17-foot containers. Grace's design included small slots at the bottom for forklifts, but Pan-Atlantic and Matson chose not to pay extra for slots because they did not use forklifts. Each company deemed the fittings it used to lift its containers the best for loading and discharging ships at top speed. Conforming to industry standards, each line felt, would mean using a system that was less than ideal for its own needs.[3]

There were two other important distinctions between standardizing rail gauges and standardizing containers. One was scope: the width of a railroad track affected only railroads, whereas the design of containers affected not just ship lines, but also railroads, truck lines, and even shippers who owned their own equipment. The other difference was timing. Railroads had been around for several decades before incompatible track gauges came to be seen as a major problem. Container shipping was brand-new, and pushing standardization before the industry developed might lock everyone into designs that would later prove undesirable. From an economic perspective, then, there was every reason to doubt the desirability of the standardization process that began in 1958. If government agencies in those days had made it a routine practice to conduct cost-benefit studies, most likely the entire process of container standardization would never have begun.[4]

These concerns were unrepresented when Marad's two expert committees held their first meetings on successive days in November 1958. Neither Pan-Atlantic nor Matson was seeking government construction subsidies, so the only two companies actually operating containerships in 1958 were not invited to join in the process of setting standards for the industry that they were creating.

Controversy arose almost immediately. After much debate, the dimension committee agreed to define a "family" of acceptable container sizes, not just a single size. It voted unanimously that 8 feet should be the standard width, despite the fact that some European railroads could not carry loads wider than 7 feet; the committee would "have to be guided mainly by domestic requirements, with the hope that foreign practice would gradually conform to our standards." Then the committee took up container heights. Some maritime industry representatives favored containers 8 feet tall. Trucking industry officials, who were observers without a vote, argued that 8½-foot-tall boxes would let customers squeeze more cargo into each container and allow room for forklifts to work inside. The committee finally agreed that containers should be no more than 8½ feet high but could be less. Length was a tougher issue still. The diversity of containers in use or on order presented a serious operational problem: while a short container could be stacked atop a longer one, its weight would not rest upon the longer one's load-bearing steel corner posts. To support a shorter container above, the bottom container would require either steel posts along its sides or thick, load-bearing walls. More posts or thicker walls, though, would increase weight and reduce interior space, making the container more costly to use. The length question was deferred.[5]

The other Marad committee, on container construction, defined its most important task as establishing maximum weights for loaded containers. Weight limits were crucial, because they would determine the lifting power required of cranes and the load that the bottom container in a stack might have to bear. The weight of empty containers, however, would not affect cranes, ships, or trucks, and the committee decided not to address it. Various other complicated issues, such as the strength of corner posts, the design of doors, and the standardization of corner fittings for lifting by cranes, were put off.[6]

The two committees appointed by Marad did not have the field to themselves. There was a competitor: the venerable American Standards Association. The association, supported by private industry, was in the business of setting standards, dealing with subjects as diverse as the size of screw threads and the construction of plaster walls. The work was vital but also mind-numbing; the engineers on a typical American Standards Association committee would study technical reports, hear the views and interests of the firms concerned, and eventually recommend standards that individual companies could abide by if they wished. To deal with containers, the association created Materials Handling Sectional Committee 5—MH-5, to all concerned—in July 1958. MH-5, in turn, organized itself into subcommittees, which were instructed to develop specifications that would "permit optimum interchange among carriers and also be compatible with domestic pallet containers and cargo containers, and foreign carriers."[7]

The MH-5 committee's first act was to ask the Marad committees to withdraw from the scene. The maritime industry alone should not be making decisions about standardization, MH-5 officials argued; the process should involve other affected industries, and should include foreign organizations so that the standards might eventually apply globally. The Marad committees refused to wait for a decade-long international process. They carried on over the winter of 1959, debating maximum weights, lifting methods, and the pros and cons of requiring steel posts every eight feet along container walls rather than just at the corners. The MH-5 subcommittees, involving many of the same participants, went to work on the same issues. The MH-5 subcommittee on dimensions quickly reached a consensus that all pairs of lengths in use or about to be used—12 and 24 feet, 17 and 35 feet, 20 and 40 feet—would be considered "standard." The subcommittee rejected only a proposal to endorse 10-foot containers, because members thought them too small to be efficient, and, in any case, none were planned.[8]

The MH-5 process was dominated by trailer manufacturers, truck lines, and railroads. These interests wanted to reach a decision on container sizes quickly, because once standard dimensions were approved, the domestic use of containers was expected to burgeon. The specifics mattered less: within the limits set by state laws, trucks and railroads could accommodate almost any length and weight. The maritime interests that were influential in the Marad committees, in contrast, cared greatly about the specifics. A ship built with cells for 27-foot containers could not easily be redesigned to carry 35-foot containers. Most ships then carrying containers had shipboard cranes built to handle a particular size, and they would have to be converted to handle other sizes. Large containers might prove impossible to fill with the available freight, but smaller ones would increase costs by requiring more lifts at the dock. Some lines had made large investments that could be rendered worthless if their containers were deemed "nonstandard." Maritime executives were especially concerned that Marad would deny financial help and perhaps even government cargoes to "nonstandard" operators. Bull Line, which carried containers 15 feet long and 6 feet 10 inches high on its breakbulk ships to Puerto Rico, begged to be left alone, because it had no desire to interchange containers with other companies. Other lines urged the government to let the market sort things out as the container industry matured. When the Marad committee on dimensions reviewed the MH-5 subcommittee's six proposed "standard" lengths in April 1959, it split. The deciding vote in favor of the MH-5 standards came from Marad itself, which was in a hurry to get standards, any standards, into place.[9]

The Marad committee also changed its mind about height. The previous November it had voted to make 8½ feet the maximum height for containers, but it ruled now for 8 feet. The change stemmed from concern that an 8½-foot-high container would violate highway height limits in some eastern states—a problem that was real for trucks hauling containers on standard trailers, but one that did not affect trucks pulling the specially designed chassis used by Pan-Atlantic and Matson. A lower height limit would benefit eastern truckers at the expense of ship lines: an 8-foot-high container held 6 percent less cargo than an 8½-foot-high container of the same length, and would be less attractive to shippers. On height standards as on length standards, the committee split, with the government once again casting a vote that would determine how private transportation companies would invest. The new standards were promptly tested by Daniel K. Ludwig's American Hawaiian Steamship Company, which wanted to build a ship carrying 30-foot-long containers. The Federal Maritime Board would not approve federal mortgage insurance for a ship fitted for nonstandard containers, so American Hawaiian asked the committee to declare 30-foot containers "standard." The committee rejected the request 3 to 2, with Marad once more casting the deciding vote. Federal aid was not forthcoming, and the ship was never built.[10]

The sister Marad committee, dealing with container construction and fittings, worked more smoothly. Members readily agreed that each container should be able to carry the weight of five fully loaded containers atop it, with the weight to be carried on the corner posts rather than on container walls. All containers should be designed to be lifted by spreader bars or hooks engaging the top corners. Rings on top for lifting by hooks or slots underneath for forklifts would be acceptable, but not mandatory. Those decisions gave engineers the basic criteria to use in designing new containers. The committee also recommended that each ship be designed with various sizes of steel cells so that it could carry multiple sizes of containers. With that, the two Marad committees scheduled no further meetings.[11]

Meanwhile, yet another player entered the standards business. The National Defense Transportation Association, representing companies that handled military cargo, decided that it, too, would study container dimensions. The effort's chief proponent was a brash entrepreneur named Morris Forgash, who had built the United States Freight Company into a $175-million-a-year business over two decades by picking up small lots of cargo from various shippers, consolidating them into truck trailers or containers, and shipping the trailers cross-country by rail. The outspoken Forgash impelled his committee to reach consensus quickly. By late summer of 1959, it had agreed unanimously that "standard" containers would be 20 feet or 40 feet long, 8 feet wide, and 8 feet high. The other lengths approved by the MH-5 and Marad committees, and the 8½-foot-high boxes supported by some truckers and most ship lines, would not be acceptable for military freight—a decision Forgash's committee was able to reach only because no one from the maritime industry was involved. No matter: individual companies' preferences, Forgash asserted, would have to yield to the need for uniformity. "Even if we reach the goal slowly, we must have a goal," he said. "Otherwise, obsolescence will overtake us all if each man is his own engineer."[12]

With the MH-5 subcommittee and the Marad dimensions committee having adopted one set of "standard" sizes, and with the National Defense Transportation Association having approved another, the wheeling and dealing began at the American Standards Association. Under the ASA's normal procedures, the February 1959 subcommittee recommendation to designate six "standard" sizes would have been sent for a mail ballot among all participating organizations. The vote never occurred. Instead, insiders set to work to change the recommendations.

A task force of the dimensions subcommittee convened on September 16, 1959, and its chairman, E. B. Ogden, announced that it was desirable to revisit the question of container length. All but two eastern states now permitted 40-foot trailers, Ogden said, so the length limit that had justified 35-foot boxes no longer existed. In the West, eight states had increased their length limits to permit trucks to pull two trailers of 27 feet each, rather than 24 feet apiece. Ogden, whose Consolidated Freightways was the country's largest truck line, urged the committee to approve 27-foot containers as a regional standard size for the West, to reduce costs for trucking companies.

Then Herbert Hall, the chair of the entire MH-5 process, intervened. Hall was a retired engineer at Aluminum Company of America, which made aluminum sheets used to manufacture containers. He had sparked the entire standardization process with a presentation to an engineering society in 1957. Hall knew little about the economics of using containers, but he was fascinated by the concept of an arithmetic relationship—preferred numbers, he called it—among sizes. He believed that making containers in 10-, 20-, 30-, and 40-foot lengths would create flexibility. A shipper could put freight for a single customer in the most suitable size rather than wasting space inside a full 40-foot container. A truck equipped to handle a 40-foot container could equally well pick up two 20-foot containers (their precise length was 19 feet 10.5 inches, to make it easy to fit two together in a 40-foot space), or one 20-foot container and two 10-footers. Trains and ships would be able to handle combinations of smaller boxes in the same way. Hall's enthusiasm was not shared by railroads and ship lines, because loading a train or ship with four 10-foot containers would cost four times as much as loading a single 40-footer. Hall reminded the task force that a higher body, the ASA's Standards Review Board, would have to approve any proposed standards, and he opined that it would not accept the 12-foot, 17-foot, 24-foot, and 35-foot containers that the MH-5 subcommittee had endorsed. The 10-, 20-, and 40-foot lengths Hall favored were promptly approved, while the other lengths were deleted from the list of "standard" sizes. Those recommendations, along with the proposed 27-foot standard for the West and several standards for container construction, were sent to member organizations for a vote late in 1959.[13]

The standards Hall wanted stood to have huge implications for the transport sector. No ships or containers then in use or in design would fit into the container system of the future. Pan-Atlantic and Matson would face an unwelcome choice. If they agreed to use only 10-foot, 20-foot, and 40-foot containers, they would be forced to write off tens of millions of dollars of investment, much of it undertaken within the previous two years, and to shift to container sizes that they deemed inefficient for their own purposes. If Pan-Atlantic and Matson declined to adopt the standards, they would forfeit eligibility for government ship-construction subsidies, while their competitors would be able to build "standard" containerships partially at government expense. Either way, the latecomers to containerization would gain at the expense of the pioneers. Individual companies did not vote in the MH-5 committee, but companies' interests were so disparate that more than a dozen of the industry organizations that did have voting rights failed to reach internal consensus. The proposed 27-foot regional standard was defeated, but the recommendation for Hall's "modular" lengths met with large numbers of abstentions.[14]

Matters were so confused that Hall decided to organize a revote. This time, the questions about container construction were left off the ballot, which now had only a single question: should the association establish standard nominal dimensions 8 feet wide, 8 feet high, and 10, 20, 30, and 40 feet long? The 30-foot container had not been debated in the various task forces and subcommittees, but Hall added it in order to have "a definite relationship between the capacities of adjacent sizes"; the fact that it appealed to Europeans worried about moving big containers through narrow city streets was an added attraction. Many steamship organizations abstained once again because of internal divisions, and again Marad backed the proposal. No vote count was released, but Hall, as chairman, decided that the 10-foot multiples had won sufficient support. On April 14, 1961, 10-, 20-, 30-, and 40-foot boxes were declared to be the only standard containers. The Federal Maritime Board promptly announced that only containerships designed for those sizes could receive construction subsidies.[15]

The standards wars were by no means over. In fact, they had barely begun. At American urging, the International Standards Organization (ISO), which then had thirty-seven nations as members, agreed to study containers. At the time, only very small containers were being shipped across borders, but bigger ones obviously were on the way. The ISO project was meant to establish worldwide guidelines before firms made large financial commitments. Delegates from eleven countries, and observers from fifteen more, came to New York in September 1961 to start the process. Most were appointed by their governments, with the United States, represented by the American Standards Association, being an exception. The United States, as the convener of the meeting, held the chair.[16]

ISO's practice, wherever possible, was to decide how a product must perform rather than how it should be made. This meant that ISO Technical Committee 104 (TC104) would focus on making containers easily interchangeable, not on the details of construction. TC104 was thus able to avoid prolonged debate between proponents of steel containers, popular in Europe, and advocates of the aluminum containers more common in America. No standard would dictate aluminum or steel. TC104 established three working groups and began what would inevitably be a slow-moving process, with many interests involved. The American Standards Association's MH-5 subcommittees continued work on other domestic standards, with the hope that whatever they agreed would later be accepted by ISO. Many leading U.S. transport engineers were involved simultaneously in both groups.[17]

The wrangling over container sizes, which had consumed three years in the United States, was now repeated at the international level. By 1962, much of Europe was allowing larger vehicles than was America, so the new American standard sizes, 8 feet high, 8 feet wide, and 10, 20, 30, or 40 feet long, faced no technical obstacles. Economic interests were another story. Many continental European railroads owned fleets of much smaller containers, made for 8 or 10 cubic meters of freight rather than the 72.5 cubic meter volume of a 40-foot container. The Europeans wanted their containers recognized as standard. The British, Japanese, and North American delegations were all opposed, because the European containers were slightly wider than 8 feet. A compromise was struck in April 1963. Smaller containers, including the European railroad sizes and American 5-foot and 6 2/3-foot boxes, would be recognized as "Series 2" containers. In 1964, these smaller sizes, along with 10-, 20-, 30-, and 40-foot containers, were formally adopted as ISO standards. Not a single container owned by the two leading containership operators, Sea-Land Service (the former Pan-Atlantic) and Matson, conformed to the new "standard" dimensions.[18]

While one set of ISO subcommittees and task forces was hashing out dimensions, other groups of experts were seeking common ground concerning strength requirements and lifting standards. In both North America and Europe, small containers were often moved with forklifts, and some had eyes on the top through which longshoremen or railroad workers could insert hooks connected to winches. The larger containers introduced in North America had steel fittings at each corner, which were welded to the corner post, to a top or bottom rail running the length of the container, and to cross-members running across the front or back end. The corner fittings were cast with holes, through which the containers could be lifted, locked to a chassis, or connected to one another. These castings were simple to make, costing about five dollars apiece in 1961.[19]

The problem came with the lifting and locking devices that fit into the holes. Pan-Atlantic, the first out of the gate, had applied for a patent on its particular system, which used conical lugs that could slip through the oblong holes of its corner fittings and automatically lock into place; a double-headed device to hold two containers together could be secured with the twist of a handle. Pan-Atlantic threatened to bring suit against anyone infringing on its design, forcing other ship lines and trailer manufacturers to develop their own locks and corner fittings. This meant that, even if container sizes were standardized, Sea-Land's cranes would not be able to lift Grace's containers, and Sea-Land containers could never ride on Matson chassis. Railroads that carried the containers of various ship lines needed complicated systems of chains and locks to secure all of the different containers, because one simple locking system would not work for all. Agreeing on a standard corner fitting thus was crucial to making containers readily interchangeable. The obstacle was that every company had financial reasons to favor its own fitting. Adopting some other design would require it to install new fittings on every container, to buy new lifting and locking devices, and to pay a license fee to the patent holder.

An MH-5 task force had tried, and failed, to come up with a new design compatible with all existing corner fittings in 1961. Inevitably, the question arose: could any of the patented corner fittings serve as the U.S. standard? It could, Hall advised at an MH-5 meeting in December 1961, so long as it was in widespread use and was available to all for a nominal royalty. The task force chairman, Keith Tantlinger, had designed the Sea-Land fitting while working for Malcom McLean in 1955. He was now chief engineer at Fruehauf Trailer Company, and he offered royalty-free use of Fruehauf's newest design, in which a steel lug slipped through the hole in a corner fitting and locked into place with a pin. Strick Trailers, a Fruehauf competitor, objected that the Fruehauf design was not good for coupling containers together, and, besides, it had not been proven in actual use. Strick's own design, however, was mired in a patent dispute and could not be offered as a standard. National Castings Company threatened a lawsuit unless any new standard was compatible with its own system, which used lugs designed to spread apart when they passed through the hole in the corner fitting.

The technical differences between these systems were important, especially for ship lines. Containerships were hugely capital-intensive, and the industry's viability depended upon minimizing port time and maximizing the time that each vessel was under way, earning revenue. The ship lines thus had special concern about "gathering," the tendency of the lugs of the lifting device to position themselves in the holes in the corner fittings. If a fitting was poor at gathering when a crane lowered its spreader to pick up a container, the crane operator often had to raise the spreader and lower it a second time. Matson chief engineer Les Harlander calculated that if gathering difficulties added just one second to the average time required to lift a container, his company would lose four thousand dollars per ship per year. After a full day of debate, the subcommittee voted on the Fruehauf design and split badly. There was no ringing endorsement of a national standard.[20]

More meetings through 1962 failed to break the deadlock. Finally, Fred Muller, an engineer serving as the MH-5 committee's secretary, offered a thought: since the Sea-Land corner fitting was working smoothly with the world's largest fleet of containers, perhaps the company would be willing to release its patent rights. Tantlinger made an appointment with Malcom McLean. McLean had no reason to be fond of the American Standards Association, which only recently had excluded Sea-Land's 35-foot containers from its list of standard sizes. Nonetheless, he understood that common technology would stimulate the growth of containerization. On January 29, 1963, Sea-Land released its patents, so that the MH-5 committee could use them as the basis for a standard corner fitting and twist lock.[21]

Agreement on a single design proved elusive. Various trailer manufacturers were still pushing their own products. Numerous ship lines and railroads had started to buy containers, albeit in small numbers, and they employed a wide variety of lifting systems. Lack of consensus meant that the U.S. delegates did not have an official design to offer when the ISO container committee met in Germany in October 1964. The Americans promoted the Sea-Land fitting as the basis for a potential international standard, with Tantlinger distributing half-size ceramic models to show other delegates what it looked like, but no design was put to a vote.[22]

Back home, the engineers' debate over the stresses and tolerances of corner fittings flared into a bitter commercial dispute. The National Castings corner fitting, an elongated box with two rectangular holes in the long side and a large square opening on the top, had been adopted by more container owners than had any other. One big company, Grace Line, had modern container cranes that operated on the National Castings system. Smaller lines that carried containers along with mixed freight in their breakbulk ships liked the National Castings fitting because the large openings let them use old-fashioned hooks for lifting and lowering. Changing to a different system would be expensive; Grace Line estimated the cost of replacing the corner fittings on its containers and the lifting frames on its cranes to be $750,000. National Castings sought wider support by agreeing to royalty-free use of its designs, although only for containers to be carried on American ships. The company persuaded the Maritime Administration that it should support the National Castings fitting as the international standard rather than a fitting based on the Sea-Land design.[23]

Four of the leading steamship lines, Sea-Land, Matson, Alaska Steamship, and American President Lines, fought back, because adoption of the National Castings fitting would have required them to change all of their containers. Instead, they proposed a minor change to the fitting that the MH-5 committee was designing based on the Sea-Land patent. If the hole on the top of the fitting were moved by half an inch, they estimated, 10,000 containers—about 80 percent of all large containers used by U.S. railroads and ship lines other than Sea-Land—would be "reasonably compatible" with Sea-Land's. The fitting they recommended, they said, would cost less than half as much as the National Castings fitting ($42.24 versus $97.90) and weigh barely half as much (124 pounds versus 236). As the battle grew intense, the politics of standardization suddenly changed. National Castings Company was sold and abandoned efforts to promote its corner fitting. Marad, which had favored National Castings, reversed course and urged ship lines to accept whatever MH-5 agreed upon. Finally, an unusual decision came from the top. The American Standards Association's Standards Review Board ignored the fact that the specialists on its MH-5 committee were still debating the finer details of corner fittings. On September 16, 1965, it approved a modified version of the Sea-Land fitting as the U.S. standard, just in time for the next meeting of the ISO container committee in The Hague.[24]

The sixty-one ISO delegates were offered two competing designs when they convened in the Dutch seat of government on September 19. The United States presented the modified Sea-Land corner fitting as the new U.S. standard, and the National Castings fitting was put forth as the British standard. The British quickly agreed that the American favorite was superior. Only one roadblock remained. ISO rules required that the documents supporting proposed standards had to be distributed four months in advance of a meeting. The MH-5 committee had made its recommendation only a few days earlier, and no technical documents were ready. The ISO committee voted unanimously to waive the four-month rule. Three high-ranking corporate executives—Tantlinger, Harlander, and Eugene Hinden of Strick Trailers—then retreated to a railcar factory in nearby Utrecht, where they worked with Dutch draftsmen for forty-eight hours nonstop to produce the requisite drawings. On September 24, 1965, the ISO delegates approved the American design as the international standard for corner fittings.[25]

The new era of freight transportation finally seemed to have arrived. In principle, land and sea carriers would soon be able to handle one another's containers. Container leasing companies could expand their fleets in the knowledge that many carriers would be prepared to lease their equipment, and shippers could make use of containers without wedding themselves to a single ship line. "Projects awaiting the outcome of the fitting question are already underway," a trade publication trumpeted within a few months of the vote in The Hague. "Container-handling hardware can now be designed with more certainty, and an increasing number of products designed to load and carry containers will be marketed."[26]

The cart, however, had gotten ahead of the horse: the ISO container committee had agreed on what the corner fitting should look like without defining all of the loads and stresses it should be able to withstand. Starting in the autumn of 1965, dozens of ship lines and leasing companies began ordering containers with fittings based on the design that had worked for Sea-Land's operations but had never been tested under other conditions. The ISO committee had yet to set maximum container weights, for example. No one could say how thick the steel in the fitting should be, because it was not clear how much weight it might have to hold. Sea-Land's cranes lifted by connecting to the tops of the fittings in the top corners of a container; it was uncertain how the fittings would perform if a container were lifted from the fittings in the bottom corners. Railroads in Europe had different coupling systems from those in the United States, meaning that the cars in a train banged against one another with greater force, and the Sea-Land fittings and locks had never been subjected to such conditions. And what if five or six containers were stacked on the deck of a ship? In high seas, the stack of containers might tilt as much as 30 or 40 degrees away from vertical. Would the newly approved corner fittings and the twist locks connecting the containers survive such stresses?

Through 1966, engineers around the world tested the new fittings and found a variety of shortcomings. As an extra check, a container was put through emergency tests in Detroit, just ahead of another meeting of the ISO committee. It failed, the fittings on the bottom of the test container giving way under heavy loads. When TC104 convened in London in January 1967, it was faced with the uncomfortable fact that the corner fittings it had approved in 1965 were deficient. Nine engineers were named to an ad hoc panel and told to solve the problems quickly. They agreed on the tests that fittings would have to pass, and then two engineers, one British, one American, were sent to a hotel room with their slide rules and told to redesign the fitting so that it could pass the tests. Requiring thicker steel in the walls of each fitting, they calculated, would solve most of the problems. No existing container complied with their "ad hoc" design. Over the bitter complaints of many ship lines that had encountered no problems with their own containers, ISO approved the "ad hoc" design at a meeting in Moscow in June 1967. The thousands of boxes that had been built since ISO first approved corner fittings in 1965 had to have new fittings welded into place, at a cost that reached into the millions of dollars.[27]

The process of standardization was proceeding nicely. The economic benefit of standardization, however, was still not clear. Containers of 10, 20, 30, and 40 feet had become American and international standards, but the neat arithmetic relationship among the "standard" sizes did not translate into demand from shippers or ship lines. Not a single ship line was using 30-foot containers. Only a handful of 10-foot containers had been purchased, and the main carrier using them soon concluded that it would not buy more. As for 20-foot containers, land carriers hated them. Ship lines "have designed, especially in their 20-foot equipment, a highly efficient port to port container without due consideration of how the box would move efficiently from port to customers," an executive of the New York Central Railroad complained. So far as truck lines were concerned, the bigger the container, the more freight could be transported per hour of driver labor. Trucking companies' preference was revealed by the truck trailers they chose to buy, almost none of which had 20-foot bodies. Hall's notion of coupling two 20-foot containers together on a single trailer proved to be impractical, because if each container was filled to its weight limit, the combined weight would violate highway regulations in every state. Towing two 20-foot containers in tandem was impractical as well, because the same truck could move more weight by pulling two 24-footers or, in many states, two 27-footers.[28]

The most powerful evidence against the international standards came from the marketplace. Despite the U.S. government's pressure on carriers to use "standard" sizes, nonstandard containers continued to dominate. Sea-Land's 35-foot containers and Matson's 24-footers, all a nonstandard 8 feet 6 inches high, accounted for two-thirds of all containers owned by U.S. ship lines in 1965. Only

16 percent of the containers in service complied with the standards for length, and a good number of those were not of standard 8-foot height. Standard containers clearly were not taking the industry by storm. The large ones were too hard to fill—too few companies shipped enough freight between two locations to require an entire 40-foot container—and small ones required too much handling. As Matson executive vice president Norman Scott explained, "In the economics of transportation, there is no magic in mathematical symmetry."[29]

Their business success notwithstanding, Sea-Land and Matson had reason to worry about the drive for standard-size containers. Both companies had raised tens of millions of dollars of private capital to buy equipment and convert their ships to carry containers, and so far neither had sought federal construction subsidies. That situation was now changing. By 1965, both Sea-Land and Matson were preparing to expand internationally, and they might want subsidies to build new ships. In addition, Marad dispensed other types of aid. It gave operating subsidies to U.S. ship lines sailing international routes, to compensate for the requirement that they employ only high-wage American seamen, and it enforced regulations giving U.S.-flag vessels "preference" to carry government cargo overseas. If Marad were to limit those subsidies only to companies adhering to the "voluntary" MH-5 standards, Sea-Land and Matson would be at a serious competitive disadvantage. Executives from the two companies met in Washington and decided to join forces to fight the U.S. government.[30]

They started back at the American Standards Association. The association's MH-5 committee had been quiescent, but in the fall of 1965, with the ISO beginning to adopt international standards for containers, the MH-5 committee named a new subcommittee to look at "demountable containers"—the sort that could be moved among ships, trains, and trucks. The chairman was Matson chief engineer Harlander, and now, in contrast to 1961, Sea-Land officials were prominent participants. At the first meeting, at the Flying Carpet Motel in Pittsburgh, Harlander surrendered the chair and made an appeal for Matson's 24-foot container size to be accepted as standard. He was followed by Sea-Land's chief engineer, Ron Katims, who called for the subcommittee to recognize 35-foot containers as well. Sea-Land's containers, the subcommittee was told, tended to hit weight limits long before they were filled to physical capacity, so 40-foot containers would not in practice hold more freight than 35-footers. With the longer size, however, Sea-Land would not be able to fit as many containers on each ship, forfeiting almost 1,800 tons of freight capacity per vessel. Harlander then called for the subcommittee to endorse 8½-foot-high containers as well. Marad's representative asked that all three questions be tabled.[31]

When discussions resumed in early 1966, the subcommittee agreed to increase the "standard" height for containers to 8½ feet, but it split on whether to recommend a change in policy to make 24-foot and 35-foot containers "standard." It bounced the entire issue up to the full MH-5 committee. The MH-5 committee itself then split. The dogged Hall, still pushing the standardization process along despite failing health, remained convinced that all approved sizes should be mathematically related. The various maritime associations on the committee, most of whose members had by then adopted 20-foot or 40-foot containers, had little incentive to cast a vote that might force them to share government subsidies with Sea-Land and Matson. Five trucking associations, whose members picked up and delivered containers for Sea-Land and Matson, submitted votes by telegraph in favor of the two additional sizes, but their votes were disallowed. Almost all of the government representatives in attendance abstained. With 15 no votes, 5 yes votes, and 54 voters abstaining or absent, the MH-5 committee had no consensus for anything. A revote the following year found the split persisting, with 24 participating organizations favoring 24-foot containers and 28 against them.[32]

Faced with the prospect of competing against subsidized competitors while being excluded from subsidies themselves, Sea-Land and Matson turned to Congress. Their lobbyists drafted legislation in 1967 to prohibit the government from using the sizes of containers or shipboard container cells as a basis for awarding subsidies or freight. Representatives and senators were soon delving into the obscure details of containerization. Other ship lines urged that the government push adoption of standard containers so that any company could handle others' containers. "The key to automation is the existence of a standardized product," British steamship executive G. E. Prior-Palmer testified. Sea-Land and Matson, competitors charged, were disrupting the effort to make containers compatible around the world. Of 107 container-carrying ships under construction in September 1967, all but six, commissioned by Sea-Land and Matson, were designed around standard sizes. Marad concurred, arguing that Sea-Land and Matson should accept the standards adopted by everyone else. Sea-Land could add five feet to each of its 25,000 containers and 9,000 chassis and alter all of its ships and cranes for about $35 million, acting Marad chief J. W. Gulick testified, and Matson, a much smaller company, could switch from 24-foot containers to 20-foot containers at a cost of only $9 million.[33]

Sea-Land and Matson, which had invested a combined $300 million in containerization, were less concerned about the cost of conversion than about the inefficiency of doing business with equipment ill-suited to their needs. Matson president Stanley Powell testified that using 20-foot containers instead of 24-footers would raise his company's operating costs by $500,000 per ship per year in service to the Far East, and would increase costs for trucks picking up and delivering containers as well. Malcom McLean followed, armed with a consultant's study showing that switching from 35- to 40-foot containers in Sea-Land's Puerto Rico service would reduce revenues by 7 percent and costs hardly at all. "I don't care what size container is adopted as a standard," he affirmed. "If the marketplace can find one that moves cheaper, that is the way the marketplace will dictate it and we want to be flexible enough to follow the marketplace."[34]

The Senate passed their legislation, but Matson sensed that a compromise would be needed to get the bill through the House. On the spur of the moment, Powell told a House committee that Matson wanted Marad to subsidize two ships with a radically new feature, adjustable steel cells for container stowage. The ships would initially carry only 24-foot containers, but if market requirements changed, the frames could be adjusted so 20-foot containers could be carried in the same space. This new feature, Powell said, would add only $65,000 to the $13 million cost. No such design existed; the entire scheme, cost estimate and all, had been drawn up on the floor of a hotel room the previous night. No matter: Congress ordered Marad not to discriminate against companies using nonstandard containers, Matson was granted its construction subsidy—and, when the company decided years later to switch from 24-foot containers to 40-foot containers, the adjustable cells conceived to satisfy a congressional committee made the shift cheap and easy.[35]

Two controversies remained. The MH-5 committee undertook a futile effort to make containers compatible with airplanes as well as with ships, trucks, and trains. The requirements were not easy to reconcile: air containers needed to be stronger than maritime containers, and they required smooth bottoms to travel on conveyor belts rather than corner fittings for lifting by cranes. After months of studies, it dawned on the engineers that shippers paying a premium for the speed of air freight would be unlikely to want their cargo carried in ships, and a separate standard was developed for air containers. Railroads raised a more serious problem, contending that containers needed heavier end walls. End walls bore no great loads when the containers were on ships, but the braking of a train could cause the end of a container to bump up against the end of the flatcar. Railroads in North America demanded end walls twice as strong as those needed by ship lines, to reduce the potential for damage claims. European railroads were even more concerned, because differences in couplings caused more forceful contact between railcars in Europe. Maritime interests resisted stronger end walls, which meant more weight and higher manufacturing costs. With the TC104 committee on their side, the railroads won the day, but not without cost; by one estimate, the requirement for stronger end walls added one hundred dollars to the cost of manufacturing a standard container.[36]

By 1970, as the International Standards Organization prepared to publish the first full draft of its painstakingly negotiated standards, the bitter battles among competing economic interests were finally winding down. In hindsight, the process can be faulted in almost every particular. It led to corner fittings that were too weak and needed redesign. Several newly approved container sizes were uneconomic and were soon abandoned. The standards for end walls may have been excessive, and the standards for lashing containers together on deck never quite added up. No one would declare that all of the subcommittees and task forces came up with an optimal result.

Yet after 1966, as truckers, ship lines, railroads, container manufacturers, and governments reached compromises on issue after issue, a fundamental change could be seen in the shipping world. The plethora of container shapes and sizes that had blocked the development of containerization in 1965 gave way to the standard sizes approved internationally. Leasing companies began to feel confident investing large sums in containers and moved into the field in a big way, soon owning more boxes than the ship lines themselves. Aside from Sea-Land, which still used mainly 35-foot containers, and Matson, which was gradually reducing its fleet of 24-foot containers, almost all of the world's major ship lines were using compatible containers. Finally, it was becoming possible to fill a container with freight in Kansas City with a high degree of confidence that almost any trucks, trains, ports, and ships would be able to move it smoothly all the way to Kuala Lumpur. International container shipping could now become a reality.[37]

## Chapter 8

# Takeoff

**The *Ideal-X* and the *Hawaiian Merchant*** were small-scale demonstrations of the container's potential. The *Gateway City*, in 1957, and the *Hawaiian Citizen*, in 1960, offered powerful examples of the efficiency that container shipping could achieve once specialized ships and equipment were brought to bear. Yet in 1962, six years after it arrived on the scene, container shipping remained a very fragile business. In the East, it accounted for 8 percent of the general freight passing through the Port of New York but hardly any elsewhere, save Sea-Land's bases in Jacksonville, Houston, and Puerto Rico. On the West Coast, a trifling 2 percent of general-cargo tonnage moved in containers. Most goods were still moved as they had been for decades, as loose freight in trucks, boxcars, or the holds of breakbulk ships. The container's economic impact was almost nil.[1]

The leaders of the nation's maritime industry were by no means unanimous that the container was the future. The steamship business was as tradition-bound as any in the country. Many of its most prominent executives were men who reveled in the romance of sea and salt air. They worked within a few blocks of one another in lower Manhattan, and spent well-oiled luncheons comparing notes with their peers at haunts like India House and the Whitehall Club. For all of their earthy bluster, their businesses had survived thanks almost entirely to government coddling. On domestic routes, government policy discouraged competition among ship lines. On international routes, rates for every commodity were fixed by conferences, a polite term for cartels, and the most important cargo, military freight, was handed out among U.S.-flag carriers without the nuisance of competitive bidding. Decisions about buying, building, or selling ships, about leasing terminals, and about sailing new routes all depended upon government directives. For men who had prospered in this environment, who loved the smells of the ocean and fondly referred to their ships as "she," Malcom McLean's wholly unromantic interest in moving freight in boxes had little appeal. It was all well and good for visionaries to proclaim that containers were a "must," but the collective wisdom of the shipping industry held that they would never carry more than a tenth of the nation's foreign trade.[2]

New union agreements and progress toward standardization encouraged shipping executives to look more seriously at containerization. When they did, though, they saw a waterfront littered with costly mistakes. Malcom McLean himself had made them; the novel shipboard cranes he had installed proved to be a nightmare, breaking down frequently, with each breakdown delaying a ship. Matson, more cautious in its investments, nonetheless built two vessels to carry both bulk sugar and containers, losing the efficiency and quick turnaround of pure containerships. Luckenbach Steamship Company had embarked on a $50 million scheme to operate five containerships between the East and West coasts, only to have to abandon the plan when government aid was not forthcoming. The Erie and St. Lawrence Corporation's container service between Port Newark and Florida, inaugurated amid great fanfare in 1960, ended six months later when paper manufacturers and food processors failed to provide enough freight.[3]

What both transportation companies and shippers were slowly coming to grasp was that simply carrying ocean freight in big metal boxes was not a viable business. Yes, it produced some savings: cranes, boxes, chassis, and containerships eliminated much of the cost of loading and unloading vessels at the dock. Shippers, though, cared not about loading costs, but about the total cost of delivering their products from factory to customer. By this standard, the advantages of containerization were less apparent. If a wholesaler was sending, say, three tons of water pumps from Cleveland to Puerto Rico, the pumps would have to be trucked to Sea-Land's warehouse in Newark, removed from the truck, and consolidated into a container along with twenty or twenty-five tons of goods from other shippers. Upon arrival in Puerto Rico the contents would have to be removed from the container, sorted, and loaded into trucks for final delivery. There was only a limited amount of traffic, involving fully loaded containers going from one shipper to one recipient over water, for which containerization indisputably made economic sense.[4]

Most big shippers had no pressing need to use coastal shipping services, whether containerized or not. They used ocean freight for exporting or importing—but only a handful of containers were being carried on international ships. Most freight shipments were domestic, going cross-country by truck or train. Not until container technology affected land-based transportation costs would the container revolution take firm hold.[5]

Up through the end of World War II, trains had been the way that most companies moved their goods. Railways' freight revenues were nine times those of intercity truck lines in 1945, when more than 400,000 carloads of manufactured goods as well as most of the nation's coal and grain were shipped by rail. The 1950s, though, were the decade of the truck. Better roads, including widespread construction of expressways, permitted larger trucks carrying heavier loads at higher speeds. The use of 40-foot trailers on superhighways instead of 28-foot trailers on congested two-lane roads led to large productivity gains that helped truckers take business from railroads. Trucking companies' intercity revenues doubled during the 1950s, and growth would have been even higher if trucks owned by or operating under contract to manufacturers and retailers were included in the count. Meanwhile, railroad freight revenues were flat. By 1963, most manufactured goods, except automobiles, moved by truck.[6]

The railroads' greatest challenge came in the smallest but most lucrative part of their business, the handling of shipments too small to fill an entire boxcar from origin to destination. Less-than-carload shipments might vary in size from a few barrels of solvent to ten thousand pounds of nuts and bolts. In 1946, these small shipments made up less than 2 percent of railroads' tonnage but brought in nearly 8 percent of their revenues. Handling these loads was inefficient, requiring railroad employees to move individual crates and cartons from one boxcar to another at connecting points at huge expense. Truckers went after the market with a vengeance, and nearly three-quarters of the railroads' less-than-carload business shifted to the highways within a decade.[7]

The loss of traffic that had always been theirs forced railroad executives to do some serious thinking about what their companies could still do best. The obvious answer was to concentrate on their strength—the ability to carry heavy loads over long distances at relatively low cost. One potential type of load grabbed their attention: trucks. Driving a truck from California to New York could require one hundred man-hours behind the wheel in the days before coast-to-coast expressways, plus time for meals and rest. Sending the truck trailer by train for the long-distance part of its journey could cut these labor costs while preserving trucks' greatest advantage, the ability to pick up and deliver at any location. Railroads had offered a service like this as early as 1885, when Long Island Railroad "farmers' trains" transported produce wagons to ferry landings opposite New York City; four wagons rode on each specially designed freight car, while the farmers and their horses traveled in separate cars. An updated version appeared in the early 1950s, as railroads began to chain truck trailers to flatcars. They called it "piggyback."[8]

Piggyback, like almost every innovation in transportation during that era, faced a very large obstacle: the Interstate Commerce Commission. The ICC regulated the rates and services of both trains and interstate trucks. It had quashed railroads' attempts to carry truck trailers in 1931 under its mandate to avoid unfair and destructive competition. Putting trailers on trains confounded the ICC's basic instincts, but in 1954 it finally outlined the conditions under which railroads could transport freight in trailers without submitting to regulation as motor carriers. Over time, the commission approved several "plans" that permitted piggyback without upsetting the structure of regulation. Plan I let truckers serving the general public—common carriers, in legal parlance—collect the cargo from shippers, put their trailers on a train, and split the revenue with the railroad, but only if the train was operating along a route that the truck line had authority to serve. Plan II allowed the railroad to own trailers and deal directly with shippers, but the shippers might have to use their own trucks to haul the trailers from a rail yard to the final destination. When it became clear that these conditions would not allow piggyback freight to prosper, the ICC approved other plans so that railroads could move trailers, or even flatcars, owned by freight forwarders or by shippers themselves. This was a huge relief to the railroads, whose financial woes were making it increasingly hard to come up with the money for new investments. Looser regulation opened the way for piggyback to grow.[9]

Piggyback solved a difficult operational problem for the railroads, the inefficient use of their enormous fleets of boxcars. U.S. railroads owned 723,962 boxcars in 1955 but got very little use from them. The typical boxcar spent as little as 8 percent of its life under way, earning revenue. The rest of the time it was a warehouse on wheels, waiting on sidings to be loaded, unloaded, or added to a train. The fact that piggyback flatcars could be put back to work as soon as the trailers had been removed freed the railroads from their role as unwilling providers of free storage. For shippers, on the other hand, piggyback, like containerization, initially offered few real cost advantages. Every railroad used different types of cars, so one railroad might be unable to unload a trailer from a flatcar originated on another line—a serious problem, given that no railroad spanned the entire country. Loading was cumbersome, often done "circus style": empty flatcars were lined up end to end, with metal bridges between them, and a truck backed each trailer along the decks of the cars to the last empty position. Most flatcars carried just one trailer, so making up a train could require switching and coupling a very large number of cars. Volume was too small for the railroads to justify the investments needed to make piggyback a truly efficient service. In addition to these operational deficiencies, the Teamsters union, whose members drove most intercity trucks, opposed a system that reduced the need for its members' labor, and it negotiated contracts that penalized truck lines for shipping trailers by train. Piggyback was a tiny business: although thirty-two railroads carried trailers on flatcars in 1955, total traffic amounted to only 0.4 percent of the railroads' carloadings.[10]

In July 1954, the mighty Pennsylvania Railroad began service between New York and Chicago with 50-foot flatcars carrying a single trailer apiece. Within months, its daily TrucTrains to Chicago and St. Louis, equipped with new 75-foot flatcars, were carrying hundreds of trailers each way. The Pennsy signed up 150 motor carriers to pick up and deliver the trailers, and soon had a $100-million-a-year business. It created a research-and-development department—a highly unusual step for a railroad—and charged it with improving TrucTrain. The biggest hurdle, TrucTrain's managers decided, was that the Pennsy could not transfer loaded cars to many connecting railroads. In November 1955, TrucTrain was incorporated as Trailer Train Company, and other railroads were invited to buy in. The idea was simple: instead of each railroad's operating its own small trailer business, Trailer Train would handle truck trailers nationwide. It would own the flatcars, collect revenue from truck lines, and pay the railroads to haul its cars over their tracks. At the end of the year, any profit would be divided among the railroads that had become shareholders. Trailer Train started small, operating barely 500 flatcars in 1956. Other railroads quickly joined the enterprise, allowing Trailer Train to gain economies of scale beyond the reach of individual railroads. By 1957, the company was buying a fleet of 85-foot flatcars, enabling it to boost efficiency by carrying two of the new 40-foot truck trailers on a single car.[11]

Three major railroads stood apart, unconvinced that the complications of loading trailers aboard flatcars were worth the trouble. In 1957, the New York Central, the Pennsylvania's direct competitor, developed a service called Flexi-Van. Flexi-Van used containers—special truck trailers that could be separated from their undercarriages through the removal of four pins. It carried them on flatcars with turntables that swung 90 degrees. A truck would back up against the side of the flatcar, the driver would release the pins to detach the trailer, and the detached trailer, with no wheels, would slide from its undercarriage along rails built into the turntable. When half of the container was atop the turntable, the driver would engage an extra wheel beneath the truck tractor to move it sideways, into a position parallel to the freight car, moving the container and turntable along with it. The driver would then release the container from the truck and push the turntable the rest of the way into position. The procedure made Flexi-Van containers much easier to load than full trailers, and made it possible to load or remove a single container without disturbing any other part of the train. Flexi-Van moved at passenger-train speeds, delivering containers from Chicago to New York in seventeen hours.[12]

In the Midwest, the Missouri Pacific Railroad took an entirely different approach. The Missouri Pacific's trucking operation used detachable truck bodies with hooks on the top. A driver would bring his truck alongside the train, beneath a wheeled crane wide enough to straddle both the train and the truck. The driver removed some pins to separate the trailer body from the undercarriage and then operated the crane himself to lift the container onto the railcar, with the entire operation taking less than ten minutes. The Southern Railway also chose containers rather than trailers as the best way to handle freight between the South and New England, where it had customers but no rail lines. By striking arrangements with truckers to haul the trailers between its terminus in Washington, D.C., and points further north, the Southern overcame its inability to send conventional trailers on flatcars through the low-ceilinged tunnels in Baltimore and New York. None of these three railroads, of course, could interchange containers with each other, much less with the railroads participating in Trailer Train. As with ship lines, so too with railroads: by the late 1950s, the drive to simplify freight handling had led to incompatible solutions.[13]

The railroads' desire to expand piggyback left the ICC in a quandary. In the early days of piggyback, the railroads' rates, like truck rates, had been based on the commodity being carried. Piggyback rates for any commodity were about the same as truck rates and a bit higher than rates for shipping in boxcars. That suited the regulators fine, because it let the railroads pick up a little traffic without shaking up the freight business. Pan-Atlantic's combined water-road rates along the Atlantic coast were 5 to 7.5 percent below railroads' boxcar rates, also in line with ICC precedents that let water carriers charge less for slower service. But then, late in 1957, the railroads tried to cut some piggyback rates to compete better against Pan-Atlantic and also Seatrain Lines, which carried loaded railcars aboard ships. Predictably, Pan-Atlantic and Seatrain objected that lower railroad rates would drive them out of business.[14]

As the ICC was trying to figure out how to help the railroads without hurting the ship lines, Congress intervened—with conflicting instructions. Congress wanted to "breathe into our whole system of transportation some new competition," Florida Senator George Smathers explained. But while it wanted the economy to benefit from lower rates and new services, Congress also wanted to protect transportation companies and their workers. The result was the Transportation Act of 1958. In a single remarkable sentence, the law ordered the ICC not to keep any carrier's rates high to protect another mode of transportation, while also directing it to block unfair or destructive competition. The commission, it seemed, could no longer use high rail rates to protect ship lines or truckers—but at the same time it was to make sure that the ship lines and truckers were not driven out of business. In confusion, the ICC told the railroads that their piggyback rates should be about 6 percent higher than Pan-Atlantic's ship-truck rates. The ICC was decisively reversed in the courts, which gave the railroads the freedom to lower piggyback rates so long as the rates covered all of their costs.[15]

The court rulings permitting lower rates made the economics of piggyback freight compelling. Trucking companies' costs were still lowest for short journeys, and the rates charged to shippers were correspondingly lower. Over longer distances, though, trucks' total cost per mile declined only slightly, because drivers' pay and fuel, the most important costs, increased along with distance. Railroads'

Table 4
Cost of Moving 20,000 Pounds of Freight, 1959

| *Distance* | *Truck* | *Rail Boxcar* | *Flatcar with One Trailer** |
|---|---|---|---|
| 500 miles | $244.47 | $206.67 | $236.59 |
| 1,000 miles | $445.86 | $337.11 | $404.14 |
| 1,500 miles | $647.13 | $467.56 | $571.69 |

*Note*: Rail costs are "fully distributed," including overhead and profit margin.
*Source*: Kenneth Holcomb. See n. 8.
* Cost per trailer, assuming that each contained 20,000 pounds of freight.

total costs per mile, on the other hand, declined sharply with distance; once the trailers or containers were loaded on board, running the train came cheap. For distances exceeding five hundred miles, piggyback freight clearly was cheaper to provide than traditional truck service. Even private truckers, who had contracts with particular shippers, could not match the cost of trailer-on-flatcar service over long distances.[16]

The railroads were in the happy situation of being able to pass their lower costs on to customers and still earn better profits than they did carrying freight the traditional way, in boxcars. Freight forwarders took advantage of the rate difference, arranging to consolidate smaller shipments into full carloads, for which they could demand lower rail rates. Manufacturers such as General Electric and Eastman Kodak quickly discovered that there was money to be saved by organizing their production so they could fill trailers or containers and ship them to a single recipient by train, rather than sending a few cases or crates by truck. By 1967, three-quarters of all manufactured goods (excluding coal and petroleum products) left the factory in shipments of at least thirty thousand pounds. Processed food, fresh meats, iron and steel products, soaps, and beer made the switch to piggyback first, but large quantities of everything from oranges to wallboard were soon riding on rails for the first time in two decades.[17]

To be sure, there were still some regulatory oddities. The ICC let railroads carry trailers at a flat rate per mile so long as the cargo was mixed, but if a trailer contained more than a certain percentage of any single commodity, the shipper would have to pay the specific rate for that commodity. Big shippers, though, were accustomed to such regulatory impediments. They saw not only that piggyback could save money, but that lower transport costs would let them sell their goods in cities that had always been too expensive to ship to. As the railroads increased train speeds, the time needed to deliver a truck trailer from Chicago to California fell from five days to three. Goods spent less time in transit, so inventory costs fell as well. The number of piggyback carloadings doubled between 1958 and 1960, then doubled again by 1965. Flexi-Van provided an astonishing 14 percent of all revenue earned by the New York Central in 1964. Trailer Train, which had less than $1 million in revenue in 1956, became a $50 million business in 1965 and owned 28,000 freight cars.[18]

International trade was not remotely a consideration when American railroads began their aggressive pursuit of trailer traffic in the mid-1950s. Yet the potential for linking piggyback freight and container shipping was apparent from the earliest days. Most piggyback loads were truck trailers, complete with wheels, that would never travel by ship. About 10 percent of piggyback shipments, however, involved trailers detached from wheels, an increasing number of which complied with the standards for container sizes and lifting methods that the American Standards Association had been developing since 1959. Standard containers were already flowing freely to and from Canada, where railroads had embraced piggyback even more eagerly than in the United States.[19]

It was the indefatigable Morris Forgash who finally got regular intercontinental container service under way. Starting in 1960, his United States Freight Company began moving containers from the United States to Japan, using U.S. railroads, Japanese truckers, and the mixed-cargo ships of States Marine Lines. A year later, the New York Central, equipped with 5,000 new Flexi-Van containers, began similar services to Japan and Europe. United States Lines, the largest U.S. line handling general cargo, carried Southern Railway containers on experimental voyages to Europe. The U.S. Army, supporting hundreds of thousands of troops in Europe, began trials sending 40-foot containers across the ocean.[20]

These first international efforts were small in scale. Malcom McLean wanted to sail to Europe in 1961, and his staff dissuaded him: the company was not ready for such a large venture. No ship line was sailing fully containerized ships to Asia or Europe, so the containers were lodged in a few cells built into one of the holds of a breakbulk ship or carried with a load of mixed cargo. Most of the cargo on these vessels was traditional freight that had to be handled one piece at a time, so loading and unloading took almost as long as for voyages without containers. Shippers saved no money by using containers internationally, because the conferences that set ocean freight rates gave them no preference: the rate for a single container holding 20 tons of auto parts was roughly the same as the rate for 20 tons of auto parts shipped in dozens of wooden crates. The containers were usually returned across the ocean empty, and that cost, too, had to be reflected in the rates. Aside from less theft, from the shipper's viewpoint the only real attraction of these early international container services was reduced paperwork. Rather than arranging and paying for each stage of the journey separately, as they always had, shippers could ask a freight forwarder to quote a single rate for the entire land-sea-land shipment from America to Asia, and could pay for it with a single check.[21]

Viewed at the start of 1965, the balance on containerization's first nine years was positive but unspectacular. In New York, container tonnage had hit a plateau, and the International Longshoremen's Association remained vociferously opposed to its growth. On the West Coast, even after rapid growth, only about 8 percent of general cargo was moving in containers. Some railroads were using containers that in theory could be interchanged with ship lines, but in practice rail-ship container traffic was negligible. The trucking companies that used demountable containers did so mainly under contracts with Sea-Land and Matson; otherwise, truckers overwhelmingly preferred trailers that were permanently attached to their wheels and could not easily be loaded aboard ships. Container shipping looked to be a viable enough business, producing $94 million of revenue for Sea-Land in 1964, but it was a niche business. The way most manufacturers, wholesalers, and retailers moved their goods had hardly changed.[22]

Behind the scenes, though, the prerequisites for the container revolution were falling into place. Dock labor costs were poised to fall massively thanks to union agreements on both coasts. International agreements were in place on standards for container sizes and lifting methods, even if few containers yet met those standards. Wharves designed for container handling were on the way. Manufacturers had learned to organize their factories so that they could save money by shipping large loads in single units to take advantage of containerization. Railroads, truckers, and freight forwarders had grown familiar with switching trailers and containers from one conveyance to another to move what was now being called "intermodal" freight. Regulators were cautiously encouraging competition so that carriers could pass some of the cost savings from containerization on to their customers. Only one crucial ingredient was missing: ships.

The ships that had launched the container era all had been leftovers, dating to World War II. As of 1965, every containership in Sea-Land's fleet was at least two decades old, and Matson's youngest had been launched in 1946. These older vessels, obtained from the U.S. government's fleet at bargain-basement prices, were slow and small, but they allowed the container pioneers to get under way without huge amounts of capital. As other companies tested containerization in the early 1960s, they generally used converted World War II freighters as well. The cost of building brand-new vessels was too great for many companies to handle, even with government subsidies, and the risks of guessing wrong about future trends in cargo handling were extremely high.[23]

No one was more aware that the world was about to change than Malcom McLean. He was already fully committed to the container. Outracing any potential competitors, Sea-Land had converted seven vessels into containerships between 1961 and 1963. The converted ships allowed it to open West Coast service in 1962 and then to buy Alaska Freight Lines in 1964, but at the cost of increasing its debt from $8.5 million to $60 million in just two years. By 1964, far greater financing needs loomed as Sea-Land began to look at Europe. Once word of McLean's next destination got out, the big American companies would surely enter the transatlantic container trade, and the European ship lines were bound to join in as well. To stay in the lead, McLean had no choice but to roll the dice once again. He did it by arranging two more extraordinary financial transactions in 1965.[24]

The first was with Daniel K. Ludwig, a man who had much in common with Malcom McLean. Ludwig, born in 1897, had entered the shipping business at age nineteen by transporting molasses around the Great Lakes. Like McLean, he ran his business with a legendary focus on costs; according to one famous story, he bought a tanker called the *Anahuac* and decided to keep the name because "it would have cost $50 to paint it out." By the 1950s his National Bulk Carriers was the largest American-owned ship line, and Ludwig was one of the world's wealthiest men. His holdings included American-Hawaiian Steamship Company, a shell since it stopped operating ships in 1953. Ludwig had watched McLean's container venture carefully, and in January 1961 American-Hawaiian suddenly applied for $100 million in federal subsidies to build ten enormous high-speed ships and open an intercoastal service through the Panama Canal. Sea-Land promptly announced its own entry into the intercoastal route and spent the next year successfully blocking Ludwig's bid for subsidies. Ludwig pared his subsidy request to three ships, to be powered by nuclear reactors, but then decided that the best way to profit from container shipping was to invest in Sea-Land instead. In early 1965, when McLean Industries' shares were trading for $13 apiece, the company issued one million shares of stock to American-Hawaiian for $8.50 per share, and Ludwig joined the company's board of directors. It was the first act of what would prove to be a long-running collaboration between Ludwig and Malcom McLean.[25]

The second deal involved Litton Industries. Litton, founded in the 1930s to make radio tubes, had been reshaped during the 1950s into a new type of company, a "conglomerate." Among its far-flung holdings was the Ingalls Shipyard in Pascagoula, Mississippi. Litton, like the other conglomerates of the day, was bent on fast growth, and it was eager for Ingalls to diversify away from naval contracts into commercial work. McLean needed ships but had no money; Litton was rolling in money but desperate to keep its shipyard busy.

Negotiations led to the creation of a company called Litton Leasing. On November 5, 1964, Sea-Land sold Litton nine containerships for $28 million, using the proceeds to help pay off $35 million in bank loans. Litton immediately leased the ships back to Sea-Land. Litton took other ships formerly belonging to the Waterman operation, which McLean was selling off. It widened them, lengthened them, and installed container cells to meet Sea-Land's specifications. For an annual rent that came to $14.6 million, the cash-strapped ship line was able to add eighteen containerships to its fleet in just four years. For good measure, Litton agreed to swap its convertible stock for 800,000 McLean Industries shares, immediately strengthening Sea-Land's stretched balance sheet by $6.8 million.[26]

Sea-Land's quick moves to acquire a huge new fleet opened the floodgates. During eight short weeks in the late summer of 1965, no fewer than twenty-six containership projects hit the headlines. Each required $8 to $10 million to convert a ship and another $1–$2 million for chassis and containers. For an industry notoriously tight with its cash, a total investment of a quarter billion dollars to enter a trade that might not generate business was almost incomprehensible. Companies that had watched containerization from a distance for years, curious but noncommittal, now felt that they had to put up real money or be swept away in the flood. Not all of them were eager. When Sea-Land threw a party at the Rotterdam Hilton to introduce its service to Dutch shippers at the start of 1966, its guests booed, and the head of Holland-America Line, itself preparing to carry containers, told a Sea-Land executive, "You come back with the next ship and take all the containers home."[27]

The established carriers' great fear was that containers would force down freight rates. Four conferences, one covering traffic each way between North America and northern Europe and the other two dealing with cargo between North America and the British Isles, set the rates for every commodity in their trades. They naturally had no provisions for containers.

Joining the conferences was not essential for Sea-Land, although belonging would surely smooth its way in dealings with European governments and ports. For United States Lines, already a conference member, conference rules on containers were critical. McLean ordered his representatives to seek admission to the eastbound and westbound North Atlantic Continental Freight Conferences, and to do so without causing fights. After Sea-Land proclaimed that it had no desire to start a rate war, the doors were opened. Sea-Land and U.S. Lines put forth two proposed rules: moving containers between piers and warehouses should be cheap, and ocean freight rates should include the use of steamship lines' containers and chassis, so that shippers would not have to pay extra. Their European competitors, contemplating container services themselves, accepted both requests. "We didn't ask for any big concessions," a Sea-Land executive recalled. "We just asked them to accommodate the container. And, you know, I think that was a big mistake they made, but they did." A European shipping executive remembered things differently: "They're afraid of Sea-Land," he said, "but they would rather keep an eye on them from the inside than have them on the outside and wonder what they're doing." Whatever the case, Sea-Land was able to enter the North Atlantic trade not as an outcast, but as a member of the club.[28]

Moore-McCormack Lines, a subsidized American carrier sailing between the East Coast and Scandinavia, opened the first transatlantic container service in March 1966, using combination ships that carried truck trailers, containers, and mixed freight. U.S. Lines, also a subsidized operator, followed almost immediately, carrying 40 20-foot containers along with mixed freight in its holds. In April, after signing agreements with 325 European truck lines to deliver its cargo to places like Basel and Munich, Sea-Land, with no government operating subsidy, began service on a totally different scale. Weekly sailings from Newark and Baltimore to Rotterdam and Bremerhaven carried 226 of its 35-foot containers.

All three carriers reported stunning efficiencies. Three medium-size containerships could handle as much transatlantic freight as six breakbulk ships, with only half the capital cost and two-thirds of the operating cost, a consultant reported. U.S. Lines found that at Port Elizabeth, one longshore gang with one crane could load as much in a ten-hour container operation as ten gangs handling conventional breakbulk freight. Moore-McCormack pegged the cost of loading containerized cargo at Port Elizabeth as $2.00 to $2.50 per ton, versus $16.00 per ton for conventional freight.[29]

Two types of freight appear to have filled those first containers crossing the Atlantic: whiskey on the westbound run, military goods on the voyage to Europe. Liquor exporters had long complained of huge losses to theft on the docks, and convincing them to use containers was not a hard sell. Among Sea-Land's first ports of call was Grangemouth, in Scotland, where it picked up Scotch whiskey. Sea-Land won the business with a tank container, made of stainless steel, designed to let exporters ship their whiskey in bulk for bottling in the United States. Two tank containers would fit neatly into a standard container cell on a Sea-Land ship, putting an end to the pilferage that had plagued the whiskey trade from time immemorial.

The military role was even more crucial. As a U.S.-flag carrier, Sea-Land was entitled to carry a portion of the freight for the quarter million U.S. soldiers in West Germany, and the military, determined to push containerization, channeled cargo Sea-Land's way. According to industry rumor, more than 90 percent of the cargo on Sea-Land's first transatlantic voyage was military freight. Military demand all but assured that Sea-Land's first voyages would be profitable, and gave it an advantage that foreign carriers could not match. When the navy finally overcame the objections of the breakbulk carriers and put European military shipping contracts out for competitive bids in the summer of 1966, Sea-Land underpriced every American competitor and won all the traffic it could handle.[30]

Authoritative numbers on shipments during that first year of transatlantic container service do not exist. The vast majority of containers going to and from Europe flowed through New York harbor, and port data offer the best guide to the magnitude of the new trade. The port's container tonnage, 1.95 million long tons in 1965, soared to 2.6 million tons in 1966, even though hardly any containers were carried during the first 10 weeks of the year. Faced with this enormous flow of cargo, more U.S. companies, two groups of British carriers, and a consortium of Continental ship lines all raced to enter the container trade. "In 1966, commitments by ship operators and ports to containers passed the point of no return," a consultant judged.[31]

Only 3 ship lines were offering international container service from the United States in the spring of 1966. By June 1967, one researcher counted 60 companies offering container service to Europe, Asia, and even Latin America (although only a handful used special containerships). More than 50,000 containers—enough to hold half a million tons of freight—crossed the ocean in the second half of 1967. Many carriers placed orders for brand-new containerships, designed to maximize the load of 8-foot-wide, 8-foot-high boxes that were quickly becoming the industry standard. In 1967, as the Port Authority touted a study showing that 75 percent of the Port of New York's general cargo could move in containers, 64 container vessels were under construction by 12 ship lines. Kerry St. Johnston, head of the British consortium Overseas Containers Ltd., warned that so much new capacity would lead to rampant rate cutting, not a happy prospect for ship lines that were in the process of investing hundreds of millions of dollars in container equipment and ships.[32]

The new, fully containerized vessels began to come on-line in 1968. Ten containerships per week sailed the North Atlantic that year, carrying a total of 200,000 20-foot containers holding 1.7 million tons of freight. The European companies whose containerships had not yet been finished coped as best they could, piling containers on the decks of their breakbulk ships. They were able to provide their customers with some semblance of container shipping, but with none of the efficiencies enjoyed by companies using fully containerized ships and high-speed cranes. "The costs were horrendous," remembered Karl Heinz Sager, the head of the German ship line Hapag-Lloyd.[33]

So far as shippers were concerned, the only reason not to join the rush to container freight was the shortage of containers. Although U.S.-flag ship lines added 13,000 containers between September 1966 and December 1967, and European ship lines bought thousands more, empty boxes could be in short supply. Otherwise, the cost savings were compelling, even with the conferences controlling transatlantic rates. Chas. Bruning Co., a maker of office machines near Chicago, found that it could get its equipment delivered to inland points in Europe in an average of twelve days. In addition to cheaper ocean freight, Bruning saved money by eliminating special export packaging, damage, and theft, and got a 25 percent discount on its insurance. So much traffic shifted so quickly that three years after containerships first sailed to Europe, only two American companies were still operating breakbulk ships across the North Atlantic, making a combined total of three sailings per month.[34]

The surge in transatlantic container traffic, coming at a time when American factories were running hard to meet the demands of a wartime economy, offered a golden opportunity for U.S. railroads to regain their place at the heart of the domestic transportation system. Their business in conventionally packaged export cargo was dying. Thousands of containers were passing through New Jersey and Baltimore every week, many of them going to or from factories in the industrial heartland of the upper Midwest. This huge scale offered no advantage to truckers, because, no matter how many boxes were being handled, one truck could pull only one 40-foot box. Scale could bring real savings aboard trains, giving the railroads a way to recover some of the export traffic they were losing.

The European railroads saw things that way. The Europeans had been trying to make a business with containers since the 1920s, and they were eager to strike deals with the ship lines. Almost as soon as transatlantic container shipping began, they offered flat per-container rates to draw traffic. In 1967, the French National Railway charged a flat 572 francs to carry a loaded 40-foot container from Bremen, in North Germany, to Basel, in Switzerland, while the German Federal Railway charged $241 to take any 40-foot box from Bremen to Munich. In Britain, using dedicated trains to carry containers to and from the port at Felixstowe had been part of Sea-Land's plan from the very beginning, and British Rail was an eager collaborator.[35]

American railroads, especially those in the East, were notably less enthusiastic. They feared that containers would draw shipments away from boxcars, bringing in less revenue. Most of them had already built ramps to load and unload truck trailers under the auspices of Trailer Train, and at a time when they were financially pressed, they had no desire to lay out additional money for overhead cranes and storage yards to handle containers. The New York Central had a successful operation with its unique Flexi-Van service, and it feared that maritime containers would siphon off Flexi-Van's customers. The railroads could not refuse to handle containers, but they could provide such poor service that no customer would want to ship them. In February 1966, the Pennsylvania Railroad delivered a flatcar holding two 20-foot containers to a Caterpillar Tractor warehouse in York, Pennsylvania. The containers were loaded while sitting on the car on Caterpillar's siding, and the railroad charged for the haul to New Jersey as if Caterpillar had simply loaded its parts into boxcars. The shipment was billed as an experiment for American Export Lines, and the railroads were rooting for it to fail. "[W]e hope that the high cost of loading, blocking and bracing, and the unloading of the containers plus the loss of the containers for 2 weeks with no offsetting per diem revenue, will discourage their pursuing this avenue further," a New York Central official wrote his counterpart at the Pennsylvania.[36]

The eastern railroads commissioned a study, which urged them to act quickly to attract container traffic. The railroads chose to do the opposite. They agreed on a new rate structure to discourage containers, providing that any container weighing more than five hundred pounds would be charged on the basis of the weight and the contents, rather than receiving the lowest full-carload rate. In addition, they insisted upon charging ship lines for carrying empty containers from the port to customers inland—hardly a policy that encouraged shippers to use rail for the land portion of international shipments. If those measures were not enough to deter container business, some railroads simply drove it away. In the spring of 1967, when Whirlpool Corporation asked the New York Central to move containers of refrigerators from an Indiana factory to the New Jersey docks, the railroad advised Whirlpool to ship its refrigerators in boxcars and put them into containers at the port; Whirlpool shipped by truck instead. Matson's plan to ship containers of Hawaiian pineapple cross-country by train met with similar hostility, because the rate for transporting the containers between Chicago and New Jersey was far below the standard per-ton rate for carrying canned goods. "It is extremely important that we defeat the proposal," a New York Central executive wrote.[37]

Malcom McLean had a different vision. For him, railroads, trucks, and ship lines were in the same business—moving freight. He wanted to turn Sea-Land's intrepid sales force loose to locate manufacturers exporting to Europe from Little Rock and Milwaukee. In 1966, as Sea-Land's transatlantic service was getting under way, McLean Industries offered an audacious proposal to build railroad yards in Chicago and St. Louis, at its own expense. Freight forwarders owned by McLean Industries would collect freight from shippers, consolidate it into McLean-owned containers, and load the containers aboard McLean-owned railcars, specially designed by the Pullman Company to carry containers stacked two high. The Pennsylvania Railroad would pull McLean's all-container train straight to a rail yard Sea-Land would build by the docks at Elizabeth, arriving in time to meet a Europe-bound ship, which would in turn connect with trucks and trains on a European dock. For the first time, a shipper a thousand miles from the sea would be able to buy not just international transportation but tightly scheduled international transportation. A seller could tell its customers when the goods were to arrive, with a reasonable likelihood that the schedule would be met.[38]

The economic advantages of this truck-train-ship combination seemed overwhelming. Trucks would do the short-haul work for which they were best suited. Trains would handle the long land haul, where their costs were lowest. Shippers' costs for the domestic leg of their international shipment would fall by half. The Pennsylvania was intrigued by the plan, the New York Central and the Baltimore and Ohio opposed. But as the Pennsylvania and the New York Central announced plans to merge, McLean's ambitions were scuttled. The railroads made the minimum counteroffer that the ICC would allow: they would carry Sea-Land's container cars—mixed in with other cars on their regular slow freights.[39]

Once again, Malcom McLean was ahead of his time—but with the railroads, he lacked the power to turn his vision into reality. Farsighted rail executives, such as Trailer Train president James P. Newell, realized that attempts to preserve high boxcar freight rates were bound to fail; Newell estimated that railroads could save 30 percent of their train and engine costs by running the sort of unit trains McLean had in mind. "Take those savings and divide them between the railroad and the shipper," Newell advised. But in 1967 and 1968, the railroads couldn't be bothered. Their piggyback traffic was booming, up 30 percent in three years, in an economy white-hot from the Vietnam War. Their mind-set, reinforced by a century of regulation, did not encourage them to conquer new business. They were content to leave the land side of the new container shipping business to trucks.[40]

Chapter 9

# Vietnam

**In the winter of** 1965, the United States government began a rapid buildup of military forces in Vietnam. In the process, it created what may have been the greatest logistical mess in the history of the U.S. armed forces. The resolution of that mess represented containerization's coming of age.[1]

Few places on earth were less suited to supporting a modern military force than South Vietnam in early 1965. The entire country, seven hundred miles long from north to south, had a single deepwater port, one railroad line that was largely inoperative, and a fragmentary highway system, mostly unpaved. The tasks of providing civilian aid and supplying the U.S. military "advisers" who had worked in Vietnam since the late 1950s—there were 23,300 of them at the start of 1965—were already overwhelming; by 1964, the small U.S. port detachment in Saigon was working twelve-hour shifts, seven days a week, to prevent a backlog of ships. The various American forces in the country had sixteen different logistical systems, a situation that led to endless competition for basic resources such as delivery trucks and warehouse space. There was no central system for keeping track of arriving cargo, and the navy's Military Sea Transportation Service (MSTS), which was responsible for chartering merchant ships to haul supplies to Vietnam, did not even have an office in the country. So far as Washington was concerned, the entire operation in Vietnam was predicated on the assumption that all troops would be withdrawn in 1965. This political fig leaf meant that spending on docks, warehouses, and other permanent infrastructure was hard to justify.[2]

The logistical challenges were well known when President Lyndon Johnson ordered 65,000 U.S. soldiers and marines to Vietnam, along with several air force squadrons, in April 1965. Being aware of the problems, though, was not the same as resolving them. By June, when U.S. troop strength in-country had reached 59,900, the supply chain was already hopelessly tangled. Ships coming from California anchored outside Vietnamese harbors, but getting their cargo safely ashore was almost impossible: the harbors were so shallow that oceangoing ships could not reach the piers. Instead, a barge or a landing ship tank (LST), an amphibious vehicle longer than a football field but with a very shallow draft, would serve as a ferry. The barge or LST would tie up to the larger ship, whose crew would off-load the cargo painstakingly, often by placing crates or boxes into nets that were lowered by rope. The process was so slow that barges carrying ammunition from ships near Nha Be required ten to thirty days to make a single round trip to shore. At Qui Nhon, LSTs brought cargo directly onto the beach and lowered their huge ramps to let trucks and forklifts inside, but unloading them still took eight days. At Da Nang, oceangoing ships had to drop their cargo into lighters four miles out at sea. Coastal ships with less than a five-meter draft could reach the dock, but the port was repeatedly thrown into chaos when they arrived without advance notice. Storms, common during the summer monsoon, could bring the intricate unloading process to a halt.[3]

The situation in Saigon was even worse. Vietnam's only deepwater port, located on the Saigon River forty-five miles from the South China Sea, was a major bottleneck. Tonnage rose by half during 1965, and the port was simply overwhelmed. There were no cranes and few forklifts, leaving almost everything to be handled by muscle power. Ships carrying military cargoes, commercial cargoes, U.S. foreign aid, and food relief shipments competed for one of only ten berths.

Once a vessel unloaded, its cargo often sat for days on the dock. Military recipients often did not know that they had freight coming. Commercial importers were accustomed to leaving their goods at the port as long as possible to put off the payment of customs duties. Cargo theft, much of it orchestrated by South Vietnamese generals, was so widespread that U.S. military police rode shotgun on the trucks taking cargo from the docks to military warehouses. Long port delays worsened the shortage of U.S.-flag ships that had forced the MSTS to activate the rustiest merchant vessels in the government-owned reserve fleet. "Military cargo requirement [*sic*] as of this date have been met only by accepting delivery of the cargo at dates later than desired," the agency's acting commander admitted in May 1965. Lacking warehouse space, army and air force commanders treated cargo ships as floating warehouses, making shipping problems worse. "Saigon just became a burying ground," a high-ranking naval officer recalled. "Ships would move up the river and they would stay, and stay, and stay, not be offloaded. The Army would argue that the press of war was such that they couldn't get the stuff ashore. Air Force didn't bother to argue, the ship was there, period, we've got it and when we are ready we'll let it go."[4]

Confusing everything was the decision by the Joint Chiefs of Staff to run a "push" supply system. In contrast to a "pull" system, in which units in the field would request the supplies they needed, the push system required supply experts back in the United States to decide what to send. The Army Materiel Command shipped more than one million automatic resupply packets, providing equipment and spare parts based on assumptions about how much a normal unit in the field would require. Supply depots in California made similar judgments about needs for food, clothing, communications gear, and building supplies. The supply experts "had never had grease under their fingernails," a top army general groused, and from a distance of thousands of miles they had no actual knowledge of the rapidly changing situation in the field. Nor were they familiar with Vietnam.[5]

In terms of getting supplies to the field as quickly as possible, the push system was a success. Spending by the Army Materiel Command, the agency that bought the army's weaponry, soared from $7.4

billion in fiscal year 1965 to $14.3 billion the following year as ammunition, weapons, building materials, and vehicles were pumped into Vietnam. What finally arrived there, though, was always unexpected and often unneeded or unwanted. Food supplies flooded in, then were suddenly cut off when it became clear that there was far too much on hand. Conex boxes, the five-ton steel containers favored by the military, would arrive with mixed loads of weapons, boots, fatigues, and assorted odds and ends, leaving quartermasters without enough of any one item to outfit their units. Troops on the ground often ran short of provisions and essential supplies.[6]

A month before the Joint Chiefs had given final approval for the troop buildup, William Westmoreland, the U.S. military commander, and James S. Killen, head of the U.S. foreign aid mission, had agreed that the best way to keep Vietnam supplied was to expand the port at Da Nang, a small city 430 miles north of Saigon. The concept was that Da Nang could receive ships arriving directly from the United States, diverting traffic from Saigon. This plan could not be executed quickly; Da Nang had shallow water and no cargo-handling equipment, and the main landing ramps for LSTs were in the middle of a major street. In April 1965, Westmoreland recommended that the United States instead focus on developing Cam Ranh Bay, 300 miles south of Da Nang, as a "second major deep water port and logistics complex." Defense secretary Robert McNamara assented in May, and army engineers quickly arrived to begin work on an airfield. Construction of piers, warehouses, and a huge maintenance complex was to follow. The logistical units that had been assigned to smaller ports were soon shifted to Cam Ranh Bay. In July, Westmoreland created a wholly new unit, the First Logistical Command, with responsibility for port operations, supply, and maintenance across all of South Vietnam, including the new Cam Ranh operation.[7]

Cam Ranh Bay was the largest natural harbor on the Vietnamese coast, but it was not an easy place to build a logistical complex. It had no infrastructure, and the shifting sands along the shore were hospitable neither to earthmoving equipment nor to standard construction techniques. Aside from the harbor, the location had one important feature: there was no South Vietnamese facility at the site. The dismal performance of the Vietnamese-run Saigon port preoccupied U.S. officials at the highest level, so much so that Ambassador Henry Cabot Lodge personally discussed port problems with South Vietnamese premier Nguyen Cao Ky in July 1965. These efforts made little headway: control over the port was too lucrative for top South Vietnamese generals, who resisted U.S. proposals that a new port authority should take over. The port at Cam Ranh Bay would ease those problems by being entirely a U.S.-run operation, free of Vietnamese corruption and inefficiency. Some top U.S. policymakers even envisioned a model community surrounded by industrial parks and residential subdivisions instead of the usual bars and brothels. The fastest way to get the port up and running was to bring in a DeLong pier, a three-hundred-foot barge with holes through which pilings could be driven into the harbor floor; the barge could then be jacked up on the pilings to the desired height above the water. The navy located a DeLong pier in South Carolina, towed it through the Panama Canal and across the sea to Cam Ranh Bay, and anchored navy ships in the harbor to provide temporary electrical power—and the port was in operation. By December, merchant ships were arriving directly from the United States, and more DeLong piers were under construction.[8]

Yet the supply problems kept growing worse. Every month, 17,000 additional U.S. troops were landing in Vietnam. Each 830-man infantry battalion hit the beach with 451 tons of supplies and equipment, each mechanized battalion 1,119 tons. Feeding, clothing, and arming the troops after arrival was using every ship the MSTS could lay hands on. By Thanksgiving 1965, 45 ships were being worked in Vietnamese ports—and 75 more, loaded with food, weapons, and ammunition, were holding off the coast or in the Philippines, where they were sent to avoid the higher pay to which merchant seamen were entitled while their ships were in Vietnamese waters. "Ten first class ports in CONUS [the continental United States] are shipping material to SVN [South Vietnam] as fast as they can—we have four second-class ports to receive it," the head of the military's trucking branch complained. When the defense secretary and the chairman of the Joint Chiefs of Staff visited Vietnam in November 1965, they got an earful about logistical problems. "Our ports are jammed with ships and cargo," the head of the First Logistical Command told them. *Life* magazine ran photos of Saigon port congestion in December, and a visiting congressman advised Westmoreland to place more emphasis on the ports. The logistical mess in Vietnam was starting to become a political embarrassment at home.[9]

Washington demanded solutions. Under heavy pressure, the South Vietnamese government agreed in late 1965 that the United States could build a new deepwater port, appropriately called Newport, in Saigon, so it could move military freight away from the downtown docks. The Pentagon simplified the supply chain by overruling navy objections and making the U.S. Army responsible for supplying all allied forces in Vietnam, including the famously independent Marine Corps. And, on orders directly from the secretary of defense, the MSTS hired a private company, Alaska Barge and Transport Co., to take charge of coastal shipping. Alaska Barge earned its living delivering cargo to remote ports in Alaska, and its boss managed to persuade McNamara that the company could straighten out logistics in Vietnam. Alaska Barge quickly began building docks, and it replaced the erratic service of Vietnamese coastal ships with a barge shuttle service to move supplies up and down the coast. "We couldn't have gotten along without it," recalled the commander of the MSTS. Alaska Barge's success left an impression on military officials used to doing things in war zones the military way: perhaps there were other jobs in Vietnam that private companies could do better than uniformed troops.[10]

It was not the excess of cargo alone that caused the port logjam in Vietnam. Aside from fuel, every bit of cargo shipped to Vietnam, military or civilian, arrived in the holds of breakbulk ships. Unloading meant lifting individual items out of the hold and placing them on the dock or, even worse, into a shallow-draft vessel that would ferry them to shore, where they would have to be unloaded a second time. Many ships made multiple stops, and if a ship had been stowed poorly in Oakland or Seattle, some cargo would have to be unloaded and then reloaded for delivery to the next port. Often, cargo could not be identified once it was finally on the dock, complicating efforts to get it to the troops who were supposed to receive it. After surveying the situation, a military study team recommended basic changes in shipping procedures in November 1965. Logistics officers in the United States should send full shiploads to individual Vietnamese ports, rather than having a ship make several port calls, so the vessel could return to America as quickly as possible. Vessels should be loaded for ease of discharge. Cargo for different consignees in Vietnam should be kept separate as much as possible, to minimize sorting at the dock. The first recommendation on the committee's list was most notable of all: all shipments should come in "unitized packaging."[11]

To military logistics experts, "unitized packaging" meant above all the ubiquitous five-ton Conex boxes that were carried with other cargo in the holds of breakbulk ships. Palletization, in which individual items were wrapped on a wooden pallet and moved on and off ships as a single piece, had come into commercial use in the early 1950s, and by late 1965 the Sharpe Army Depot, the main supply base in California, was promoting it for military cargo. McNamara, however, knew that the commercial world had moved far beyond small containers and wooden pallets. Leading shipping executives were invited to Washington, where they were shown film clips of sailors lowering cargo nets by rope and asked for advice. When Malcom McLean saw the film, a colleague recalled, "[h]e got obsessed with the idea of putting containerships into Vietnam. He was back and forth to Washington, talking to people, and they told him there isn't anything you can do in Vietnam."[12]

McLean finally got the ear of Frank Besson, the four-star general who headed army supply operations. At Christmas 1965, Besson agreed that McLean could take a look at the situation in Vietnam. McLean phoned his engineering chief, Ron Katims, and consulting engineer Robert Campbell, both just arrived in Europe to plan the start of Sea-Land's European service, and told them to meet his Pan Am flight in Paris the next morning. A day later, with their wool suits and overcoats, the trio was in steamy Saigon. They visited Da Nang and Cam Ranh Bay, got rounds of military briefings, and crossed paths with a delegation from the International Longshoremen's Association, which had arrived in Saigon on December 16. McLean's team concluded that containerization would solve much of the logistical confusion in Vietnam. He received the prompt endorsement of ILA president Teddy Gleason—the same Teddy Gleason who had fended off containerization in New York for the better part of a decade. On departing Vietnam in late January, Gleason urged the government to lease as many containerships as it could find.[13]

The military command was of two minds when it came to such a radical change in procedures. On the one hand, political pressure to bring in private-sector expertise was intense. At a top-level conference in Honolulu in January 1966, the Joint Chiefs announced a new policy "to contract for civilians to perform duties which they could accomplish, such as the operation of port facilities." On the other hand, no one in the military had experience with containerization. The MSTS had never leased a containership or run a supply exercise involving containers. The Defense Department's initial request for proposals to run a "containership service" to Vietnam involved only 7-foot Conex boxes carried as breakbulk cargo, not the far larger aluminum containers that could be handled by high-speed cranes and lifted directly onto truck chassis for delivery. The numerous port construction projects under way by early 1966, including the deepwater complex at Cam Ranh Bay, the Newport development in Saigon, and the new piers at Da Nang and other ports, all were proceeding as conventional breakbulk facilities. Container shipping, no matter how important in the commercial world, was not something that the military knew how to do.[14]

Through the winter of 1966, McLean struggled to convince the Pentagon that containerization could solve its logistical problems in Vietnam. In April, he finally obtained a foothold. Equipment Rental Inc., a new division of McLean Industries, was awarded an army contract to run a trucking operation at the Saigon piers. The contract had nothing to do with containers, but McLean was so eager to show what his company could do that Equipment Rental began moving freight two months ahead of schedule. In May, Besson asked the MSTS to give Sea-Land a contract to run three containerships between Oakland and the Japanese island of Okinawa, a major staging point for Vietnam. Sea-Land was to deliver 476 35-foot containers every 12 days. At the same time, with conventional carriers unable to supply additional vessels, the MSTS solicited bids for containerships to sail directly from the United States to Vietnam. Sea-Land would have to compete for the business—but as the largest containership operator by far, the only one sailing across the Pacific, and the only one with a handy supply of vessels equipped with the shipboard cranes needed to unload in Vietnam, it had the edge. When several competitors sought to submit a joint proposal, the MSTS refused to accept their bid. Things finally seemed to be going McLean's way.[15]

But Vietnam was still not ready for container shipping. The First Logistical Command, the agency in charge of the military's port, warehousing, and trucking operations in Vietnam, was notably unenthusiastic. The port delays facing military ships had eased during the first half of 1966, from an average of 6.9 days in February to 5.3 in July, and port congestion problems had less urgency for the forces on the ground. No cranes or storage areas for containers were available, and none were under construction. With Besson's Washington-based Army Materiel Command pushing container shipping aggressively, and with Sea-Land showing in Okinawa that container service required only half as many ships and one-sixteenth as much labor as breakbulk service, Westmoreland ordered the First Log to reassess its opposition. In July 1966, that command finally conceded that a containership service was desirable, but no sooner than October 1967. Sea-Land won a contract to open a containership service to Subic Bay, the huge U.S. naval base in the Philippines, but the bidding for service to Vietnam was postponed.[16]

It took yet another logistical breakdown in Vietnam to put an end to bureaucratic resistance. The improvements of the first half of 1966 were being reversed by August under the mounting flood of supplies and war matériel: shipments from California to Vietnam would be 55 percent higher in the year starting July 1, 1966, than they had been in the year just ended. Ships were once again arriving by surprise, tying up ports, while the lack of basic supplies was so severe that the air force had to airlift half a million tons of meat from Okinawa because there was none in-country. Nonmilitary cargoes, including U.S. aid, were taking two weeks or more to unload, and the backlog of military cargo was rising even as the army was grappling with such perplexing supply problems as a surplus of "ecclesiastical items, with greatly limited demand experience," and evacuating nonessential candelabras and crucifixes. Unloading at Saigon was hampered by resistance from Vietnamese longshoremen, who feared that the U.S. military would take over the docks and eliminate civilian jobs. When McNamara visited Vietnam in October 1966, the secretary of defense spent much of his time on port problems. "McNamara Acts to End Port Tie Up," the military newspaper headlined.[17]

In that environment, the MSTS reopened the bidding for containership service to Vietnam on October 14. Three companies were interested, but Sea-Land upstaged the competition by offering to provide not only containers but chassis, trucks, and terminals. To the government's surprise, Sea-Land proposed a fixed price per ton rather than the customary markup over its costs. After much negotiation, Sea-Land was awarded a $70 million contract in March 1967 to provide seven ships. Three of its largest vessels were to start sailing between Oakland and Seattle and Cam Ranh Bay by August. Sea-Land would install shore-based cranes there to handle the traffic. Three small vessels with shipboard cranes would cover routes between the West Coast and Da Nang starting in June, four months earlier than the First Logistical Command had deemed possible. The seventh ship was to serve as an interport shuttle within Vietnam. Sea-Land agreed to furnish refrigerated containers, to unload its own ships, and to deliver the containers with its own trucks and chassis to any point within thirty miles of its piers.[18]

Almost overnight, Cam Ranh Bay was turned into a large containerport. One of the DeLong piers was redesigned to support massive container cranes, and South Korean welders worked in intense heat inside the pier to reinforce the wooden deck. Crane rails were installed on the deck, while Sea-Land assembled two cranes in the Philippines from a patchwork of parts. In June, two barges, loaded with the partially built cranes, trucks to haul containers, campers for workers to live in, and even a sewage disposal plant, were floated across the South China Sea from the Philippines to Cam Ranh Bay. Then the realities of construction in the middle of a war zone intervened. The Da Nang operation got under way on August 1, a few weeks late, as the first containership to serve Vietnam, the *Bienville*, arrived from Oakland and unloaded its 226 containers in fifteen hours. The containerport at Cam Ranh Bay, though, did not see its first ship until November 1967, three months behind schedule. When it finally arrived, the *Oakland*, 685 feet long, delivered 609 35-foot containers—as much cargo as could be carried on ten average breakbulk ships hauling military freight to Vietnam.[19]

Another huge containership brought 600 or so containers to Cam Ranh Bay every two weeks. One-fifth of them were typically refrigerated units filled with meat, produce, and even ice cream. The remainder held almost every type of military supply except ammunition, which was not approved for shipment by container. Sea-Land's trucks carried about half the food to nearby bases, and the rest was transshipped to Saigon or other coastal ports on the rusty vessel Sea-Land used as a feeder ship. Sea-Land's state-of-the-art computer system at Cam Ranh Bay used punch cards to keep track of every container from loading in the United States to arrival in Vietnam to its return to America. Supplies flowed in, and the cargo backlog dissipated. "The port congestion problem was solved," the army's history of 1967 declared triumphantly. The seven Sea-Land containerships, MSTS Commander Lawson Ramage estimated, moved as much cargo as twenty conventional vessels, going far to alleviate the chronic shortage of merchant shipping.[20]

Just like commercial shippers, military shippers needed to learn how to use the container to best advantage. At first, they treated it simply like a big, empty box. Much of the initial load on Sea-Land's Okinawa service consisted of the smaller steel Conex containers. Four Conex boxes were loaded into each 35-foot Sea-Land container, meaning that a quarter of the weight inside the big container was nothing more than the steel of the smaller boxes. Logistics officers were uncertain how to make efficient use of the 45,000-pound load limit of an unrefrigerated container while utilizing all of its 2,088 cubic feet, so containers were frequently shipped underweight or with only half of their volume utilized. On the runs to Okinawa and to Subic Bay in the Philippines, for which the government had guaranteed a minimum number of containers, "It has become common practice . . . to take cargo that should go break-bulk and stuff it into Sea-Land's containers to fulfill the guaranteed minimums," an MSTS administrator complained. Military record keeping often was inadequate to take advantage of the container's efficiencies: in early 1968, after MSTS started shipping from California to Hawaii in Matson's containers, it found repeatedly that the containers listed on ships' cargo manifests were not necessarily the ones arriving in Honolulu.[21]

Such adjustments notwithstanding, in 1968, the first full year of container operations, one-fifth of all military cargo in the Pacific was shipped in containers. The containerized share of nonpetroleum cargo was probably closer to two-fifths. On-time performance was spotty—the shipboard cranes on the vessels serving Da Nang were continual problems, and repair often required delaying the ship—but Sea-Land managed to deliver between 1,230 and 1,320 containers to Cam Ranh Bay every month in its first year of operation. By June 1968, the army's supply operation in Da Nang was asking the MSTS for more container capacity. In October, Sea-Land added a fourth large C-4 containership to its Vietnam fleet and expanded its delivery area to fourteen interior locations. With other ship lines clamoring to enter the market, the Joint Chiefs sought to double container service to Vietnam at the end of 1968, only to run up against the fact that Sea-Land controlled the only deepwater container piers on the coast. Sea-Land itself offered to revise its contract and increase service. "Potential for multi million dollar cost reduction . . . is visualized," reported the army's assistant chief of staff for logistics.[22] Indeed, evidence of lower costs and reduced damage was already mounting impressively. McLean estimated in

1967 that loading a containership, sailing it to Vietnam, and unloading it there cost about half as much per ton as carrying the same cargo in a navy-owned merchant ship, not counting the reduction in loss and damage. Looking back from 1970, Besson calculated that the armed forces could have saved $882 million in shipping, inventory, port, and storage costs between 1965 and 1968 if they had adopted containerization when the buildup began.[23]

The military, hesitant to adopt container technology, now became its greatest advocate, and containerization became a tool for reform. MSTS chief Ramage warned in October 1968 that the potential of the container system could not be fully realized until supply agencies and military traffic managers revamped their procedures. The army instructed its depots to stop combining shipments that would need to be sorted in Vietnam and to abide by the Three Cs: one container, one customer, one commodity. In 1968, McNamara named Besson, containerization's most ardent military supporter, to head the Joint Logistics Review Board that would evaluate supply-system performance in Vietnam. Besson, with the support of McNamara's successor, Melvin Laird, used the opportunity to push for a more centralized system of military logistics, run by the army and built around land-sea transportation of intermodal containers. The venerable Conex container, designed for the days when the cranes on oceangoing ships could lift only five tons, would be phased out. The army bought its first commercial-size containers, twenty feet long, able to hold six and a half times as much freight as Conex boxes, and fully compatible with the newest containerships and cranes.[24]

Once the commitment to containerization was made, the transition was swift. Half of military cargo going to Europe was containerized by 1970. Military engineers drafted plans for portable terminals to unload containers in underdeveloped locations on short notice. The army and the navy tested containerization of ammunition by loading containers at factories and shipping them aboard a dedicated vessel to combat units in Vietnam; containers were a perfectly safe way to transport ammunition, the studies concluded, although artillery shells were so heavy that shipping them in containers longer than twenty feet left a lot of empty space. "Containerization cannot be considered just another means of transportation," Besson told Congress in 1970. "The full benefits of containerization can only be derived from logistic systems designed with full use of containers in mind." It was a conclusion that shippers in the private sector were only beginning to reach.[25]

Malcom McLean's persistence in pushing containerization was vital to the U.S. war effort in Vietnam. Without it, America's ability to prosecute a large-scale war halfway around the world would have been severely limited. The U.S. military would have experienced extreme difficulty feeding, housing, and supplying the 540,000 soldiers, sailors, marines, and air force personnel who were in Vietnam by the start of 1969. Continual headlines about theft, supply shortages, and massive waste would have caused domestic support for the war to erode even faster than it did. Containerization enabled the United States to sustain a well-fed and well-equipped force through years of combat in places that would otherwise have been beyond the reach of U.S. military might.

Containerization was vital as well to the growth of Sea-Land Service. Defense Department contracts had long been life-or-death matters for U.S.-flag ship lines with international routes. Until 1966 and 1967, when military transportation agencies first put their shipping needs up for competitive bidding, the military's freight on a given route had been divided up among all the U.S.-flag lines serving that route, guaranteeing every carrier a piece of the pie. The Defense Department's involvement with container shipping had been minimal, and it had never tendered freight to Sea-Land, even on domestic routes to Puerto Rico and Alaska, because the military was not equipped to use 35-foot containers. Vietnam broke the barrier. From almost nothing in 1965, Sea-Land's Defense Department revenues rose to a total of $450 million between 1967 and 1973. In the peak year, fiscal 1971, $102 million of Vietnam-related contracts accounted for 30 percent of the company's sales.[26]

Like everything else Malcom McLean did, venturing into Vietnam entailed considerable risk in hopes of large reward. The cost and risk of reinforcing the pier at Cam Ranh Bay, assembling the cranes, floating equipment and vehicles across from the Philippines, and building the truck terminals were entirely Sea-Land's. The U.S. government was liable only for damage to Sea-Land's trucks and equipment caused by enemy fire. It did not furnish men or material to help Sea-Land get its operations up and running. In a place where replacement parts could not simply be ordered from a nearby distributor, the chance that something would go wrong, blowing budgets and cost calculations, was very high. McLean was running a commercial operation in a war zone, and betting that he could control costs well enough to make a profit from his fixed-price bid.[27]

The gamble paid off hugely. In return for his willingness to bear risks on the cost side, McLean negotiated contacts that assured Sea-Land's revenue. The MSTS guaranteed a minimum number of containers on each trip to Okinawa and the Philippines. To Vietnam, the rate per container was fixed, but Sea-Land's contract required the government to offer it "all of its containerizable cargo" outbound from Seattle and Oakland, leading to extremely high utilization: in 1968, 99 percent of the container slots were filled.

No figures are available on profits from the Vietnam service, but high capacity utilization must have translated into robust profitability. Each round trip from the West Coast to Cam Ranh Bay brought Sea-Land more than $20,000 per day, and each smaller vessel sailing to Danang took in about $8,000 per day, at a time when MSTS was paying $5,000 per day to lease large breakbulk ships. Sea-Land was also protected against the risk that its containers would vanish into the jungles of Vietnam. A central control office kept track of each container, and containers had to be emptied and returned within specified time limits or the unit holding them had to pay extra charges.[28]

The contracts also permitted Sea-Land to make some extra profit. The Philippines service was supposed to call at both Manila and Subic Bay, but after Sea-Land threatened to charge $500 an hour for port delays in Manila, the air force decided it could pick up its spare parts just as easily at Subic Bay; the contract remained unaltered, and Sea-Land was able to save $6,800 per trip by skipping the stop in

Manila. Sea-Land collected additional revenue any time an army unit in the field restuffed a container with material to be "retrograded" to the United States, because its MSTS contracts were westbound only. These payments for eastbound freight were pure profit and were high enough that in March 1968 the U.S. command revoked permission to retrograde freight via container because, it was explained delicately, Sea-Land's charges were "not rate-favorable."[29]

Malcom McLean was not one to pass up an opportunity for profit. Now, an obvious one awaited. He had six ships, three large and three small, sailing between the U.S. West Coast and Vietnam. Westbound, they were loaded nearly full with military freight. Eastbound, they carried little but empty containers. The rates paid by the U.S. government for the westbound haul covered all costs for the entire voyage. If Sea-Land could find freight to carry from the Pacific back to the United States, the revenue would be almost entirely profit. Thinking the situation through, McLean had another of his brainstorms: why not stop in Japan?

Japan was the world's fastest-growing economy during the 1960s: between 1960 and 1973, its industrial output quadrupled. Already the second-largest source of U.S. imports, by the late 1960s Japan was quickly moving up the ladder from apparel and transistor radios to stereo systems, cars, and industrial equipment. It took little imagination to envision the potential for container shipping. The Japanese government had used a typical industrial policy exercise to endorse containerization in 1966, when the Shipping and Shipbuilding Rationalization Council urged the Ministry of Transport to eliminate confusion and excessive competition in order to derive maximum national benefit from the new technology. The council called for container service between Japan and the U.S. West Coast to begin in 1968, with services to the U.S. East Coast, Europe, and Australia to begin by 1970. It asked the government to build container terminals initially in the Tokyo/Yokohama and Osaka/Kobe areas. The government, the council said, should require Japanese and foreign shipping lines to form consortia to operate the containerships and terminals, but should structure that cooperation to avoid undermining Japanese ship lines. If all went as planned, the council said, half of Japan's exports would be containerized by 1971, traveling on 12 huge ships carrying 1,000 containers each.[30]

The government acted with unusual speed. Delegations visited Oakland and other U.S. ports to learn how a containerport should be run. New port legislation was approved in August 1967, and Japan's first two container cranes began operation in Tokyo and Kobe by the end of the year. Matters on the land side were not quite so easy. Standard trucks in Japan hauled loads smaller than 11 tons, and in any case highway regulations barred full-size containers, except on a handful of new toll roads. The Japanese National Railway was not equipped to carry containers longer than 20 feet. The type of intermodal transportation being practiced in North America and since 1966 in Europe, with containers transferred almost seamlessly from a ship to a truck or railcars to the recipient's loading dock, would not be simple to replicate in Japan.[31]

The first to try was Matson Navigation. In February 1966, Matson won U.S. government approval to operate an unsubsidized freight service between the West Coast, Hawaii, and the Far East. The company's management had visions of fast ships racing across the Pacific with television sets and wristwatches, discharging cargo at Oakland directly to special trains that would carry it east. On the return trip, there might be military cargo for the U.S. bases in Japan and South Korea. The key assumption was that Matson would have two or three years to capture the business of Japan's leading exporters before other ship lines entered the market. Matson sent two of its C-3 breakbulk ships to a Japanese shipyard to be converted into self-unloading ships able to carry 464 containers and, in recognition of Japan's rapidly growing auto exports, 49 cars. It ordered two high-speed containerships in Germany for delivery in 1969. To encourage Japanese customers, it entered a joint venture with a Japanese ship line, Nippon Yusen Kaisha (N.Y.K. Line). In September 1967, before a single container crane was operating in the country, Matson began service to Japan.[32]

Competitors were not far behind. In January 1968, four Japanese ship lines signed leases for container berths in Oakland. In March

1968, the same month that army officers in Vietnam were ordered not to ship cargo back to the States via Sea-Land, Sea-Land announced that it would provide weekly sailings from Japan.

Like almost everything else connected with Malcom McLean, Sea-Land's entry in Japan stemmed more from instinct than from analysis. "We've got these empty ships coming back from Vietnam," former Sea-Land executive Scott Morrison recalled. "So we have a meeting, and Malcom says, 'Anybody know anybody at Mitsui?' " McLean handed around the Japanese trading company's annual report and announced that he wanted to fly to Tokyo to meet its president. Two weeks later, a huge delegation from Mitsui was touring Sea-Land's docks at Elizabeth. Malcom McLean wanted nothing to do with joint ventures, but he hired a Mitsui group company to build Sea-Land a terminal in Japan. Another Mitsui company agreed to be Sea-Land's agent, and a third agreed to handle domestic trucking within Japan. With its ship operating costs fully covered by its military contracts for Vietnam, Sea-Land was guaranteed to make money no matter how little cargo it picked up in Japan.[33]

The first Japanese containership, owned by Matson partner N.Y.K. Line, completed its maiden voyage to America in September 1968. Six weeks later, having been duly admitted to the transpacific conference, Sea-Land began six sailings a month from Yokohama to the West Coast, its ships laden with televisions and stereos produced by Japanese factories. Other Japanese carriers entered as well. The Japan–West Coast route, which had no commercial container service at all before September 1967, was suddenly crowded with ships needing to be filled. Seven different companies were competing for less than seven thousand tons of eastbound freight each month by the end of 1968, and more were about to join. The lack of business proved to be only temporary. The cargo would soon come, in a flood that no one imagined.[34]

## Chapter 10

# Ports in a Storm

**The Coastwise Steamship Company** had been created to serve the paper industry. Since the 1930s, its ships had picked up rolls of newsprint from Crown Zellerbach's mills at Port Angeles and Camas, in Washington State, and hauled them down the coast to California. The business was reliable—until the Southern Pacific and Union Pacific railroads went after the traffic. They exempted newsprint from their general rate increases in the 1950s, and then they began to lower their rates in order to capture the cargo. To compete, the ship line had to slash its tariff for newsprint from $32 to $18 per ton. By 1958, Coastwise Steamship Company was near insolvency, and the Pacific coastal trade in newsprint was dead.[1]

As with newsprint, so with cotton and oranges, chemicals and lumber. American coastal shipping withered during the 1950s in the face of a competitive onslaught by trains and especially trucks. The number of cargo ships engaged in coastwise trade, aside from tankers, fell from 66 in 1950 to 35 in 1960, and the total tonnage of active coastal ships dropped by one-third. The waterfronts that had once been vital to local economies fell into decay as the ships stopped coming. Docks were abandoned, warehouses bricked up. Over the thirteen years from 1945 to 1957, total investment on construction and modernization at all North American ports outside New York came to a meager $40 million a year.[2]

Two events tied to containerization brusquely awakened the somnolent port industry. In December 1955 came the Port of New York Authority's decision to turn 450 acres of New Jersey salt marsh into a futuristic port for containerships, a scheme utterly beyond the capability of any other port in the world. Less publicized, but even more ominous, were the changes in Malcom McLean's container service. McLean had gone to great trouble to secure rights to serve ports from Boston to Galveston, and the fully containerized ships Pan-Atlantic introduced in 1957 were given expensive onboard cranes so they could call at almost any port. The plan was for Pan-Atlantic's vessels, like traditional ships, to call at all the important towns along its routes. That plan was scrapped almost immediately, as Pan-Atlantic reshaped its service to focus on four ports—Newark, Jacksonville, Houston, and San Juan, Puerto Rico—and cut back or eliminated other stops.

These two unrelated developments—the rise of New York, the neglect of Tampa and Mobile—revealed the economics that would affect seaports as container shipping grew. For ports, capturing container traffic was going to be expensive, requiring investments out of all proportion to what had come before. For ship lines, the days when vessels meandered along the coast, calling at every port in search of cargo, would soon be over. Every stop would mean tying up an expensive containership that could generate revenue and profit only when it was on the move. Only ports that could be relied upon for large amounts of freight were worth a visit, and all others would be served by truck or barge.

By the late 1950s, the lesson for public officials already was clear. As container shipping expanded, maritime traffic would be drawn to a small number of very large ports. Many established centers of maritime commerce would no longer be needed, and ports would have to compete to be among the survivors. Most important, the scope of the investment that would be required—filling the sea to provide hundreds of acres of solid waterfront land, building enormous cranes and marshaling yards, creating off-dock infrastructure such as roads and bridges—was far beyond the ability of ship lines to finance. If they hoped to capture the jobs and tax revenues that would come with being a major transportation center, government agencies would have to be far more closely involved in financing, building, and running ports than ever before.[3]

The new economic reality was grasped first by the ports along America's West Coast. The Pacific ports were backwaters during the 1950s. Domestic maritime commerce was fading, except where there was no alternative, such as Seattle's trade with Alaska and runs between California ports and Hawaii. America's international trade was overwhelmingly oriented toward Europe; excluding petroleum and other tanker cargoes, barely 11 percent of imports and exports passed through Pacific ports in 1955. Counting petroleum and chemicals, all the West Coast ports together handled less cargo in a year than New York City alone.[4]

Above all, the Pacific ports were victims of geography. Although the port cities themselves were large and growing quickly, their hinterlands were very thinly populated. All of California beyond Los Angeles and San Francisco Bay had barely six million inhabitants in 1960, and the eight Rocky Mountain states, stretching a thousand miles to the east, had a combined population smaller than New York City's. The first important city inland of Seattle was Minneapolis, sixteen hundred miles away. While the West's industries were expanding rapidly, only Los Angeles–Long Beach had a manufacturing base that could rival the factory centers of the East and Midwest. Whereas Baltimore and Philadelphia could handle the foreign-trade needs of Pittsburgh and Chicago, the West Coast ports had no similar domestic markets, and potential trading partners across the Pacific, such as Korea, China, and the nations of Indochina, were cut off by war or politics. With cargo flows flat, except for oil, the ports had no avenue for growth. Seattle's docks saw 10 percent less cargo in 1960 than in 1950. Tacoma, mainly a lumber port a few miles south on Puget Sound, had lost one-third of its traffic over the same decade as the timber companies shifted their business to the rails. Tonnage in Portland fell by 17 percent. The only West Coast port that grew during the 1950s was Los Angeles, which invested in new wharves and warehouses in hopes of challenging the regional dominance of San Francisco.[5]

Containerization offered a chance to escape these geographic constraints. Instructed by the efforts of Matson, whose studies of container service to Hawaii envisioned the Pacific ports becoming hubs for truck pickups and deliveries in Denver and Salt Lake City, civic leaders up and down the West Coast looked afresh at their deteriorating waterfronts. The action began in San Francisco, where the state of California oversaw ninety-six outmoded piers. Many were narrow wooden docks unchanged since the 1920s, and even those that were structurally sound were not designed for large trucks. Consultants recommended a new "superterminal" big enough to handle eight oceangoing ships, to be located at Army Street south of downtown. In 1958, California voters approved $50 million of bonds for the port, a substantial amount for the time.[6]

Seattle soon followed San Francisco's lead and hired a consultant to help save its port. All of Seattle's twenty-one piers predated the end of World War II, and most had been built for sailing ships shortly after 1900. By the late 1950s, only six piers were in part-time use for general cargo, and the port's tax-advantaged foreign trade zone was so quiet that the Seattle Port Commission, a county agency, considered closing it. A local television documentary about the sorry state of the port turned the political situation around in 1959. Business leaders formed a port committee, and in July 1960 the Port Commission unveiled a $32 million construction plan that included two container terminals. The port was suddenly the center of attention: in November 1960, no fewer than seventeen candidates stood for election to the Port Commission. Voters approved a $10 million bond issue for the first stage of construction.[7]

Los Angeles, where the new Long Beach and Harbor Freeways had been designed to move trucks out of the port, kept pace. City officials tried aggressively to persuade voters of the port's economic importance, and they were rewarded with authority to issue revenue bonds—bonds that would be serviced with ship lines' lease payments—in a 1959 referendum. By 1960, Matson's two-year-old container service to Hawaii, using facilities Matson had improved at its own expense, moved seven thousand containers across the docks. It was a tiny operation compared with Sea-Land's base at Newark, but it was enough to make Los Angeles the largest containerport on the West Coast. The municipal port department, with the strong backing of city hall, promptly embarked upon a five-year, $37 million program to build wharves and cranes for containerships.[8]

The most dramatic change, though, occurred in Oakland, on the east side of San Francisco Bay. Through the start of the 1960s, Oakland was a sleepy agricultural port one-third the size of Long Beach, Seattle, or Portland, and far smaller than San Francisco. Its waterfront was lined with industries—a dog food plant, a dry ice plant, a brake shoe factory—that had long since ceased to be important port users. Oakland had almost no incoming traffic; typically, European ships would arrive at San Francisco, unload, and then sail across to Oakland to take on canned fruit, almonds, and walnuts for the voyage home. The Oakland Port Commission, a city agency, had issued its first revenue bonds in 1957 to repair a few old docks, but it had no grander plans. Then came an unexpected development. Officials in San Francisco, where Matson had based its container service to Hawaii, ignored Matson's request for a separate container terminal, because the city's port director thought container shipping a passing fad. When Matson installed the world's first land-based container crane in 1959, it was built not in San Francisco, the West's greatest maritime center, but in Alameda, a small city within plain view of the Oakland docks.[9]

Matson's operation focused the attention of Oakland port officials on container shipping. In early 1961, they learned of American-Hawaiian Steamship's application for government subsidies to build a fleet of large containerships. The vessels would run through the Panama Canal, mainly carrying fruit and vegetables from California canneries to East Coast markets. This was a natural cargo for Oakland to capture. Port director Dudley Frost and chief engineer Ben Nutter prepared two binders of facts and figures, added leather covers stamped "American-Hawaiian Steamship Company," and flew east in April 1961. Meetings with government and industry officials in Washington changed their plans. "Somebody said, 'Oh, forget those guys. They're no good. Go and see Sea-Land,'" Nutter recalled. "I said, 'Sea who?' " The cover on one of the binders was quickly replaced by one stamped "Sea-Land," and Frost and Nutter made their way to Port Newark. A Sea-Land executive stopped their presentation to inform them that it had already decided to run containerships from Newark to California. If they could offer a suitable site at a reasonable price, Sea-Land would establish its northern terminus at Oakland.[10]

Oakland had never hosted a containership, but it immediately began to promote itself as a future containerport. Nutter dreamed up a lease very different from the norm of so many cents per ton: Sea-Land would pay a minimum fee high enough to cover the cost of building its terminal and would pay more as its tonnage rose, but beyond a certain point there would be no additional charge. That "mini-max" provision gave Sea-Land an incentive to pump cargo through Oakland, because once its tonnage exceeded the upper limit, its average port cost per ton would plummet. Oakland spent $600,000 to upgrade two berths, and the federal government agreed to deepen the harbor from 30 to 35 feet to permit larger containerships in the future. In September 1962, Sea-Land's *Elizabethport*, the largest freighter in the world, steamed through the Panama Canal to call at Long Beach and Oakland.[11]

The West Coast ports already had more than doubled their annual investment over the span of two years, and the competitive battle had barely begun. Oakland, with two railroad yards right next to the port, appeared to have the edge. Los Angeles countered with another bond issue in 1962, this time for $14 million. Then the adjacent port of Long Beach reemerged. Long Beach had struggled through the 1950s after the pumping of oil from beneath the harbor caused the harbor floor to subside and docks to collapse. When the mess was finally cleaned up, the city-owned port found itself with a deeper harbor than Los Angeles. It snagged Sea-Land's southern California terminus in 1962 and dedicated its oil revenues to constructing a 310-acre landfill. The twin ports were soon waging a rate war that left Los Angeles, which had no oil and needed to turn a profit, at a disadvantage. Los Angeles and San Francisco asked the Federal Maritime Commission, the regulator, to block Sea-Land's deals at Long Beach and Oakland on the ground that they involved unfair subsidies, to no avail. Up the coast, Seattle, where two container terminals were under construction, announced a $30 million program to build more in August 1962, even though the seasonal Alaska Steamship operation was the only container business in the port. The ports were suddenly full of activity. The flow of nonmilitary cargo, flat for a decade, rose by one-third between 1962 and 1965.[12]

Then Oakland raised the bar. The port's ambitions centered on an area known as the Outer Harbor, bisected by an embankment that had once carried passenger trains to their terminus at a ferry landing. The Port Commission was out of money after expanding the Oakland airport, but the Bay Area Rapid Transit District, which began designing its regional rail system in 1963, came to its rescue. In return for permission to tunnel beneath port property, the rail agency agreed to clear abandoned buildings along the embankment, construct a 9,100-foot dike, and fill the enclosed area with dirt excavated in tunnel construction. Oakland designed an enormous terminal for the 140-acre site, with 12 berths, wharves 78 feet wide—wide enough to erect cranes that would straddle the dockside railroad tracks—and the ability to accommodate ships of almost any length. The outermost dock was barely a mile from 60-foot water depth in San Francisco Bay, assuring that the port would remain accessible as ships became larger.

Oakland delegations visited Japan, in 1963, and Europe, in 1964, and learned that several ship lines were interested in containerization. None of them was prepared to sign a contract that would allow the port to sell revenue bonds, but a timely $10 million grant from the federal Economic Development Administration, intended to generate jobs in the depressed city, supplied construction funds. A new terminal was rushed into construction, without a tenant, before new environmental regulations took effect. With that project under way, Sea-Land decided in 1965 that it needed more space, and signed a contract for a twenty-six-acre terminal with two big shoreside cranes. A few months later, Matson, hitherto a purely domestic company, announced that Oakland's new landfill would be its base for an unsubsidized service carrying containers between the Pacific Coast and Asia.[13]

Behind this frenzied expansion of long-neglected ports was the emergence of an entirely new line of thought about economic growth. Manufacturing was almost universally regarded as the bedrock of a healthy local economy in the 1960s, and much of the value of a port, aside from jobs on the docks, was that transportation-conscious manufacturers would locate nearby. As early as 1966, though, public officials in Seattle were sensing that their remote city, with little industry, might be able to develop a new economy based on distribution rather than on factories. The lack of population close at hand would be no obstacle; Seattle could become not merely a local port for western Washington but the center of a distribution network stretching from Asia to the U.S. Midwest. "Commodity distribution has grown out of the dependent sector to link production and consumption," port planner Ting-Li Cho wrote presciently. "It has become an independent sector that, in return, determines the economy of production and consumption." Much the same message, with opposite implications for the local economy, was transmitted that same year to San Francisco officials by consulting firm Arthur D. Little. A great proportion of the city's wholesaling, trucking, and warehousing business would soon relocate to be near the emerging port facilities on the eastern side of San Francisco Bay, Little warned, because it no longer needed to be close to the other business activities in San Francisco.[14]

Fortified by increasing confidence in their economic prospects, Seattle, Oakland, Los Angeles, and Long Beach were in a constant state of construction. Sea-Land opened container service from Seattle to Alaska in 1964, a few days before the devastating Alaska earthquake created huge demand for construction supplies and relief shipments. The U.S. buildup in Vietnam sent a flood of aid cargoes through Los Angeles and Long Beach. Oakland's nonmilitary container shipments, just 365,085 tons in 1965, quadrupled to 1.5 million tons in 1968 and doubled again to 3 million tons in 1969, as Japanese and

European ship lines began pushing containers through the port. By now, nearly 60 percent of Oakland's cargo was moving in containers. Los Angeles attracted four Japanese lines to a new terminal. Long Beach, anticipating that the shallower Los Angeles harbor would soon force ship lines to look elsewhere, began building space to handle ten ships at a time at three new terminals, including a hundred-acre site for Sea-Land alone. Seattle began no fewer than three new terminals with no tenants in place, driven by a new imperative: if the supply of terminal space was not adequate to meet the demand for container shipping, the ships might go somewhere else.[15]

Two traditional maritime centers stood aside from the frenzy. Portland, which handled nearly as much cargo as Seattle during the 1950s, could not muster the money or resources to build a containerport. The consequences were severe. Seattle's foreign trade more than doubled between 1963 and 1972, but Portland's barely grew. Once Japanese containerships began to call at Seattle, in 1970, Portland found itself receiving Japanese goods by truck from Seattle rather than by ship from Yokohama. San Francisco faced more fundamental problems, because the city's location, on a congested peninsula with direct rail access only to the south, was ill-suited to handling freight to and from points east. City officials managed to get dredging under way in 1968 after wresting control of their port away from the state, but actual construction of container piers was delayed so long that even American President Lines, whose predecessor companies had made their home in the city for more than a century, finally decamped for Oakland. The city's plans kept changing, even as Seattle, Oakland, Los Angeles, and Long Beach opened one huge purpose-built container terminal after another. In 1969, a Swedish ship line's massive neon sign, a fixture of the San Francisco waterfront for decades, was moved across to the Oakland side of the bay, the glowing words "JOHNSON LINE" offering San Franciscans a nightly reminder that their city's time as a major port was over.[16]

The explosion of port construction on the Pacific coast, starting in the 1950s, had no counterpart on the other side of the country. After Grace Line abandoned its ill-fated container service to Venezuela,

Sea-Land was left as the only company using dedicated ships to move containers in the East. The many other lines that advertised container service handled boxes along with mixed freight and did not need special cranes or storage yards. More important, the extraordinary enthusiasm for containerization shown by Pacific Coast ports was little in evidence on the Atlantic, except at Sea-Land's home port in New Jersey. The West Coast ports that embraced containerization, save for Los Angeles, were withering in the late 1950s, and they saw salvation in the new technology. The ports on the Atlantic and the Gulf of Mexico had a steadier flow of cargo: as late as 1966, nine of the ten largest maritime routes for U.S. international trade passed through ports on the East Coast or the Gulf, and only one touched the West Coast. The eastern ports had less to gain from containerization, and, outside of New York, their eagerness to invest millions of dollars of public money was correspondingly less acute.[17]

The Port of New York Authority docks in Newark and Elizabeth were expanding without cease as container shipping became an international business. By 1965, half a dozen ship lines announced plans to launch container services from New Jersey docks to Europe in 1966, and dozens of new ships were on order. The embrace of containerization was not repeated up and down the coast. The obstacles were the same almost everywhere: labor and money.

On the labor front, the New York Shipping Association and the ILA had negotiated smaller gang sizes and a guaranteed annual income for displaced longshoremen in 1964, but union locals at other ports, save Philadelphia, had not. Unlike New York, where the efforts of the Waterfront Commission had eliminated casual dockworkers in the 1950s, most other East and Gulf Coast ports had large numbers of part-time longshoremen well into the 1970s. Boston longshoremen worked an average of one and a half days per week, New Orleans longshoremen two days. If containers were allowed, these part-time jobs were likely to vanish as the industry shifted to a permanent, full-time workforce. The ILA had seen how container shipping had decimated its membership rolls in New York, and it was loath to tolerate it in other ports until income guarantees were in place.[18]

Interunion disputes were a problem as well. In Boston, the Massachusetts Port Authority spent $1.1 million to build a container crane in 1966 so that Sea-Land ships could call on their way between New York and Europe, but the terminal was kept closed first by a dispute between the ILA and port employers and then by a dispute between the ILA and the Teamsters union. Sea-Land and its competitors soon learned that they could operate more profitably by trucking Europe-bound containers to New York and having their ships bypass Boston, and port traffic never recovered. In New York and other ports, the Teamsters objected to contracts that guaranteed ILA members the right to consolidate partial loads into containers at inland warehouses. The ILA viewed these contracts as essential to preserving its members' jobs as traditional maritime functions moved away from the waterfront, but the Teamsters viewed them as infringements on their own jurisdiction over the warehouse industry. Contests over which union's members would do the work persisted until 1970.[19]

Aside from Baltimore, most ports found the cost of building dedicated container facilities so daunting that they postponed decisions. The city of Philadelphia, short of funds, did nothing toward containerization until worried business leaders pushed for the creation of a port corporation with authority to issue bonds, in 1965. Only after a study predicted that Philadelphia would soon be losing a million tons of freight per year did the new corporation reluctantly invest in a container terminal, which opened in 1970. Miami built ramps for roll on–roll off ships but no specialized wharves for containerships. Gulf Coast ports such as Mobile decided not to invest in containerization because many of the Caribbean islands to which they exported were too small to require large containers. New Orleans, long the largest Gulf port, handled containers over the same wharves used for other types of cargo; its first purpose-built container terminal, located on a canal that subsequently proved too shallow, did not open for business until 1971. Houston, Sea-Land's original western terminus, invested sooner, and it firmly established itself as the premier containerport on the Gulf.[20]

The net result of these decisions was that a single port, the Port of New York Authority's complex at Newark and Elizabeth, dominated container shipping in the East. In 1970, only one other harbor between Maine and Texas, the Hampton Roads of Virginia, could boast even one-ninth the container capacity of the wharves on New York harbor. The emerging economics of container shipping meant that the laggards faced potentially serious consequences. The newly built containerships coming on the scene in the late 1960s carried far more cargo than the vessels they supplanted; even if the total amount of cargo grew, fewer voyages would be required. Shipowners wanted to keep their ships under way to recover the high construction cost, so they preferred that each voyage involve only one or two stops on either side of the ocean rather than four or five. Secondary ports would not see transatlantic ships but would get only feeder services to bigger ports. Once a port had slipped from the first rank, it would have a hard time climbing back: a less active port would have to spread the cost of building a capital-intensive container terminal across fewer ships, and its higher costs per ship would in turn drive away business. Ports that came late to the container game either would have to take huge risks in hopes of attracting tenants or would need to find a large ship line willing to help bear the costs of establishing a major new port of call.[21]

Some late starters did succeed in turning themselves into major containerports with relentless investment. The first containers passed through Charleston, South Carolina, in 1965, but the port had only one berth and no specialized crane to handle containers. Then, in the late 1960s, Sea-Land decided to expand its very modest Charleston operation. The state-owned port began an ambitious development program, growing from an initial fifteen-acre container terminal into three terminals covering nearly three hundred acres by the early 1980s. Charleston, with almost no container traffic in 1970, ranked eighth among continental U.S. ports by 1973 and climbed to fourth by 2000. The nearby port of Savannah, Georgia, another late starter, followed a similar trajectory after belatedly installing its first container crane in 1970. But as container shipping made the transition from the emerging technology of the early

1960s to the booming business of the early 1970s, the opportunity for ports to establish themselves as major maritime centers was diminishing rapidly. "The maintenance of a major port in every major coastal city is no longer justified," a government report declared in the early 1970s. Such long-standing ports as Boston and San Francisco, Gulfport, Mississippi, and Richmond, California, would have to find other roles in the container age.[22]

The first decade of container shipping was an American affair. Ports, railroads, governments, and trade unions around the world spent those years studying the ways that containerization had shaken freight transportation in the United States. They knew that the container had killed off thousands of jobs on the docks, rendered entire ports obsolete, and fundamentally altered decisions about business location. Even so, the speed with which the container conquered global trade routes took almost everyone by surprise. Some of the world's great port cities soon saw their ports all but disappear, while insignificant towns on little-known harbors found themselves among the great centers of maritime commerce.[23]

Nowhere was the transformation more tumultuous than in Britain. London and Liverpool were by far Britain's biggest ports in the early 1960s, but their business was remarkably close to home. Exporters and importers tended to use the nearest port in order to minimize trucking costs. About 40 percent of Britain's exports in 1964 originated within twenty-five miles of their port of export, and two-thirds of all imports traveled fewer than twenty-five miles from the port of discharge. London, a huge industrial center in its own right, and Liverpool, serving the industrial heartland in the English Midlands, each handled one-fourth of British trade, with dozens of other ports claiming small shares.[24]

Both the London and the Liverpool docks were run by local government agencies, which since the 1940s had walked a tightrope between improving operations and antagonizing the powerful Transport and General Workers Union; as one careful student aptly described the situation, the docks had been modernized in "a leisurely manner." Dozens of small stevedoring companies competed to load or unload each vessel, and the stevedores in turn hired longshoremen by the day. These transitory arrangements involving lightly capitalized companies provided no incentive to make long-term investments in automation. Although productivity gains had been sluggish since the mid-1950s, pay gains had been robust. The average full-time docker earned about 30 percent more than the average male worker in Britain in the mid-1960s, compared with a gap of about 18 percent a decade earlier.[25]

Numerous government commissions had studied ways to make the ports more efficient. An inquiry in 1966 had called for a reduction in the number of stevedore companies, in the expectation that the survivors would be larger, more professional, and better able to finance equipment for efficient cargo handling. In return, the government promised that automation would not lead to redundancies among dockers. Given time, perhaps a deal could have been struck: on both coasts of the United States, the union agreements that opened the way for containerization had taken half a decade to arrange. No time was available in Britain, because technological change was forcing its way onto the docks. In March 1966, United States Lines carried the first large containers, along with other freight, on a voyage from New York to London. The following month, Sea-Land's *Fairland* steamed across the North Atlantic to Rotterdam, Bremen, and the Scottish port of Grangemouth, carrying only containers. With barely a year's notice, Rotterdam and Bremen had lengthened docks, deepened channels, and begun installing container cranes. London had not—and *Fairland* did not bother to call.[26]

London's formidable docks, it was obvious, were not well suited for container shipping. The docks were grouped in sheltered enclosures off the Thames that were difficult even for conventional ships to navigate; large vessels had to unload into lighters nearer the mouth of the river. Labor issues aside, transferring huge cargo containers from oceangoing ships to lighters made no economic sense, and the prospect of hundreds of lorries hauling 40-foot loads through the narrow streets of East London was a nightmare. Liverpool's aging docks had similarly few attractions for container operators. The British Transport Docks Board, the government's oversight agency, turned to consultants McKinsey & Company for advice. McKinsey predicted that container shipping would quickly consolidate around a few companies using gigantic ships carrying standardized containers. Ports, it said, would need to be very large to gain economies of scale in transferring containers between ships, trains, and trucks at high speed. Containerization could cut Britain's ocean freight bill in half, McKinsey found—but only if a single huge port were to handle all cargo to and from North America and then use unit trains to link the port to other parts of the United Kingdom. A simultaneous study by consultant Arthur D. Little predicted that in 1970, ships would carry the equivalent of 1,800 20-foot containers each week from America to Britain, and 1,580 from Britain to America. Every aspect of these findings represented a threat to the power of the Transport and General Workers Union: there would be fewer ships, fewer ports, and fewer workers at each port. An important part of traditional dock work, the loading and unloading of cargo, would move to warehouses miles inland, where dockers surely would not be employed.[27]

The British Transport Docks Board and local port agencies agreed on major investments that would total £200 million ($550 million at the 1967 exchange rate) between 1965 and 1969. The largest was the Port of London Authority's £30 million ($83 million) container complex at Tilbury, a longtime port twenty miles down the Thames. Away from the traffic congestion of central London, and with populous southeast England at its doorstep, Tilbury had the potential to become Europe's premier containerport, or so the government hoped. There were to be five deepwater berths for containerships, each with twenty acres of land to store containers. Another containerport was built at Southampton, southwest of London, and the Mersey Docks and Harbour Board began a container terminal at Seaforth, north of Liverpool close to the deep water of the Irish Sea.[28]

Tilbury's opening in 1967 was accompanied by a "voluntary" severance scheme for dockworkers, funded by charges on cargo at major ports. The union soon accused employers of abusing the rules to get rid of workers, and it objected to the new government policy encouraging permanent employment rather than daily hiring on the docks. Reaching back to a tactic tried by the International Longshoremen's Association in New York a decade earlier, the union imposed a ban on containers at Tilbury from January 1968.[29]

The Transport and General Workers Union was powerful, but not omnipotent. It had never bothered to sign up members at the tiny port of Felixstowe, located on a North Sea estuary ninety miles northeast of London. Felixstowe, one of the hundreds of towns on Britain's coasts, had two docks owned by the Felixstowe Railway and Dock Company, a private company controlled by an importer of grain and palm oil. The docks had been destroyed in the storms of 1953, and by 1959 the only activity involved ninety permanent workers who unloaded tropical commodities into a few storage tanks and warehouses. Felixstowe had no general-cargo business to protect, no militant unions, and, because it had never employed casual dock labor, no requirement for ship lines to contribute to the national dockworker severance scheme.

In 1966, while the British government was trying to convince container ship lines to call at Tilbury, Felixstowe's owners had had the foresight to strike a private deal with Sea-Land Service. They spent £3.5 million pounds ($10 million), a fraction of the government's outlay at Tilbury, to reinforce a wharf and install a container crane. Sea-Land opened service in July 1967 with a small ship shuttling containers back and forth to Rotterdam, and soon added ships directly from the United States. In 1968, with Tilbury closed by the strike, the hitherto obscure Port of Felixstowe became Britain's largest container terminal. U.S. Lines finally was able to use Tilbury after reaching a union agreement, but the port remained closed to most other container carriers, including British lines. By 1969, Felixstowe, with two or three North Atlantic sailings each week and several feeder services running across the North Sea to Rotterdam, handled 1.9 million tons of general cargo, every bit of it in containers.[30]

Tilbury's prolonged closure hit hard at the two consortia of British carriers that had planned to launch container services there, one across the North Atlantic and the other to Australia. They struck back in the traditional way: by trying to stifle their competitors. Sea-Land's request to join the conferences that set rates between the United States and Britain was rejected until the company filed an antitrust suit in British courts. When a small U.S. company, Container Marine Lines, offered Scottish distillers a through rate covering shipments between their plants and U.S. ports, including land transportation from Scotland to Felixstowe, the conference objected that a through rate would lead to "regulatory disintegration." Only the threat that the U.S. Federal Maritime Commission would curb the conference's right to set rates forced it to accept more competition.[31]

Felixstowe's fortune came directly at London's expense. London's port had been busy through the mid-1960s. The shift to container shipping boosted average tonnage per man-hour 66 percent in just four years. The abrupt fall in costs at other ports drove the London docks to collapse. As Tilbury opened, the famous East India Dock closed without warning in 1967. As Felixstowe burgeoned, the St. Katherine Docks, adjoining the Tower of London, were shut in 1968. The nearby London Docks followed immediately, and the Surrey Docks, across the river, closed in 1970. Of the 144 wharves that had operated in London at the start of 1967, 70 closed by the end of 1971, and almost all of the rest followed soon after. The number of dockworkers fell from 24,000 to 16,000 in less than five years. Factories and warehouses, with no further need to be near the Thames, began to flee, taking their import-and-export business elsewhere, and the waterfront communities tied to the port began to disintegrate.[32]

The Transport and General Workers Union finally lifted its ban on handling containers at Tilbury after twenty-seven months, in April 1970. No sooner did the port reopen than it shut again, as the union waged a three-week nationwide dock strike to protest the stevedores' preference for employing permanent, skilled workers to run their expensive equipment rather than hiring day labor. Nationally, dockers won a 7 percent pay raise, but a special agreement in London allowed containerization in return for double pay. Tilbury was able to open for container shipping at long last. But the delay took a toll. By the time it reopened, greater London had lost its place as the maritime center of Europe.[33]

The new center was Rotterdam, in the Netherlands. A port since the 1400s, Rotterdam had been demolished by German bombing in 1940. The old port had been modest, with two-thirds of the cargo being off-loaded into barges because deepwater ships could not reach the docks. The ruins provided Dutch planners a clean canvas on which to build a modern port starting in the 1950s along the river Maas. Road, rail, and barge connections to Germany helped Rotterdam prosper as the two countries joined in the European Common Market. By 1962, its vast imports pushed Rotterdam ahead of New York as the world's largest port by tonnage. Rotterdam set aside land for containers early on, and Dutch longshoremen, unlike their British counterparts, posed no objections when containerships began calling in 1966. During two and a half years of union-induced delay in Britain, Rotterdam spent $60 million to build the European Container Terminus, with ten berths and room for more. Traffic that had once fed through London to other British ports was now transshipped at Rotterdam, which was on its way to becoming the largest container center in the world.[34]

In Liverpool, meanwhile, the Mersey Docks and Harbour Board had become a financial disaster, its condition worsened by the diversion of cargo to the containerports. Parliament was forced to approve an emergency bailout in 1971. With Felixstowe now seen as a model, the national government took over the city's docks. An infusion of government loans and grants paid off redundant dockers and financed completion of the new pier complex at Seaforth, including three terminals for containers. As the Royal Seaforth Docks opened in 1972, ten of Liverpool's historic piers, some of them two centuries old, were abandoned for good. The great maritime center of the British Empire, the cosmopolitan city whose cotton trade fueled the Industrial Revolution and whose Cunard and White Star steamers dominated the North Atlantic, fell into an economic stupor that would last for three decades.

The container contributed to a fundamental shift in the geography of British ports. In the precontainer era, London and Liverpool had dominated Britain's international trade, their docks and warehouses filled with goods headed to or from factories located nearby. The two ports each loaded one-quarter of Britain's exports, with no other port handling more than 5 percent. The container stripped Liverpool of its competitive advantages. Its costs per ton of cargo were too high, and it was on the wrong side of an island that was reorienting its trade toward continental Europe. In 1970, only 8 percent of Britain's rapidly growing container traffic moved through Liverpool, and the port's share of all British maritime trade in manufactured goods was falling toward 10 percent. Within five years, the exodus of port-related manufacturing would leave the city's economy devastated.[35]

Britain's accession to the European Economic Community in 1973 reoriented its trade to Europe, favoring London and other southern ports over northern and western ports such as Liverpool and Glasgow. Even so, London continued to struggle. "From being the world's largest port after Rotterdam and New York, London has been overhauled by Antwerp, Hamburg, and Le Havre," the British shipping magazine *Fairplay* warned in 1975. "If the present situation is allowed to continue she will slip still further down the 'big league' and face the grim prospect of being relegated to the role of feeder port to the continent." Meanwhile, Felixstowe surged. In 1968, the new containerport had handled all of 18,252 loaded containers. By 1974, 137,850 loaded boxes passed through the port, which was well on its way to becoming the major port for British trade with North America. As containerization's economies of scale begin to take hold, more than 40 percent of all container movements in British harbors would soon occur in a single port, Felixstowe, whose traffic at the dawn of the container age had been too small even to merit statistical mention.[36]

The preparations for container shipping in the United States and Europe provided Asian governments a lesson. In the United States, ports responded to containerization with no overriding rhyme or reason; cities such as New York and San Francisco squandered tax money on wharves and cranes that had little chance of recouping the initial investment, even as cities that might have become important containerports, such as Philadelphia, failed to invest in time. In Britain, the government was so terrified of the waterfront unions that it took few steps to prepare for the container era until the first ships were already in port. In continental Europe, the ports that had the foresight to plan for container shipping, notably Rotterdam, Antwerp, and Bremen, were the first to capture the traffic. Along Asia's Pacific rim, it seemed apparent that containerization would require major change, and change had to be planned.[37]

Time seemed to be on the Asians' side. The shift from breakbulk shipping to container shipping had greatly reduced the cost of loading or unloading a vessel, but it made no difference at all in the operating cost once the vessel left port. This meant that the benefits from switching to container shipping were greatest on shorter routes, for which the savings in cargo handling and port time came to a very large proportion of the total cost of a voyage. The experts reckoned that there was less money to be saved on long-distance routes involving weeks at sea, such as those from the United States to Japan or Britain to Australia. Some even argued that containerization was infeasible in the Pacific and Australian trades, because expensive ships would be tied up for too long and because returning empty containers across seven thousand miles of ocean would prove impossibly expensive.[38]

The race to put containerships on the North Atlantic by the winter of 1966 drew attention in Asia. In early 1966, with Sea-Land preparing to deliver containers to a U.S. base on the Japanese island of Okinawa, a council established by the Japanese transport ministry was issuing directives to promote container services. The transport ministry soon came up with a plan to build twenty-two containership berths in Tokyo and in Kobe, near Osaka, while Sea-Land developed docks in Yokohama. The Australian Maritime Services Board quickly scrapped plans to build conventional wharves at Sydney and invited bids for construction of a container terminal in September 1966, although no international ship line had yet expressed interest in providing containership service to Sydney. The first fully containerized ship to serve the Far East, belonging to

TABLE 5
Largest Containerports by Tonnage, 1969

| | *Container Cargo (metric tons)* |
|---|---|
| New York/New Jersey | 4,000,800 |
| Oakland | 3,001,000 |
| Rotterdam | 2,043,131 |
| Sydney | 1,589,000 |
| Los Angeles | 1,316,000 |
| Antwerp | 1,300,000 |
| Yokohama | 1,262,000 |
| Melbourne | 1,134,200 |
| Felixstowe | 925,000 |
| Bremen/Bremerhaven | 822,100 |

*Source*: Bremer Ausschuß für Wirtschaftsforschung, *Container Facilities and Traffic* (1971).

Matson, sailed from Tokyo to San Francisco in September 1967, and large-scale container shipping arrived the following year. International containerships came to Australia in 1969, and Sydney, Yokohama, and Melbourne quickly vaulted to the top rank among the world's containerports.[39]

Other governments were not far behind. Taiwan's national port program began planning container terminals at five different ports. The Hong Kong Container Committee, appointed by the British colonial government in August 1966, looked at other developments in the western Pacific and issued a warning that December: "[U]nless a container terminal is available in Hong Kong to serve these ships the trading position of the Colony will be affected detrimentally." And no government anywhere was more aggressive in preparing for the container age than Singapore's.[40]

Singapore was a new country in the late 1960s, having been ejected from Malaysia in 1965 amid armed conflict between Malaysia and Indonesia. Its port was more significant as a military base than as a shipping hub. The British had 35,000 soldiers and sailors on the 226-square-mile island, and 25,000 civilians worked at the bases and the naval shipyards. The commercial port comprised a handful of wharves and Singapore Roads, the anchorage offshore where cargo was transferred from one small trading vessel to another. The amount of general cargo actually crossing the docks was about one-fifth of that handled in New York. The Port of Singapore Authority had been created in 1964 to take responsibility for most of Singapore's docks, but it had little to work with. The initial value of all of its assets, including apartment complexes and office buildings as well as docks and warehouses, was less than $50 million.[41]

Immediately upon independence, the new government launched a crash effort to build the economy by drawing foreign investment, especially in manufacturing. Amid a general government crackdown on dissent, the Port of Singapore Authority was able to slash the size of longshore gangs from twenty-seven to twenty-three, institute a second shift, and boost by half the amount of cargo handled per man-hour. It put forth a plan in 1965 to build four berths for conventional ships at a site known as the East Lagoon, which had a breakwater but no major docks. Within months, the plan was scrapped. The containerships that were about to cross the Atlantic had captured the interest of port officials. They announced in 1966 that instead of conventional berths, they would build a port for containers.[42]

Singapore's strategy was to use containers to become the commercial hub of Southeast Asia. With a $15 million World Bank loan covering nearly half the cost, the port authority began work on a terminal at which long-distance vessels from Japan, North America, and Europe could hand off containers to smaller ships serving regional ports. Construction started in 1967, the same year that the first containers—3,100 of them, mainly empties—were deposited on the island's docks. When the British announced in 1968 that their bases and naval shipyards would close within three years, the government countered with even more ambitious plans to build ships, develop industry, and expand the port. "It may be necessary to embark on further construction depending on the build-up of shipping and container traffic," the Port of Singapore Authority advised, even though its first container project was barely under way.[43]

When large-scale container shipping finally arrived in Pacific ports beyond Japan, in 1970, the question of whether it would be viable on long-distance routes quickly became laughable. The $36 million East Lagoon complex opened in June 1972, three months ahead of schedule, cementing Singapore's reputation as an island of efficiency. As the only port in the region with docks long enough for 900-foot containerships, Singapore became a major transshipment point, with third-generation ships handing off containers to smaller vessels that shuttled them to Thailand, Malaysia, Indonesia, and the Philippines. With longshore gangs reduced to only fifteen men and with steep charges on boxes left in the new 120-acre container yard for more than three days, the port ran as smoothly as any in the world.[44]

Singapore's containerport grew beyond all expectations. In 1971, before the new terminal opened, the Port of Singapore Authority forecast 190,000 containers after a decade in operation. Instead, it handled over a million boxes in 1982 and was the world's sixth-largest containerport. By 1986, Singapore had more container traffic than all the ports of France combined. In 1996, more containers passed through Singapore than through Japan. In 2005 Singapore became the world's largest port for general cargo, pulling ahead of Hong Kong, and some 5,000 international companies were using the island-state as a warehousing and distribution hub—testimony to the power of transportation to reshape the flow of trade.[45]

Chapter 11

# Boom and Bust

**n January 10, 1969,** the maritime world was shaken by an unexpected piece of news. Malcom McLean, the father of container shipping, was selling out. Once again, his timing was impeccable.

Three years earlier, at the start of 1966, container shipping had been an infant industry. Only two ship lines, Sea-Land Service and Matson Navigation, moved containers in any quantity. Both served only U.S. domestic traffic, using old ships originally built for a very different sort of business. Almost none of the world's international trade was containerized, and no port outside the United States had the ability to load containers aboard ships except by having longshoremen clamber atop each box and attach hooks at each corner. Most of the world's manufactured goods and foodstuffs moved as they had for a hundred years, painstakingly loaded piece by piece into the holds of breakbulk ships. A leading maritime executive could still hold the opinion, voiced in 1966, "I do not think the time for the all-container ship is now nor in the next decade."[1]

Fast-forward three years and the world had changed. The equivalent of 3,400 20-foot containers of commercial imports or exports passed through U.S. ports each week during 1968, up from zero in

1965.* Rotterdam, Bremen, Antwerp, Felixstowe, Glasgow, Montreal, Yokohama, Kobe, Saigon, and Cam Ranh Bay all had modern facilities for handling containers. Revenues at McLean's Sea-Land Service, whose 31 ships made it far and away the largest container operator in the world, had mushroomed from $102 million in 1965 to $227 million in 1968 as Sea-Land expanded to Vietnam, Western Europe, and Japan. Container shipping had turned into a rip-roaring business—and an extremely expensive one. Sea-Land's debts at the end of 1968 reached $101 million, $22 million of which was payable within the year. During 1969 it was to take possession of six more rebuilt ships costing an additional $39 million, plus $32 million for containers and equipment.[2]

The financial demands would only grow, for the maritime equivalent of an arms race was under way.

The first generation of containerships, the ones that had plied the East and Gulf coasts and brought the container revolution to Puerto Rico and Hawaii, Alaska and Europe, had consisted almost entirely of older vessels, originally built for other purposes. Most of them were small, about 500 feet long, and very slow, straining to steam at 16 or 17 knots. Many of these early container vessels carried only a couple of hundred containers along with breakbulk freight, refrigerated cargo, even passengers. Only three ships in the entire world were equipped with enough container cells to hold more than 1,000 20-foot containers. The first-generation containerships had cost the ship lines almost nothing; of the 77 U.S.-flag ships equipped to carry containers at the end of 1968, 53 were relics of World War II. Most lines had no ships with container cells in their holds and desperately

* The quantity of breakbulk shipping was measured either by weight or by "measurement tons," a standard method for converting volume into tonnage, and these conventions were initially applied to container cargo. The capacity of containerships and cranes, however, was determined by the quantity of containers rather than their weight, and by the mid-1960s ports and ship lines began to emphasize the number of containers they handled. Raw numbers proved problematic, because they failed to distinguish between empty containers and full ones, and between large containers and small ones. In 1968, the Maritime Administration began to report container traffic in standardized 20-foot equivalent units, or TEUs. A 40-foot container represents 2 TEUs, and one of Matson's 24-foot boxes registered as 1.2 TEUs.

tried to meet customers' demands by packing containers into conventional breakbulk ships. Breakbulk ships, however, were hard to service with high-speed container cranes. Each time a container was to be moved, longshoremen would have to climb atop the box, attach hooks at the corners, and then remove the hooks once the container had been lifted. With none of the operating efficiencies of cellular containerships, most carriers were losing money on every container they carried.[3]

The second generation of containerships was of a totally different order. Sixteen of these newly built ships were at sea by the end of 1969, and another 50 were under construction. These vessels were designed from the start to work smoothly with dockside container cranes. They were large, they were fast, and they came with very high price tags.

The first of these new vessels was *American Lancer*, owned by United States Lines, Sea-Land's biggest competitor across the North Atlantic. The *Lancer*, which made its maiden voyage between Newark and Rotterdam, London, and Hamburg in May 1968, was far bigger than any containership on the seas. It could carry 1,210 20-foot containers at a speed of 23 knots—half again as fast as the reconstructed ships in Sea-Land's fleet. In August 1968 U.S. Lines asked the Maritime Administration for a $95 million subsidy to build six more such behemoths. Other American, European, and Japanese companies raced to place their orders. Almost always, a ship was designed with a specific route in mind. Atlantic vessels usually held 1,000–1,200 containers, because too large a ship meant too much port time after a relatively short voyage. Ships meant for the Asia trades were typically larger, carrying 1,300–1,600 20-foot containers, because the relatively long ocean voyages from Europe or America to Japan generated enough additional revenue to cover the added construction cost.[4]

The expense of building and equipping these second-generation containerships staggered even the largest ship lines. Between 1967 and the end of 1972, a consultant would later calculate, the total cost of containerization around the world would come to near $10 billion—an amount close to $40 billion in 2005 dollars. Individual

European ship lines had no prospect of raising financing of this magnitude: the total after-tax profit of all thirty-seven British steamship companies in 1966 came to less than £6 million. With few alternatives, the British formed consortia such as Overseas Containers Ltd., whose members shared the $185 million cost of building six ships, and containers to go with them, between 1967 and 1969. The smaller Belgian, French, and Scandinavian carriers sought strength in numbers as well. If four ship lines joined forces, each building one or two vessels, in combination they might have enough ships to be significant players.[5]

The American carriers were slightly more prosperous, thanks to government subsidies and military shipments, but they were hardly rolling in money. Sea-Land generated a total of $30 million of profit from 1965 through 1967, almost all of it on domestic routes. The largest American ship line, United States Lines, earned $4 million of profit over those three years. The Americans were not forced into joint ventures, though, because they had an option that the Europeans did not. The American conglomerates that aspired to remake the business world in the late 1960s spotted opportunity in the traditionally low-profit maritime industry, and they wanted to be in on the container boom. Litton Industries, of course, had invested in Sea-Land. Walter Kidde & Co. opened its wallet to buy United States Lines in January 1969. City Investing, another conglomerate, won a bidding war for Moore-McCormack Lines until the ship line reported a big loss for 1968 and the deal fell apart. The "cold, pragmatical thinking" of conglomerates threatened the industry, a maritime executive complained in 1968. "Such conglomerates, the newcomers, assign no value to the romance of the sea or the traditions of the railroads and the highways. They are strictly readers of the bottom lines of financial reports."[6]

No conglomerate chieftain was a more avid reader of financial reports than Malcom McLean. He knew what the cost of competition was going to be, and he knew that Sea-Land, its balance sheet stretched to the limit, had no hope of borrowing the money. His previous conglomerate backer, Litton, which held 10 percent of Sea-Land's shares, was tapped out. McLean turned toward an entirely unexpected source of funds: R. J. Reynolds Industries. Reynolds, based in Winston-Salem, North Carolina, was the nation's largest tobacco company. Its cigarette business threw off cash by the bucketful, and its managers were using that cash to turn the company into a conglomerate. U.S. cigarette consumption had fallen in 1968, and impending restrictions on marketing—the government would ban cigarette advertising on American television at the start of 1971—boded ill for its core business. The ship line's huge investment needs would provide Reynolds with a convenient shelter from corporate income tax. An added inducement was McLean's status as a local genius—he had moved McLean Trucking to Winston-Salem after World War II and had lived there for a decade. Reynolds offered $530 million, with McLean Industries' shareholders free to choose between Reynolds securities and $50 per share in cash. Litton Industries cashed out at a huge profit, as did Daniel K. Ludwig: the $8.5 million that Ludwig had invested in Sea-Land in 1965 now was worth $50 million. Many Sea-Land executives, shocked by word of their company's sale, instantly became very wealthy men.[7]

If anyone doubted McLean's timing, his wisdom soon became clear. In October 1968, he had commissioned designs for an entirely new kind of containership, the SL-7. The SL-7 was meant to leave U.S. Lines' new *Lancer* looking as outdated as a Liberty Ship. It would be nearly a thousand feet long, just a few feet shorter than the famed *Queen Mary*. Its capacity was 1,096 of Sea-Land's 35-foot containers, equivalent to more than 1,900 20-foot boxes—a far greater load than any other ship afloat. Its most striking characteristic, though, was its speed. The SL-7 would travel at 33 knots, twice as fast as any ship in the Sea-Land fleet. It would be fast enough to sail around the world in 56 days, so fast that a fleet of eight ships could provide a round-the-world sailing from each major port each week. U.S. Lines boasted that the *Lancer* and its sister ships could deliver a container from Newark to Rotterdam in 6½ days. The SL-7s would do it in 4½ days and could cross the Pacific from Oakland to Yokohama in just 5½ days. Only one commercial ship ever built, the venerable passenger liner *United States*, was fast enough to keep up with it.[8]

More was at work than megalomania. McLean, once more, had conceived a way to gain a strategic edge. He planned to deploy the new ships in the Pacific. Sea-Land was a conference carrier in the Pacific, charging the same rates as its competitors. The SL-7s' faster transit time would help Sea-Land attract cargo, and other carriers, bound by the conference agreement, would not be able to drop their rates in response. In the summer of 1969, the Sea-Land division of R. J. Reynolds Industries made its plans for the SL-7 public, ordering eight vessels from European shipyards. The price tag was $32 million per ship. Containers and other equipment would bring the total cost of the SL-7s to $435 million. For McLean Industries, even if it could have raised the money, spending nearly half a billion dollars on ships would have been a bet-the-company gamble. For R. J. Reynolds, it was almost spare change. The tobacco giant was so cash rich that in 1970 it purchased a petroleum company, American Independent Oil Company, to provide a cheap source of fuel for Sea-Land's expanding fleet.[9]

The first stage of the container boom occurred entirely on the North Atlantic. The second happened on the Pacific. Matson sailed the first fully containerized ship from Japan in September 1967, in what it thought would be a partnership with Japanese ship lines. Having learned the business, the Japanese soon left Matson behind and began their own container service to California in September 1968. Sea-Land, using containerships headed home from Vietnam, began carrying 35-foot boxes from Yokohama and Kobe the following month. If there had been any doubt about whether Japanese exporters would adopt containerization, it was settled quickly. Within a year, container tonnage between Japan and California was two-thirds that across the North Atlantic. The impact on trade flows was immediate. Japanese seaborne exports, 27.1 million metric tons in 1967, rose to 30.3 million with the start of containerization late in 1968 and then soared to 40.6 million tons in 1969, the first full year of container service to California. The value of Japanese exports to the United States leaped 21 percent in 1969 alone.[10]

Cars, which did not travel in containers, accounted for part of the export surge. Much of the increased trade, however, was stimulated by containerization. Within three years, nearly one-third of Japanese exports to the United States were containerized, as were half of Japan's exports to Australia.[11]

Electronics manufacturers had been among the first Japanese exporters to see the advantages of shipping their fragile, theft-prone products in containers. Electronics exports had been on the rise since the early 1960s, but the lower freight rates, inventory costs, and insurance losses from container shipping helped turn Japanese products into everyday items in the United States, and soon in Western Europe. Exports of televisions climbed from 3.5 million sets in 1968 to 6.2 million in 1971. Shipments of tape recorders went from 10.5 million to 20.2 million units over the same three years. Containerization even gave new life to Japanese clothing and textile plants. Rising wages had put an end to the growth of Japan's apparel exports in 1967, but the drop in shipping costs briefly made it viable for Japanese clothing manufacturers to sell in America again.[12]

In 1969, as United States Lines was preparing to add eight fast containerships to its U.S.-Japan service, the Japanese government put shipping at the center of its economic development strategy. Its new five-year plan called for a 50 percent expansion of Japan's merchant fleet, including tankers and ore carriers as well as containerships. The government offered $440 million to help Japanese ship lines begin container service to New York, the Pacific Northwest, and Southeast Asia, using Japanese-built ships. The subsidies were an incredible bargain. A ship line needed to put up only 5 percent of the cost of building its new vessel. The government development bank provided most of the funds. No payments were due for three years, after which the ship line was to repay the loan over ten years at an interest rate of 5.5 percent—a lower rate than the Japanese government paid to borrow the money that its development bank lent out. The remainder of the construction cost came from commercial banks, with the government paying 2 percentage points of interest. With such giveaway terms, Japanese ship lines had no fewer than 158 vessels on order or under construction by the end of 1970, all in Japanese shipyards.[13]

Hong Kong received its first visit from a fully containerized ship in July 1969, even before its container terminal was ready. The following year, as Sea-Land opened container service to South Korea and Matson began biweekly visits to Taiwan, Hong Kong, and the Philippines, container capacity on transpacific routes reached nearly a quarter million units in 73 vessels. Other new services linked Australia to Europe, North America, and Japan. Regular sailings with fully containerized ships between Europe and the Far East began in 1971.[14]

Shipyards around the world were choked with new orders. East Asia's ports, with years to prepare themselves, were ready and waiting as the new vessels came on line in 1971 and 1972. Trade soared, as a story similar to Japan's was repeated along the Pacific Rim. Oceanborne exports from South Korea, 2.9 million tons in 1969, reached 6 million tons in 1973. Korean exports to the United States trebled over those three years as lower shipping costs made its garments competitive in the U.S. market. Hong Kong followed much the same course. Before it filled ninety-five acres of harbor to build a containerport, the colony's shipping had been so primitive that oceangoing ships anchored far out in the harbor, and small boats shuttled imports and exports back and forth to shore. With the new terminal allowing containerships to collect cargo straight from the dock, Hong Kong's shipments of clothing, plastic goods, and small electronics rose from 3 million tons in 1970 to 3.8 million in 1972, and the value of its foreign trade rose 35 percent.

Exports from Taiwan, $1.4 billion in 1970, were $4.3 billion by 1973, and imports more than doubled. The pattern in Singapore was much the same. In Australia, the opening of container traffic coincided with a surge in manufactured exports and a dramatic shift away from traditional exports such as meat, ore, and greasy wool. The volume of exports other than minerals or farm products rose 16 percent annually from 1966–67 to 1969–70. Prior to 1968, the value of Australia's industrial exports typically came to less than half its exports of grain and meat. By 1970, most of Australia's general-merchandise trade was already moving in containers, and factory exports nearly matched farm exports. In the process, Australia left behind its past as a resource-based economy and began to develop a much more balanced economic structure.[15]

The container cannot claim sole credit for this burst of international commerce, but it is entitled to a share. A 1972 study by McKinsey & Company, an international consulting firm, laid out some of the ways in which containerization stimulated trade between Europe and Australia, where containers came into use on mixed vessels in 1967 and fully containerized ships opened service in 1969. Previously, Australia-bound ships had spent weeks calling at any of eleven European ports before starting the southbound voyage. Containerships collected cargo only at the huge containerports at Tilbury, Hamburg, and Rotterdam, whose enormous size kept the cost of handling each container low. Previously, shipments took a minimum of 70 days to get from Hamburg to Sydney, with each additional port call adding to the time; containerships offered a transit time of 34 days, eliminating at least 36 days' worth of carrying costs. Insurance claims for Europe-Australia service were running 85 percent lower than in the days of breakbulk freight. Packaging costs were much lower, and ocean shipping rates themselves had dropped. The total savings from containerization were so great that traditional ships abandoned the Australia route almost immediately.[16]

The breakneck construction of new containerships transformed the world's merchant fleet. In 1967, 50 American-owned vessels, most of them built during World War II and rebuilt during the 1950s or 1960s, accounted for all but a handful of the fully containerized ships in operation around the world. From 1968 through 1975, no fewer than 406 containerships entered service. Most of the new vessels were at least twice as large as any that had been on the scene in 1967. Beyond these fully containerized vessels, ship lines added more than 200 partially containerized ships, with container cells built into some of their holds but not others, and almost 300 roll on–roll off ships to serve routes that lacked the volume to justify containerships. With these hundreds of new vessels, container shipping was coming into full flower.[17]

The U.S. merchant fleet changed almost overnight. In 1968, there were still 615 general-cargo freighters flying the U.S. flag.

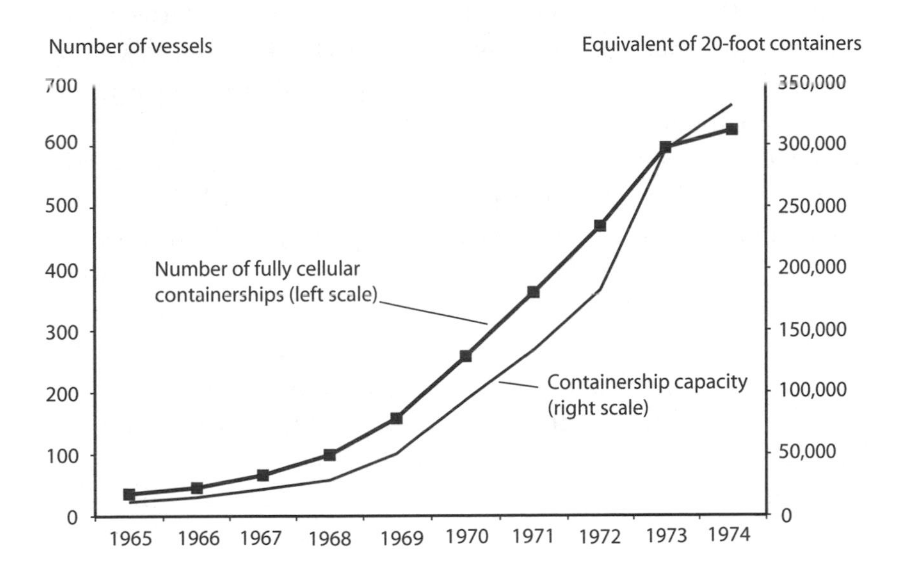


**The containership boom. Source: UNCTAD.**

Within the next six years, more than half of those vessels had left American-flag service, either cast off to the tenuous ship lines of poor countries or sold for scrap. Replacing them were fewer but much larger and faster ships. The American seamen's unions were wont to cite the diminished fleet as a sign of maritime weakness, but the truth is that the few dozen new containerships could carry far more cargo than the hundreds of rust buckets they supplanted. Even as the U.S.-flag fleet shrank nearly by half, the number of vessels able to carry more than 15,000 tons of cargo rose from 49 in 1968 to 119 in 1974. New steam-turbine engines helped boost the average speed of large U.S.-flag freighters from just 17 knots in 1968 to 21 knots in 1974. The difference was enough to cut a full day off a transatlantic crossing.[18]

The launch of so many vessels resulted in a quantum jump in capacity. The basic economics of containerization dictated as much. Once a ship line had made the decision to introduce containerships on a particular route, other carriers in the trade normally followed swiftly lest they be left behind. The capital-intensive nature of container shipping put a premium on size; quite unlike breakbulk shipping, in which an owner of "tramp" ships could eke out a profit picking up freight wherever it could be found, a container line needed enough ships, containers, and chassis to run a high-frequency service between major ports on a regular schedule. When a ship line decided to enter a trade, it had to do so in a large way—which meant that on every major route, several competitors were entering with several vessels apiece. Capacity on the largest international routes increased fourteen times over between 1968 and 1974. Between the United States and northern Europe, where only a handful of small containers had moved prior to 1966, there were enough new ships to carry nearly a million boxes a year by 1974. The containership route between Japan and the U.S. Atlantic coast, opened only in 1970, was served by thirty vessels in 1973.[19]

Demand, robust through it was, could not possibly keep up with this explosion of supply. The result was a new and painful experience for the shipping industry: a rate war.

Overcapacity was an old story in ocean shipping. The flow of cargo had always been volatile, based on economic growth, changes in tariffs and trade restrictions, and political factors such as wars and embargoes. In the 1950s and 1960s, though, a temporary imbalance between the amount of space on breakbulk ships and the amount of general cargo usually was not a fatal problem. The war-surplus ships that filled most merchant fleets had been acquired for little or nothing, so shipowners were not saddled with huge mortgage payments. Their main expenses—cargo handling, fees for the use of docks, pay for crews, fuel—were operating costs. If business was bad, the shipowner could lay the vessel up and most of the costs would go away.

The economics of container shipping were fundamentally different. The huge sums borrowed to buy ships, containers, and chassis required regular payments of interest and principal. State-of-the-art container terminals meant either debt service, if a ship line had borrowed to build its own terminal, or rent, if the terminal was leased from a port agency. Those fixed costs accounted for up to three-quarters of the total cost of running a container operation, and they had to be paid no matter how much cargo was available. No company could afford to lay up a containership just because there was too little cargo. So long as each voyage collected enough revenue to cover operating costs, the ship had to keep moving. In container shipping, quite unlike breakbulk, overcapacity would not diminish as owners temporarily idled their ships. Instead, rates would fall as carriers struggled to win every available box, and overcapacity would persist until the demand for shipping space eventually caught up with the supply.[20]

Overcapacity preoccupied everyone connected with containerization. "Now that standardized containers have been introduced, the rush to 'get on the bandwagon' will probably lead to substantial overexpansion," a study for the British government warned in 1967. By one early estimate, 5 ships carrying 1,200 containers each, sailing at 25 knots, could move all of the U.S.-UK trade that could be containerized. By another, just 25 ships could handle the entire general-cargo trade between Europe and North America. A third estimate foresaw that the 5 ships ordered by the American carrier Farrell Line would be adequate for all of Australia's exports to the United States. With hundreds of containerships on order, experts projected that half the available container slots across both the Atlantic and the Pacific would go unused by 1974. In the North Atlantic, "by the early 1970s there will be excess container capacity," a study for the U.S. government predicted in 1968.[21]

The wolf was at the door even sooner than anticipated. In early 1967, less than a year after fully cellular containerships entered the trade, the North Atlantic conferences cut rates for containers by 10 percent—an action that a leading U.S. shipping executive termed "a disaster." That was only the beginning. With too many ships chasing too little cargo, the long-standing structure of ocean freight rates began to fall apart.[22]

Prices for international shipping, unlike domestic shipping, usually were not set by government regulators. Instead, rate setting was the realm of liner conferences, voluntary cartels of the operators on each route. No fewer than 110 different conferences set rates on routes to or from the United States, and similar conferences governed routes elsewhere in the world. The conference members negotiated a rate schedule among themselves and often assigned each member ship line a percentage of the total traffic. All shippers using conference carriers were supposed to pay the official rates, with no special deals, although cheating was common; "rebating," secretly refunding part of a shipper's payments, was a widespread if illegal practice. Conferences in trades serving the United States were required to publish their rates and to be "open"—that is, to accept new lines that wished to join—but many other routes around the world featured confidential rates and "closed" conferences that excluded newcomers. On most routes, governments did not require ship lines to be conference members—but if a carrier began operating as an "independent," it was likely to find the conference letting its members slash rates and add capacity in order to destroy the intruder. Most of the time, all carriers had an interest in going along with the system.[23]

The conferences structured their rates very much as railroads did. There was a separate rate for each commodity, or sometimes two rates, one measured by weight and one by volume. For breakbulk shipping, there was logic behind this: some commodities were more complicated to load than others, some took more shipboard space and some less, and different rates were a way to recognize the differing costs. Applied to containers, the commodity-based system made no sense at all: a ship line's cost to move a 40-foot container of bicycle tires was identical to its cost for a 40-foot container of table lamps. When containers appeared, though, the conferences, dominated by companies that still sailed breakbulk ships, relied on the tried-and-true system of commodity-based rates. On the North Atlantic, the rate per ton for a product shipped in a container was the same as if it were shipped breakbulk, with a discount of 5 to 10 percent for a full container of a single commodity. Rates for mixed freight made even less sense. When the Europe-Australia conference set container rates in 1967, a year before containership service opened, it decreed that each commodity in a mixed container would be charged the per-ton rate for that commodity. The only way to find the correct rate was to open the container and weigh every single item inside.[24]

This economically illogical system could not last. Ship lines had no reason to care what was inside the containers they carried, and with rampant excess capacity they were willing to accept any payment that exceeded their cost to carry the container. By early 1967, Waterman Steamship, Malcom McLean's former company, switched to a flat rate for shipments from the United States to southern Europe: $400 for a shipper-owned 20-foot container, $800 for a 40-foot container, regardless of the contents. Waterman did not yet have any containerships and its rate structure had no imitators, but its move reveals the pressure on prices. Carriers began threatening to leave their conferences unless rates came down. The conferences struggled vainly to keep the rate structure intact. In the summer of 1969, the transatlantic conference system blew apart. Eight lines formed a new conference with the aim of leaving commodity-based rates behind and establishing rates that made sense in the world of containers.[25]

As the artificially high rate structure collapsed, ship lines faced profit squeeze. Restructuring was the only way out. In July 1969, barely three years after container shipping had become an international business, West Germany's two biggest shipping companies agreed to merge as Hapag-Lloyd, a huge new player in the North Atlantic. Three months later, Malcom McLean responded in kind. McLean had always preferred consolidation to competition; had the U.S. government not blocked him, he would have acquired Sea-Land's sole East Coast competitor, Seatrain Lines, in 1959, and its main competitor to Puerto Rico, Bull Line, in 1962. Now, on Sea-Land's behalf, he committed $1.2 billion of R. J. Reynolds's money to an audacious deal with United States Lines. U.S. Lines was in the midst of building 16 containerships, all able to carry more than 1,000 containers and to steam faster than 20 knots. It would soon have the greatest containership capacity of any line. Sea-Land proposed to lease that entire fleet, all 16 vessels, for 20 years. U.S. Lines would surrender its status as a subsidized carrier, which would allow

Sea-Land to deploy the ships wherever it wanted, without government approval. A major competitor would be out of the game, and Sea-Land would become by far the largest ship line on both the Atlantic and the Pacific.[26]

Competitors cried foul—but they reacted promptly. In early 1970, Grace Line was merged into Prudential Lines. Matson surrendered its international ambitions, selling its ships in 1970 and giving up its efforts to turn Honolulu into a hub for commerce across the Pacific. Moore-McCormack Lines sold its four newest freighters and exited the North Atlantic. Two British carriers, Ben Line and Ellerman Line, joined forces on the UK–Far East route, and three Scandinavian companies combined their ships to create a single international carrier called Scanservice.

Those shifts were far from enough to stabilize the industry. In the Australia trade, Overseas Containers Ltd. lost $36 million between 1969 and 1971. Hapag-Lloyd suffered losses in 1969, 1970, and again in 1971. On the North Atlantic, where one-third of containership capacity was unutilized, American Export Isbrandtsen Line lost so much money in 1970 and 1971 that its parent company's shares were suspended from trading on the New York Stock Exchange and its president was forced out. U.S. Lines, operating in both the Atlantic and the Pacific, lost $14 million in 1970 and as much again the following year. Even Sea-Land had a difficult passage after the U.S. government blocked its efforts to combine with U.S. Lines, its profit falling from $39 million in 1969 to $21 million in 1970 and barely $12 million in 1971. R. J. Reynolds, like the other conglomerates that had invested in ship lines, was learning that container shipping was not the gold mine it had imagined.[27]

In desperation, the leading carriers on important routes tried an old-fashioned solution: reducing competition. Five competitors in the Europe–Far East trade, two British, two Japanese, and the German Hapag-Lloyd, combined their Pacific interests in an alliance called TRIO. Among them, the companies agreed to build nineteen large ships, with each company allocated a number of container slots on each ship. A second Europe-Pacific consortium soon followed, with the Swedish carriers and the Dutch company Nedlloyd merging their Asian operations into a company called ScanDutch. Those two alliances drastically cut the number of competitors between Europe and Japan, helping stabilize rates. An even more powerful cartel, the North Atlantic Pool Agreement, was born in June 1971. The pool agreement, strongly backed by six European governments, combined the efforts of what had been fifteen separate ship lines from six countries. It spelled out exactly what percentage of the total cargo each company would carry. All of the members agreed to charge identical rates, and revenues from North America–Europe service were to be shared. The cartel managed finally to put a floor under rates. "Without the pool, a lot of us would go under," one executive admitted in 1972.[28]

Economic growth around the world picked up in 1972, and with it the flow of trade. Container tonnage nearly doubled from 1971 to 1973, and as carriers finally found enough cargo to fill their ships, they earned profits once more. But the shipping industry that survived the carnage of containerization's first rate cycle was quite different from the one that had existed in 1967. Far fewer independent companies were left, and they had no illusions about the future. Rate wars would obviously be a permanent feature of the container shipping industry, recurring every time the world economy turned down or ship lines expanded their fleets. Shippers would pay according to the distance their containers traveled, regardless of the weight or the nature of the contents, and in difficult times rates would dip so low that carriers would barely cover their operating costs. Ship lines would be under constant pressure to build bigger ships and faster cranes to reduce the cost of handling each container, because at some point overcapacity would return, and when rates collapsed the carrier with the lowest cost would have the best chance of survival.[29]

The next collapse was not long in coming.

The years 1972 and 1973, as it turned out, represented a peaceful interlude in an economically turbulent decade. Industrial production rose 18 percent in the United States, 19 percent in Canada, 22 percent in Japan, 12 percent in Europe. International trade grew strongly enough to transform the glut of container shipping into a shortage, despite the launch of 143 containerships in just two years. The sharp rise in oil prices that began in 1973 proved initially to be an unexpected blessing for the maritime industry, giving containerships, which transported more cargo per barrel of oil, a further cost advantage over the remaining breakbulk ships. The amount of containerized ocean cargo around the world rose 40 percent in 1973 alone. Companies ordered their ships to reduce speed in order to conserve fuel, cutting the number of voyages they could make over the course of a year, which further tightened the market. Freight rates soared, as conferences pushed through hundreds of rate increases and added surcharges to cover exchange-rate movements, higher fuel costs, and port delays. "Many shippers, faced with rate increases of over 15 per cent plus surcharges, must have found their freight bills increased by as much as 25 to 30 percent," a United Nations report declared.[30]

The boom lasted into 1974, when a weaker dollar drove exports from U.S. factories up 42 percent in a single year. Rate increases, along with the various agreements around the world to limit capacity, pool revenues, or join forces, finally worked magic on the shipping industry's bottom line. Sea-Land reported a healthy $142 million profit, up from $16 million in 1973. Even U.S. Lines, which had seen little besides red ink out of its sixteen new containerships, posted a $16 million profit for 1974. Judged the head of Atlantic Container Line, "If an operator can't make it on the North Atlantic now, he will never make it."[31]

The oil crisis, though, ended up devastating the shipping industry. The world economy tumbled into recession in the second half of 1974 as central banks tightened monetary policy to counteract the inflationary consequences of dearer oil. Industrial production collapsed, and with it the flow of trade. World exports of manufactured goods fell in 1975 for the first time since the war, and the amount of seaborne trade dropped 6 percent. Even as trade flows diminished, shipyards kept delivering new containerships—and every new ship weakened the ship lines' ability to hold rates up. Containerships from the Soviet Union joined the competition in both the Atlantic and the Pacific outside the conference structure, pressuring rates further. The shipping conferences were forced to roll back or eliminate surcharges six hundred times between 1974 and 1976.[32]

The second crisis of container shipping was made worse by the carriers' own choices. The hundreds of containerships built in the first half of the 1970s had been designed for the world of the late 1960s. High speed was important because of the closure of the Suez Canal in the 1967 Arab-Israeli war, which forced ship traffic between Europe and Asia and Australia to take a much longer route around the tip of Africa. High fuel consumption—the inevitable result of high speed—did not much matter, because oil was cheap. The world of the mid-1970s was totally different. The price of fuel quadrupled. On the North Atlantic, fuel went from one-fourth of operating costs in 1972 to half in 1975. On the Europe–Far East route, the unexpectedly fast reopening of the Suez Canal in June 1975 eliminated the reason for having fuel-guzzling high-speed ships to sail around Africa. Many carriers were stuck with the wrong vessels for the times.[33]

Prominent among them was the Sea-Land division of R. J. Reynolds Industries. Malcom McLean, acting, per usual, on intuition rather than cautious analysis, had overridden objections from Sea-Land's board to move ahead with the SL-7 in 1968, and Reynolds had agreed to build eight of the ships when it bought the ship line in 1969. The costliest merchant ships ever built were also the thirstiest, each burning five hundred tons of fuel per day. At full speed, they consumed three times as much fuel per container as competitors' vessels. When the price of bunker fuel jumped from $22 per ton to $70 within a matter of months, the SL-7s became a crushing burden. Although R. J. Reynolds boasted to its shareholders that the SL-7s "provide the fastest container shipping service in the world," the ships consistently missed their ambitious schedules and could not make money.[34]

A settling of scores was inevitable. McLean, unhappy with Reynolds's bureaucratic ways, began selling his stock in 1975 and left the board in 1977. Reynolds, frustrated with its inability to control the extraordinary volatility of the steamship business, reorganized Sea-Land to put the ship line under tighter corporate control. The changes did not help. In 1980, Reynolds finally took a $150 million loss on the SL-7s, which had been in service for less than eight years, and dumped them on the U.S. navy for rebuilding as fast supply ships. Four years later, it got out of the shipping business altogether and spun off Sea-Land as an independent company. As R. J. Reynolds's new management explained to investment analysts, "investors who might be interested in owning RJR stock were not the type who ordinarily would be interested in a capital-intensive, cyclical transportation company."[35]

Quite so. For R. J. Reynolds, and for the other corporations that had chased fast growth by buying into container shipping in the late 1960s, their investments brought little but disappointment. Sea-Land and its competitors were not at all like Polaroid or Xerox, companies whose proprietary technology and constant stream of innovations provided inordinately high profits for decades. Ship lines' end product was basically a commodity. Just like farmers and steelmakers, they would always be hostage to external forces, their prices and profit margins depending mainly on economic growth and on their competitors' decisions to build new ships. The go-go years were over. By 1976, less than a decade after container shipping became an international business, the *Financial Times* could declare that "the revolutionary impact of containerization, the biggest advance in freight movement for generations, has largely worked itself out."[36]

Except that the *Financial Times* got it wrong. The revolutionary impact of containerization, as it turned out, was yet to come.

Chapter 12

# The Bigness Complex

**Malcom McLean sold** his stock and quietly left the board of R. J. Reynolds Industries in February 1977. By all accounts, the marriage had not been a happy one. McLean was frustrated by the tobacco giant's bureaucracy and bewildered by its repeated changes of strategy. Most of all, though, he was restless. "I am a builder, and they are runners," McLean explained. "You cannot put a builder in with a bunch of runners. You just throw them out of kilter."[1]

After giving up day-to-day responsibility for Sea-Land Service in 1970, he had spent $9 million to buy Pinehurst, the famed golf resort in central North Carolina, not far from his birthplace in Maxton. He acquired a small life insurance company, an estate in Alabama, a trading company. Then, in 1973, he started First Colony Farms on 440,000 acres in the swamps of eastern North Carolina. Modeled after his friend Daniel Ludwig's plantation in the Amazon, First Colony was probably the largest agricultural development in U.S. history. McLean spent millions draining wetlands to start a massive peat-harvesting operation, then built a plant to turn the peat into methanol. Nearby he planned the world's largest hog farm, where the hogs would be raised mechanically to slaughter weight and then shipped to a slaughterhouse he would build on-site. The peat scheme, though, was blocked by one of the earliest environmentalist campaigns, and the hog farm, able to raise 100,000 animals a year, never made money. When he got an offer for the hog farm in 1977, McLean sold—for $12 million plus 40 percent of the profits for twenty years—and looked around for something new.[2]

In October 1977, he found it. To the surprise of almost everyone, he arranged to buy United States Lines.

U.S. Lines was not exactly a prize. It had long since been supplanted by Sea-Land as the largest American-flag ship line, and its owner, the conglomerate Walter Kidde & Co., had been trying to unload it almost since the day it purchased the company in 1969. Its famous flagship, the luxury liner *United States*, had been sold off to the U.S. government. U.S. Lines had lost money through most of the 1970s. Nonetheless, McLean spotted value. For an investment of $160 million, of which $50 million went to pay off debts, he got thirty ships; $50 million in cash; a huge new terminal on Staten Island, in New York Harbor; and an important network of routes to Europe and Asia. U.S. Lines, unlike Sea-Land, was entitled to operating subsidies from the U.S. government on its international routes. The subsidies were a curse as well as a blessing: the ship line was assured a source of revenue, but the Maritime Administration got to dictate where and how often its ships sailed.

In 1978, as his new ship line was eking out a very modest profit, McLean hatched an audacious plan. U.S. Lines would build a series of enormous containerships, half again as large as anything else on the sea, and send them around the world. The timing was right, because shipbuilders' order books were shrinking after the headlong expansion of the 1970s and construction prices were falling. A round-the-world route, McLean thought, would solve one of the industry's inherent problems, the imbalanced flow of freight that left some ships sailing full in one direction and half-empty in the other. The new vessels would have the lowest construction cost per container slot of any vessel in the world and the lowest operating costs per container as well. U.S. Lines would achieve what it took to succeed in container shipping: scale.

Scale was the holy grail of the maritime industry by the late 1970s. Bigger ships lowered the cost of carrying each container. Bigger ports with bigger cranes lowered the cost of handling each ship. Bigger containers—the 20-foot box, shippers' favorite in the early 1970s, was yielding to the 40-footer—cut down on crane movements and reduced the time needed to turn a vessel around in port, making more efficient use of capital. A virtuous circle had developed: lower costs per container permitted lower rates, which drew more freight, which supported yet more investments in order to lower unit costs even more. If ever there was a business in which economies of scale mattered, container shipping was it.

Ship lines responded to the imperative of scale by extending their reach. The old breakbulk companies had often been content to serve a single route. In 1960, no fewer than twenty-eight carriers had sailed the North Atlantic, from the mighty Cunard Line to such one-ship minnows as American Independence Line and Irish Shipping Limited. In the container age, minnows could not survive, and the truly big fish, companies such as Sea-Land, U.S. Lines, and Hapag-Lloyd, wanted to be in every major trade, either with their own ships or with an arrangement that allowed them to book space on someone else's. The more ships they had, the more ports they served, the more widely they could spread the fixed costs of their operations. The more far-flung their services, the easier it would be to find loads to fill their containers and containers to fill their ships. The broader their networks, the more effectively they could cultivate relationships with multinational manufacturers whose needs for freight transportation were worldwide.[3]

Ocean carriers added 272 containerships to their fleets between 1976 and 1979. Four times during the 1970s, worldwide container shipping capacity increased by more than 20 percent in a single year. Total cargo capacity aboard containerships, 1.9 million tons in 1970, reached 10 million in 1980, not counting the tonnage of vessels designed for a mix of containers and other freight.[4]

The quest for scale brought not just more ships but bigger ships. The *Fairland*, the first Sea-Land ship to cross the Atlantic in 1966, was only 469 feet long. The purpose-built containerships of the late 1960s were about 600 feet from stem to stern, and the fast vessels launched in 1972–73 were as much as 900 feet long and 80 feet wide, with drafts of 40 feet. At that point, containership design seemed to be approaching its limits. The locks of the Panama Canal, through which almost all traffic between Asia and the Atlantic coast of North America had to travel, are 1,000 feet long and 110 across, and bigger ships would not fit. The oil crisis, which caused so many financial problems for ship lines, unexpectedly brought relief. Shipowners decided to build slower vessels to save fuel: the average speed of newly delivered containerships dropped steadily from 25 knots in 1973 to 20 in 1984. Naval architects were no longer forced to design streamlined shapes to help achieve high speeds, and could concentrate instead on increasing payloads. Without getting much longer, vessels got much larger. The ships entering service by 1978 could hold up to 3,500 20-foot containers—more than had entered all U.S. ports combined during an average week in 1968.

These Panamax vessels—the maximum size that could fit through the Panama Canal—could haul a container at much lower cost than could their predecessors. The construction cost itself was lower, relative to capacity: a vessel to carry 3,000 containers did not require twice as much steel or twice as large an engine as a vessel to carry 1,500. Given the extent of automation on board the new vessels, a larger ship did not require a larger crew, so crew wages per container were much lower. Fuel consumption did not increase proportionately with the vessel's size. By the 1980s, new ships holding the equivalent of 4,200 20-foot containers could move a ton of cargo at 40 percent less than could a ship built for 3,000 containers and at one-third the cost of a vessel designed for 1,800.[5]

And still the vessels grew. The economies of scale were so clear, and so large, that in 1988 ship lines began buying vessels too wide to fit through the Panama Canal. These so-called Post-Panamax ships needed deeper water and longer piers than many ports could offer. They were uneconomic to run on most of the world's shipping lanes. They offered no flexibility, but they could do one thing very well. On a busy route between two large, deep harbors, such as

Hong Kong and Los Angeles or Singapore and Rotterdam, they could sail back and forth, with a brief stop at each end, moving freight more cheaply than any other vehicles ever built. By the start of the twenty-first century, ship lines were ordering vessels able to carry 10,000 20-foot containers, or 5,000 standard 40-footers, and even bigger ships were on the drawing boards.

As ships got bigger, ports got bigger. In 1970, the equivalent of 292,000 loaded 20-foot containers passed across the piers at Newark and Elizabeth, far and away the world's largest container complex. In 1980, the wharves around New York Harbor, including the new U.S. Lines terminal on Staten Island, handled seven times that many loaded boxes, even though New York's share of all U.S. container traffic had declined. Container traffic from Britain to points outside Europe, almost all of which passed through either Felixstowe or Tilbury, more than trebled in a decade, despite Britain's weak economy. Deep-sea ports from Rotterdam, Antwerp, and Hamburg to Hong Kong, Yokohama, and Kaohsiung, on Taiwan, more than doubled the number of boxes they handled in the late 1970s. More and more, the biggest ports traded largely with one another: in 1976, nearly one-quarter of all U.S. containerized foreign trade went through Kobe, Japan, or Rotterdam, in the Netherlands, and another quarter went through just five Asian or European ports.[6]

The ceaseless expansion of port capacity was driven by the same force as the ceaseless increase in ship capacity, the demand for lower cost per box. New ships sold for as much as $60 million apiece in the late 1970s, despite the depression in shipbuilding. To cover their mortgage payments, ship lines had to maximize the time that their vessels were under way, filled with revenue-generating cargo, and minimize the time spent in port. The equation was simple: the bigger the port, the bigger the vessels it could handle and the faster it could empty them, reload them, and send them back out to sea. Bigger ports were likely to have deeper berths, more and faster cranes, better technology to keep track of all the boxes, and better road and rail services to move freight in and out. The more boxes a port was equipped to handle, the lower its cost per box was likely to be. As one study concluded bluntly, "Size matters."[7]

Size mattered, but a port's location mattered less and less. Traditionally, ports had prospered from interrupting the flow of trade. Customs brokerage, wholesaling, and distribution had been concentrated in port cities, as they were in New York, because all inbound and outbound cargo made a stop there. A port usually had overwhelming financial and commercial links with the interior region that was its hinterland. Geographers, once upon a time, had designated inland points as "tributary" to a particular port.

There were no tributaries in container shipping. Containers turned ports into mere "load centers," places through which large amounts of cargo flowed with hardly a break. Each ship line organized its operations around a small number of load centers to minimize the number of stops its costly vessels would make. Customers did not care where those load centers happened to be: an Illinois manufacturer shipping machinery to Korea was indifferent as to whether its goods went by truck to Long Beach or by rail to Seattle, much less whether they entered Korea at Busan or Inchon. The ship line would make those decisions at its discretion—and it would make them based entirely on which combination of vessel operating costs, port charges, and ground transport rates would lead to the lowest total cost per box.[8]

This new maritime geography brought decidedly nontraditional trade patterns. Exports from southern France might move most cheaply through Le Havre, on the English Channel. Imports for Scotland might ride the train from southeastern England. Japanese cargo headed for San Francisco Bay might well be imported through Seattle rather than Oakland, with the ship line's saving of one day's steaming time in each direction outweighing the cost of putting some of the cargo on a train from Seattle to California. Port cities along the Gulf of Mexico increasingly did their European and Asian trade through Charleston or Los Angeles, because ship lines deemed sailing to the Gulf uneconomic. The Hampton Roads of Virginia displaced Baltimore as a major load center through no fault of Baltimore's, but because a ship line serving Europe could make four more trips per year from Hampton Roads—and when a $60 million ship was involved, those four trips could spell the difference between profit and loss.[9]

The local economic benefits of a successful port still were large. A metropolitan area with a port would have a concentration of jobs in trucking, railroading, and warehousing, would need customs brokers and freight forwarders, and would reap tax revenue from port-related businesses. Where those jobs happened to be, however, depended upon commercial considerations much more than on geography. Ports such as Seattle, with small local markets, had a realistic hope of being major container gateways, "reaping substantial economic benefits as a result," one study argued. Ports such as Tokyo and London, with populous markets close at hand, were not guaranteed to prosper. With the ship lines calling the tune, ports were forced to compete for the carriers' favor.[10]

Competition involved investment on a dizzying scale. The World Bank and the Asian Development Bank poured $1.3 billion into port projects in developing countries during the 1970s. American ports spent $2.3 billion for container-handling facilities between 1973 and 1989. Ship lines used their bargaining power to push the risks of building new berths, buying new cranes, and digging deeper channels on to government-run port agencies. The ports insisted that the ship lines sign leases, but leases often did not guarantee a flow of cargo. Ship lines could, and often did, move their load centers from one port to another as their strategies changed, making only minimal payments to the ports they left behind. In a single year, thirty ship lines shifted their service at North American ports, leaving some with a severe loss of traffic. Having the fanciest facilities did not guarantee success; Oakland spent a disproportionate share of its revenues for container terminals in the late 1970s but lost enormous market share to Long Beach nonetheless. In 1983, after yet more huge investments, an environmental lawsuit brought a dredging project to a halt, and American President Lines responded by moving much of its traffic to Seattle. Seattle, in turn, lost out in 1985, when Tacoma, a few miles south on Puget Sound, built a $44 million terminal and lured Sea-Land away.[11]

Inevitably, much of the investment in port facilities went to waste. Baltimore's new docks led to a surge of cargo in 1979 and 1980—but the port handled fewer containers in 2000 than it had two decades earlier. Taiwan's expensive containerport at Kaohsiung was a roaring success, but the government's decision to build a port at Taichung as well was a costly mistake. San Diego was one of many ports to order high-priced container cranes that then found little use. Entire technologies, such as on-dock rail, proved to be sinkholes: ports that laid train tracks on the docks, so that cranes could transfer cargo directly from ships to waiting railcars, learned that the time required to move the train forward as the crane loaded each railcar delayed ships and reduced productivity. Many of the on-dock rail lines were abandoned, but the cost of failure was largely borne by the ports.[12]

The increasing riskiness of the port business did not pass unnoticed. Government investment in ports had been crucial to the development of container shipping in the 1960s and 1970s. With the exceptions of Felixstowe and Hong Kong, every major containerport in that era was developed at public risk and expense. At the time, there had been no alternative: the undercapitalized ship lines and stevedoring companies could not possibly have financed port development on their own. As investment needs grew larger, public officials began to lose their enthusiasm for running ports. "The incremental costs are now staggering," the head of Seattle's port said in 1981. The possibility that a ship line's departure or demise could leave a public agency to pay for idle cranes and silent container yards was too great for many governments to chance.[13]

British prime minister Margaret Thatcher broke the ice by selling off twenty-one ports to a private company in 1981. Governments in other countries followed suit. Malaysia leased its container terminal at Port Klang to a private group in 1986, and ports from Mexico to Korea to New Zealand were soon in private hands. The investors included not only stevedoring and transport companies, but also leading ocean carriers. Containership lines were by now huge businesses, able to raise the large amounts of capital that ports required. As port users, they had an interest in having facilities that could handle their ships quickly. Unlike government agencies, the private port operators had no imperative to expand for the sake of local economic development; they could insist on long-term contracts, backed by banks or by collateral, to assure that they would recover whatever investments they made. Governments retreated to the role of landlords, renting out waterfront land to private companies. By the end of the twentieth century, nearly half the world's trade in containers would be passing through privately controlled ports.[14]

In 1977, container shipping reached a landmark. Containerships were put into service between South Africa and Europe, the last big maritime route still handled by breakbulk ships. Containers were by no means universal; on many less trafficked routes, especially to Africa and Latin America, traditional ships still dominated. In commercial terms, though, these were niche markets, not large opportunities. The major ocean routes had become the floating highways that Malcom McLean had envisioned. Seventeen ships with a total capacity of 20,000 20-foot containers sailed each week from the U.S. Pacific coast to Japan in 1980. From northern Europe, 23 ships set sail weekly for Atlantic and Great Lakes ports in North America, and 8 more, with a capacity of more than 15,000 containers, left Europe for Japan. Even the long route between Australia and the U.S. East Coast had an average of 2.5 containerships in each direction each week, carrying Australian meat to America and manufactured goods the other way.[15]

In their endless quest to get bigger, ship lines set their sights on a new way of linking ports that they already served: sailing around the world.

Round-the-world service had been an idea hardly worth contemplating in the days of breakbulk shipping. With slow ships and long port calls, the 39,000-mile trip from New York across the North Atlantic, through the Straits of Gibraltar and the Suez Canal, calling at Singapore and Hong Kong and Yokohama and Los Angeles and heading home through the Panama Canal, would easily have taken six months. Faster vessels and shorter port calls made a three-month voyage imaginable. In 1978, the year after McLean's departure, R. J. Reynolds ordered 12 diesel-powered vessels at a cost of $580 million and promised that Sea-Land would soon launch a "new weekly round-the-world service."[16]

The idea was not entirely insane. Most ship lines suffered from highly imbalanced traffic patterns. Sea-Land, for example, was a major carrier in the North Pacific, but Japan's huge trade surplus with the United States meant that it carried far more cargo eastbound than westbound. It had the reverse problem in the Middle East, where countries flush with petroleum revenues were importing vast quantities of manufactured goods but had little containerized freight to export. A service sailing eastbound around the world could help resolve this imbalance by allowing ships to discharge full containers at Middle Eastern ports, take on empties that had been delivered on previous voyages, and carry them onward to Japan. On the way, the vessels might stop in Singapore and Hong Kong, where they would be met by smaller ships shuttling freight from developing economies, such as India and Thailand, that did not yet trade enough to justify a container route to Japan or the United States.

Yet round-the-world service was a risky venture. Traffic flows between different pairs of ports were vastly different; a vessel that might be perfect for loads between New York and Rotterdam could easily be too large between Singapore and Hong Kong. Delays due to a storm, a dock strike, or a mechanical problem could play havoc with schedules intended to have a ship calling at each port on the same day each week. This was no minor problem: Israel's Zim Line, which sailed from America's Atlantic coast through the Suez Canal to America's West Coast—the closest thing to a round-the-world container service in 1980—arrived within one day of schedule on only 64 percent of its trips and was more than a full week late one trip in seven. If shippers were to decide that a standard point-to-point service was more likely to run on schedule than one circumnavigating the globe, round-the-world ships might find themselves hard-pressed to attract freight. In the face of these risks, Sea-Land gave up its plans to sail around the world.

Two of its major competitors did not. One was Evergreen Marine. Evergreen, founded as a tramp company by the ambitious Taiwanese entrepreneur Chang Yung-fa in 1968, had become a major operator across the Pacific and on the Far East–Europe route, undercutting conference freight rates to gain traffic. In May 1982, Evergreen ordered 16 containerships from yards in Japan and Taiwan at a cost of $1 billion and announced that it would run round-the-world services heading both east and west. The vessels, originally planned to carry 2,240 20-foot containers, were soon redesigned to hold 2,728. Chang called these vessels his "G-class" and named them accordingly: *Ever Gifted, Ever Glory, Ever Gleamy.* They would steam at 21 knots, fast enough that each of the 19 ports of call would see an Evergreen ship in each direction every 10 days. Evergreen's ships would circumnavigate the world in 81 days eastbound, 82 days going west.[17]

The other competitor in the race around the world was an equally self-confident shipping magnate, Malcom McLean. In 1982, his U.S. Lines placed orders for 14 gigantic containerships. By building at Korea's Daewoo shipyard, the company forfeited any rights to U.S. government construction subsidies but won the freedom to deploy the vessels where it chose, without government involvement. Each new ship could carry the equivalent of 4,482 20-foot boxes, half again as much as Evergreen's G-class vessels. The ships were wide, flat, and utilitarian, designed—in the words of their architect, Charles Cushing—to look "much like a big shoe box above the water." McLean's strategy was different from Chang's. His ships would circle the globe only in an eastbound direction, and they would do so slowly. McLean had learned from his mistakes with the speedy SL-7s, whose fuel bills ate up all their profits. The new ships were built for an era of expensive oil. They would conserve fuel by sailing at only 18 knots, taking longer than Evergreen's vessels to sail around the world.[18]

McLean dubbed his new vessels Econships, because their fuel economy, along with the scale economies created by their enormous size, would produce the lowest cost per container of any ships anywhere. The ships alone cost $570 million. McLean's new, publicly traded holding company, named, like its predecessor, McLean Industries, had no difficulty raising the money. The world was eager to invest with the founder of container shipping as he turned U.S. Lines into a "global bus service."[19]

The profit potential of these new services was questionable from the start. Almost the entire container shipping industry was struggling: Seatrain Lines went bankrupt in 1981; Delta Steamship and Moore-McCormack Lines collapsed into the arms of U.S. Lines in 1982; Hapag-Lloyd sold its headquarters building for cash to repay creditors; Taiwan's Orient Overseas holding was forced to restructure $2.7 billion of debt; and Sea Containers, which owned British ferry services as well as 15 containerships and a container leasing company, nearly went under. Matters grew even worse in 1984 and 1985, as Evergreen and U.S. Lines began their round-the-world services. New shipping capacity flooded onto the market. The available space for containers in the North Pacific rose 20 percent between May 1983 and May 1984, leaving ships from North America to Japan running almost half empty. *Lloyd's Shipping Economist* reported "widespread rate cutting actions by carriers in search of market share."[20]

Nor was the service quite as expected. Each port call required not just a stop at the dock but a costly, time-wasting diversion from the round-the-world route. Unless stops were severely curtailed, voyages would become impractically long. As a result, most ports were connected to the round-the-world services with feeder ships that transferred their containers at major load centers, lengthening total transit time for cargo. Evergreen's globe-girdling ships eventually stopped visiting Britain altogether, using Le Havre, in France, as their regional load center and shuttling 200,000 containers per year to ports in England, Scotland, and Ireland. McLean's Econships required too much depth for many ports, sometimes leaving cargo on the dock in order to depart on the high tide. Neither Evergreen nor U.S. Lines had faced the fact that their ships might not be the best way to move cargo; although shipping containers cross-country on double-stack trains cost more than sending them through the Panama Canal, American President Lines' ship-rail service could get a container from Japan to New York in only 14 days, a transit time neither Evergreen nor U.S. Lines could come close to matching.

On-time performance was a constant problem as well. With a ship sailing around the world, bad weather in the Bay of Biscay or a troublesome crane in Dubai could obliterate the schedules promised to customers at Yokohama and Long Beach.[21]

Disaster was not long in coming. Instead of rising from $28 to $50 a barrel, as McLean had expected, oil prices collapsed to $14 in 1985. U.S. Lines' slow, fuel-conserving ships were suddenly the wrong vessels for the market, and the oil sheikdoms of the Middle East could no longer afford the limitless quantities of imports that were supposed to keep the Econships filled with cargo. The competition was tougher, too: unlike the 1970s, when one mismanaged breakbulk company after another collapsed before the onslaught of container shipping, the players in the 1980s were professionally managed firms with no inclination to surrender. After posting a $62 million profit in 1984, McLean Industries reported a $67 million loss in 1985. The company missed an interest payment in early 1986 as McLean struggled to restructure his loans. It was to no avail. In the first nine months of 1986, McLean Industries lost $237 million on revenue of $854 million. Container terminals in Europe began demanding cash up front before allowing the Econships to load. Creditors tightened their terms. On November 24, staggering under $1.2 billion of debt, McLean Industries suspended all service and filed for bankruptcy.[22]

The collapse of United States Lines was, at the time, the largest bankruptcy in U.S. history. It was also one of the most tangled. A total of 52 ships were arrested at ports from Singapore to Greece. The seven U.S. banks that held the mortgages on the Econships scrambled to recover what they could from vessels that no other company wanted; 16 months later, the ships were sold to Sea-Land for 28 cents on the dollar. More than 10,000 containers and 5,500 chassis were dumped back on Flexivan Leasing, which had been renting them to U.S. Lines for a few dollars a day. U.S. Lines' $12 million annual lease for its new container terminal on Staten Island was annulled, leaving the Port Authority of New York and New Jersey liable for $60 million of dredging and construction work. First Colony Farms passed into the hands of its bankers, much of it to end up as a wildlife refuge. Unsecured creditors, who were owed $260 million, came away with almost nothing. Malcom McLean's shareholding, representing 88 percent of McLean Industries' common stock, was wiped out, and he and his son Malcom McLean Jr., a vice president, were ejected from the management. Thousands of people lost their jobs.[23]

"Malcom never got over the U.S. Lines bankruptcy," a longtime associate said later. He went into seclusion, shunning journalists and avoiding public appearances. His failure followed him, the knowledge that he had hurt thousands of people a constant source of shame. Yet he still was a driven man. In 1991, five years after the failure of U.S. Lines, sheer boredom led him to launch yet another shipping company at the age of seventy-seven. Former Sea-Land executives, many of them now among the shipping industry's leading lights, prevailed upon him to return at least occasionally to public view, to accept the awards and honors that were his due. On the morning of his funeral, May 30, 2001, containerships around the world sounded their whistles in his memory.[24]

Yet if the failure of United States Lines was a personal disaster for many, it was far from a catastrophe for the industry Malcom McLean had created. By 1986, the year of U.S. Lines' collapse, ports, transportation companies, and shippers around the world had invested $76 billion in order to carry freight in containers. Another $130 billion of outlays was forecast by the end of the twentieth century, on even bigger ships, ports that could turn a vessel inside of twelve hours, cranes that could move more than one box per minute. Container shipping was becoming a very big business, and as it grew, the cost of moving a containerload of cargo was steadily coming down.[25]

Chapter 13

# The Shippers' Revenge

**isappointing as they** were for investors, the first years of international container shipping introduced an entirely new dynamism to the boring old business of moving freight. The decade-long rate war, as Karl Heinz Sager of the German carrier Hapag-Lloyd commented later, "entailed tremendous losses for shipowners, but it brought on the other hand the breakthrough of the 'box' with shippers." The new container technology spread widely, and it would soon begin to penetrate deeply into the world economy.[1]

The initial effects of the container were felt mainly within the narrow confines of the maritime industry, by ship lines, port agencies, and dockworkers. Carriers staggered under the enormous costs of the transition to container shipping, and some did not survive it. Ports literally had to rebuild themselves to handle containers in quantity, taking on entirely new roles developing and financing terminals of previously unimagined scale. Longshoremen almost everywhere lost their jobs in large numbers, although in many cases their unions resisted strongly enough to win compensation for acceding to the changes that would quickly reshape work on the docks.

The sweeping changes within the maritime world initially had very few wider consequences. Ocean transportation itself accounted for only a very small share of the world economy, and, except in dockside communities, longshore work was a tiny percentage of total employment. The true importance of the revolution in freight transportation would be found not in its effect on ship lines and dockworkers, but later, as the impact of containerization resonated among the hundreds of thousands of factories and wholesalers and commodity traders and government agencies with goods to ship. For most shippers, except perhaps government agencies, the cost of transporting goods was decisive in determining what products they would make, where they would manufacture and sell them, and whether importing or exporting was worthwhile. The container would reshape the world economy only when it changed shippers' costs in a significant way.

This did not happen quickly. As late as 1975, after containership lines had been crossing oceans on regular schedules for nearly a decade, a United Nations agency could declare that "[f]ew shippers have benefited from long-term reductions in liner transport costs." A decade later, the situation would look very different.[2]

The first years of international container shipping, up through the early 1970s, brought large reductions in ship lines' costs. The most important was the saving in loading and unloading vessels, which had been the greatest single expense in precontainer days. Capital costs, although higher than for traditional ships, were not exorbitant, because old vessels refitted with cells to hold containers made up most of the container fleet. Container berths at ports cost ten times as much to build as conventional berths, but they could handle twenty times as much cargo per man-hour, so the cost per ton was lower. The early containerships were cheaper to operate than breakbulk ships on a per-ton basis, because each ship carried more freight. Adding it all up, the United Nations Conference on Trade and Development (UNCTAD) concluded in 1970 that ship lines' costs of moving freight on containerships were less than half those on conventional ships.[3]

Shippers shared part of these initial cost reductions from containerization. The complexity of commodity-by-commodity rate structures makes it hard to estimate average rates, but anecdotal evidence suggests strongly that the introduction of international container shipping immediately brought lower rates than were available on breakbulk vessels. The rate decline, however, was probably less than justified by shipowners' large savings, because the same conferences that set rates for containers also set rates for traditional shipping. Many conference members were carrying containers inefficiently on breakbulk ships until their new containerships were built. They wanted to keep container rates close to breakbulk rates in order to protect their profits, and to slow the growth of containerization until their new vessels arrived.

The result was that the early rates for containers were based not on the cost of container shipping, but on the cost of breakbulk freight. If a container held mixed freight, each item was charged only slightly less than if it were moving in a breakbulk ship. Containers filled with a single product received discounts that were larger, but not generous. At the start of service from Europe to Australia in 1969, for example, a Welsh refrigerator plant could save only 11 percent from breakbulk rates by shipping full containers of its product, and almost nothing by sending small shipments that would travel in mixed containers along with other cargo. Full refrigerated containers of Australian meat went to Britain at a fairly meager 8.65 percent discount from the breakbulk rate.[4]

From the containership operators' point of view, offering lower rates for full containers than for mixed containers made perfect sense. A box fully loaded with a single commodity, sealed at the factory and not opened until it arrived at its final destination, was the most economical sort of cargo to handle, whereas mixed freight had to be consolidated by a freight forwarder or a ship line, using expensive longshore labor. In the 1960s, however, manufacturers were not accustomed to doing business by the containerload. Often, they would produce goods as orders came in and send out each order as it was finished. A 1968 study of 235 shipments of manufactured goods between North America and Western Europe found that 40 percent weighed less than one ton and 84 percent less than ten tons. These loads were too small to fill a single container and would not have qualified for the cheapest rates.[5]

Cost structures changed dramatically with the arrival of second-generation containerships starting in 1969. The new vessels had been designed with ease of loading and unloading foremost in their architects' minds, and their cost for handling cargo was very low. Quite unlike either breakbulk ships or first-generation containerships, though, the second-generation ships came with obligations payable regardless of the business situation. Interest and principle on money borrowed to buy ships, chassis, and containers loomed large. Instead of port fees that varied with time at the dock and the amount of cargo loaded or unloaded, there were long-term leases for wharves, cranes, and marshaling yards, with rent owed even if traffic was down. Transporting empty containers back across the ocean was a burden with no corollary in the breakbulk world, and it could be heavy: more than half the 100,000 containers passing through the port of Antwerp in 1969 were empty. The computer systems to keep track of the containers and prepare loading plans for ships were a major new fixed cost.[6]

The new ships' larger sizes and higher speeds allowed them to move far more freight over the course of a year than earlier vessels. Ships purchased in the early 1970s by European lines sailing to the Far East, for example, had four times the cargo capacity of the breakbulk ships they replaced, and their higher speeds and faster port turnaround times let them make six round trips each year rather than 3⅓. Over the course of a year, each one of these new ships could carry six or seven times as much cargo as a conventional vessel. Profitability required that at least three-quarters of the container cells be filled; beyond that point, the fixed costs could be spread widely and the cost per container would be low. Profits thus depended not only on the number of vessels competing for cargo, but on the business cycle. A global recession would hit shipowners twice over: the lack of freight would cause their fixed cost per container to increase at the same time as it would weaken their ability to hold rates at profitable levels.[7]

Precisely such a lack of freight led to lower shipping rates in the early 1970s. Shipping machinery from southern Germany to New York cost one-third less by containership than by breakbulk freight, a bank study found in 1971. From whiskey distillers in Scotland to apple growers in Australia, major users of international shipping abandoned breakbulk freight as soon as regular container service was able to meet their needs. They had no reason to make this switch unless they found container shipping advantageous. Shippers' overwhelming choice—in economic terms, their "revealed preference"—is very strong evidence that containerization on a trade route lowered the cost of shipping. The willingness of ship lines to share revenues through arrangements such as the North Atlantic Pool in 1971 indicates their desperation as freight rates tumbled.[8]

Then came the oil crisis. The dramatic oil-price rises that began in 1972 and accelerated after the Yom Kippur War in October 1973 had a disproportionate impact on all transportation industries. The average price of crude oil on the world market rose from just over three dollars per barrel in 1972 to more than twelve dollars per barrel in 1974. Freight costs, whether by truck, train, or sea, rose relative to the cost of manufacturing.

The new containerships were hit especially hard. Their high speeds meant that they consumed two or three times as much fuel for a given amount of freight as the breakbulk ships they replaced. This had not been a concern at the time the fuel-guzzling vessels had been designed; in the early 1970s, fuel accounted for only 10 to 15 percent of containerships' operating costs. By 1974, though, fuel prices were a crushing burden, eventually to reach half the total cost of running a ship. The liner shipping conferences raised rates, slapped fuel surcharges and currency adjustment surcharges onto customers' bills, and repeatedly raised the surcharges as fuel costs kept rising and the dollar kept falling. The cost of container shipping on long-distance routes, on which fuel mattered most as a share of total costs, rose disproportionately. Importers and exporters responded by curtailing long-distance trade in manufactured goods much more sharply than short-distance trade. To freight users around the world, container shipping no longer seemed quite such a bargain.[9]

Determining exactly what happened to the cost of shipping from 1972 through the late 1970s poses an insurmountable challenge for the historian. Only short sea routes, such as those across the North Sea, had flat rates per container for most of that period. Elsewhere, charges were based not on the container but on the commodity inside. There is no reliable way to calculate an average cost, much less to track its change over time.[10]

Three sources other than actual freight rates have been used to estimate the trends in shipping costs. One is the cost of leasing "tramp" ships, vessels that are chartered rather than providing regularly scheduled, or "liner," service. The charter price per ton of tramp capacity rose sharply, as was widely reported in shipping publications during the 1960s and 1970s. Most tramps, however, carried grain or other bulk cargo rather than manufactured goods, so the rental cost sheds little light on the price of container shipping. As container shipping gained importance, tramps were increasingly relegated to carrying low-value freight that could not efficiently be containerized, making tramp prices of little relevance to the cost of containerized freight.[11]

A second main source of freight-cost data is the Liner Index compiled by the German Ministry of Transport. This shows that freight rates flattened out in 1966, as container shipping arrived, but then rose steeply, trebling between 1969 and 1981. The Liner Index, though, is highly problematic as a gauge of global transport costs. It tracked rates on cargoes passing through ports in northern Germany, the Netherlands, and northern Belgium, not worldwide, and its coverage included a large proportion of noncontainer shipping. Changes in the exchange rate of the German mark seem to have had a huge influence on the index's movements. It took four German marks to buy one U.S. dollar in 1966, three in 1972, and only two by 1978. For shippers who did business in dollars, ocean freight rates as measured by the Liner Index rose well below the rate of inflation during the 1970s.[12]

The third alternative is the estimate of standardized charter rates for containerships published by Hamburg ship broker Wilhelm A. N. Hansen starting in 1977. Hansen's measure, unlike the Liner Index, shows prices falling in 1978 and 1979. However, it is drawn from charters of very small containerships, the kind most likely to be available for charter. It is not clear whether it accurately reflects rates charged by operators with larger, more efficient vessels.[13]

The technical problems involved in measuring shipping rates during the 1960s and 1970s are so great that reliable measures of the container's price impact are unlikely to be developed. International shipping rates were often set in U.S. dollars, and dramatic changes in exchange rates changed shipping costs for companies in many countries independent of changes in technology. Many conferences offered discounts of as much as 20 percent to shippers that signed "loyalty agreements" promising to use only conference members' ships, so the published conference rates were not necessarily the rates that important shippers paid. Many large shippers demanded, and received, under-the-table rebates from ship lines in return for paying the published rate; although rebates on routes to the United States were illegal—Sea-Land was fined $4 million in 1977 for distributing $19 million in secret payments to customers between 1971 and 1975—the practice was common elsewhere. Rebates, of course, made the actual prices that shippers paid much lower than the prices ship lines claimed to charge.[14]

Complicating matters even further is the fact that traditional breakbulk ships remained in service long after container shipping arrived. Breakbulk ships transported more of the United States' general-cargo trade than did containerships until 1973. They remained important on routes to developing countries in Africa and Latin America well into the 1980s, because in many trades the flow of cargo was too small to justify the capital outlay for dedicated containerships and ports. Any measure of overall ocean freight costs during the first decade of international container shipping thus is capturing a large amount of breakbulk shipping. It is also capturing inflation. Consumer prices in every industrial country more than doubled during the 1970s, and it would be an extraordinary achievement indeed if containerization actually brought the nominal cost of shipping down.[15]

Trying to compute the extent to which containerization changed "average" maritime rates for shippers keeping their accounts in different currencies and moving a wide variety of goods under hundreds of conference rate structures is an exercise in futility. On balance, the evidence suggests strongly that the cost of shipping a ton of international freight began to decline as containerization became important around 1968 or 1969, and that it fell through 1972 or 1973. As fuel prices rose steeply, freight costs reversed direction, rising until 1976 or 1977. Rates on American-flag vessels other than tankers, overwhelmingly general-cargo ships, show a similar trend, with ship lines' revenues falling relative to the value of their cargo until the oil crisis brought the rate cutting to a temporary end in 1975.[16]

And what if container shipping had not taken the transportation world by storm? Dockers' pay soared during the 1970s. Productivity improvements in breakbulk shipping were minimal. The labor-intensive task of loading a breakbulk ship would have been far costlier in 1976 than it was a decade earlier. Even at the peak of oil prices in 1976, when fuel surcharges were pushing ocean freight rates sky-high, very few shippers seem to have entertained the thought of going back to breakbulk shipping.[17]

Ocean freight, of course, is not the only cost involved in transporting imports and exports. The total freight bill includes not just ship rates, but land transportation to and from the ports; packaging; storage and other port charges; damage and insurance; and the cost of money tied up in goods that are in transit. In the days of breakbulk shipping, the relative importance of these various costs depended heavily upon the details of the particular shipment. Moving a load of packaging material from the United States to Western Europe in 1968, for example, yielded $381 per ton for the ship line and only $34 for truckers or railroads. Moving a load of auto parts with long land shipments at both ends, by contrast, cost $152 per ton for land freight and only $20 for ocean freight. For the packaging materials, a change in ocean freight rates would have made a dramatic change in the total freight bill, but for the auto parts it would barely have mattered.[18]

The containerization of ocean shipping initially did not reduce the costs on land. In many countries, rates for truck lines and railroads were based upon the commodity and the distance, just like ocean freight rates. Regulations in the United States barred ship lines even from quoting a single through rate to an inland destination, much less negotiating special discounts for land transportation on behalf of their customers. Moving a container of televisions from Hiroshima to Chicago thus required the exporter to pay the standard Japanese truck rate for televisions, plus the appropriate ocean freight rate, plus the domestic U.S. truck or rail rate for electronic products, plus a payment to a freight forwarder to make all the arrangements. Land freight rates moved sharply higher during the 1970s, driven by increased fuel prices and higher wages. Shippers exporting to the United States increasingly favored routes that involved longer ocean voyages and shorter land hauls, an indication that land transport costs were increasing relative to the cost of ocean freight.[19]

Businesses near ports ignored by container operators may have ended up with disproportionately higher shipping costs in the 1970s, because their goods now had to move much longer distances over land. Seventeen different ports had handled New Zealand's international trade in breakbulk days, but containers were shipped through only four, leaving meat or wool processors to pay for getting their products to Auckland or Wellington. The same happened to industrial companies around Manchester; Britain's fifth-largest port fell into disuse in the 1970s as containerships avoided the time-consuming trip up the thirty-six-mile canal from the sea, and local customers had to cover the land costs of trading through Liverpool or Felixstowe. Manufacturers in northern New England faced the added cost of trucking their exports to New York after their traditional port, Boston, ended up with only occasional visits from containerships.[20]

Many nonfreight costs undoubtedly fell with the growth of container shipping. Packing full containers at the factory eliminated the need for custom-made wooden crates to protect merchandise from theft or damage. The container itself served as a mobile warehouse, so the traditional costs of storage in transit warehouses fell away. Cargo theft dropped sharply, and claims of damage to goods in transit fell by up to 95 percent; after insurers were persuaded that container shipping in fact had fewer property losses, premiums fell by up to 30 percent. Faster ships and reductions in the time needed to load and unload vessels at ports resulted in lower costs for inventory in shipment.[21]

As Malcom McLean had understood back in 1955, it is the sum of these costs, not just the published rate of a ship line or railroad, that matters to shippers. Ideally, we would like to trace the door-to-door cost of the same shipment over time, so that we could measure the change as containerized transportation took hold. With similar information on a hundred different consumer products and industrial goods, we might be able to assemble a reasonable index of freight costs. This task, alas, is beyond even the most intrepid investigator. Data on door-to-door shipping costs were not compiled in 1965, and they do not exist today. Even a rough estimate of how the arrival of containerization in international trade affected the cost of trade is sheer guesswork.

What we do know is that the overall cost of shipping goods internationally remained relatively high through the mid-1970s, even with containerization. One 1976 shipment studied in detail by the Maritime Administration, involving $25,000 worth of wheel rims shipped from Lansing, Michigan, to Paris, France, incurred $5,637 of freight costs—22.6 percent of the value of the cargo. The bill included $3,600 for ocean freight from Detroit to Le Havre, more than $600 in trucking costs, and over $1,300 in fees and insurance costs. With the 7 percent French import tariff added on, the wheel rims cost one-third more in France than in Michigan.[22]

At some point in the late 1970s, the trend line seems to have begun to change. Although fuel costs continued to rise, the real cost of shipping goods internationally started to fall rapidly.[23]

What happened to make shipping cheaper? And why did it start to happen around 1977 rather than with the onset of international container shipping a decade earlier? The answers have to do with a group that has received little attention in these pages: shippers. Containerization required the buyers of transportation to learn a whole new way of thinking about managing their freight costs. As they became more knowledgeable, more sophisticated, and more organized, they began to drive down the cost of shipping.

Shippers were not a major force in the days of breakbulk freight. Many governments frowned upon rate competition, supporting rate fixing by the liner conferences and, on some routes, prohibiting low-rate independent carriers altogether. Even where governments allowed nonconference lines to compete, relatively few did, because there often was not enough freight: shippers typically agreed to pledge all of their freight on a route to conference members in return for "loyalty" discounts—a pledge that strengthened the conference by making it harder for nonconference intruders to get business. Shippers, ship lines, and governments all thought of ocean ship lines in much the same way they thought of trucking companies and railroads, as providers of a public service entitled to raise their rates whenever their costs went up. "Our future depends on having strong conferences supported by strong commercial shipper bodies," an executive of a British ship line said in 1974—as if the interests of ship lines and their customers were one and the same.[24]

The huge capital requirements of container shipping left fewer ship lines on each route, strengthening the conferences and tilting the playing field against shippers. The 1971 pooling agreement on the North Atlantic, to take the most extreme example, essentially combined the efforts of fifteen lines that had once been competitors. On the Europe-Australia route, the thirteen companies that sailed between Europe and Australia in 1967 had combined into seven by 1972. As these new groupings began to curb competition, shippers reacted by working together more closely. By 1976, private-sector shippers' councils were active in thirty-five countries.[25]

It was in Australia, where farmers were almost totally dependent upon exports, that shippers began to flex their muscles. In 1971, four groups representing sheep farmers and wool buyers formed a joint organization to oppose rises in freight rates. A year later, rubber traders in Singapore responded to conference surcharges by finding a nonconference carrier to move their product to Europe for 40 percent less. Australian dairy producers signed with a nonconference carrier to save 10 percent on freight rates to Japan. By 1973, shippers' power on the East Asia–Europe route was substantial enough that the conference was forced to bargain, and the Malaysian Palm Oil Producers Association won an unprecedented two-year rate freeze. "Increases in the liner freight rates met considerable opposition from shippers in certain trades," UNCTAD reported in 1974. In 1975, the Australian Meat Board bargained for unusually deep rate reductions in return for giving four ship lines all its meat shipments to the U.S. East Coast.[26]

Shipper organizations had no legal status in the United States, and shippers were reluctant to negotiate jointly lest they be accused of violating antitrust law. The biggest shippers, however, began to exert influence on their own, even as they changed the way they worked in order to take advantage of the container.[27]

In the early days of containerization, users dealt with shipping much as they had in breakbulk days. Traffic management was decentralized, with each plant or warehouse making its own arrangements. If the company as a whole could have saved money by sending fully loaded 40-foot containers to individual customers, well, that was not the concern of the freight managers at individual locations, whose job was mainly to get the products out the door. Most shippers favored 20-foot maritime containers, which cost more per ton to ship, because they could not coordinate production of various orders well enough to fill a 40-foot box. The biggest shipper of all, the U.S. military, divided responsibilities between one agency that handled land shipping and another that dealt with sea freight, often paying extra because it had selected the wrong size container for a given load.[28]

In industry, the traffic department, housed in the back of the plant near the loading dock, would be given whatever the manufacturing department produced, with instructions to ship it. A tariff clerk, his desk piled high with the freight classification guidelines of various liner conferences, trucking conferences, and railroads, would try to describe the cargo in whatever way brought the lowest rate. An export manager would then call ship lines to select a vessel, balancing the desire for fast delivery with the need to keep from becoming too dependent on a particular carrier. With decentralized organizations and fairly primitive computer systems, even large, relatively sophisticated multinational corporations could end up paying dramatically different prices for the same type of cargo, depending upon what the tariff clerk and the export manager could accomplish. "In some cases we'd pay $1,600 for a 40-foot container in the North Atlantic, and in other cases we'd be paying $8,000 for the same container," recalled a former chemical-industry executive.[29]

Big shippers typically signed dozens of loyalty agreements covering different routes, obtaining discounts in return for commitments to ship all of their freight with conference members, and then dealt with hundreds of individual conference carriers. The result, often enough, was unsatisfactory. A loyalty agreement did not guarantee space on a ship; if the manufacturer had cargo to ship to India but no space was available on a conference member's vessel, the cargo had to wait until a conference ship had room. Sending the freight on an independent liner or a tramp ship violated the contract and would expose the shipper to a heavy fine from the conference. If the only available conference vessel was making multiple port calls before heading overseas, the cargo would have to wait while the vessel loaded other freight in each port. Managing relationships with ship lines and doling out cargo were administrative nightmares for major manufacturers, requiring large numbers of staff.[30]

As ship lines combined their forces to gain market power, manufacturers responded aggressively. The first step was to look beyond the conferences.

Nonconference carriers had always played a role in the major trades, but a small one. The biggest shippers rarely used them. Independents, as the nonconference lines were known, offered discounts of 10 to 20 percent from conference rates, but most of them were too small to provide frequent service on the routes they plied. If a shipper used an independent carrier and then required service that the independent could not provide, it would end up paying a conference line more than if it had signed an agreement with the conference in the first place. Shippers that had a highly predictable flow of cargo could handle that risk. For manufacturers, who might have a sudden need to ship an unanticipated order, sticking with the large conference carriers, even at higher cost, was the safer strategy.[31]

When containers came on the scene, the economics of container shipping were thought to work against independent lines. The costs were so high that small operators could not enter the business on a whim. Establishing a viable container operation in the United States–Asia trade, one economist estimated in 1978, would require $374 million to buy five ships plus containers, chassis, and cranes. Common sense suggested that anyone putting up that much money would join conferences in hopes of keeping rates high enough to recover costs. But in the second half of the 1970s, it turned out that the barriers to entry were not as high as they seemed. Shipbuilding costs, which had risen 400 percent from the end of 1970 through the end of 1975, began to fall as the collapse of the oil tanker market left shipyards bereft of orders. Builders slashed prices and extended loans just to keep their yards at work. Bargains on new vessels allowed traditional ship lines such as Maersk of Denmark and Evergreen Marine of Taiwan to elbow their way into container shipping. Maersk and Evergreen operated as independents on most routes, with rates far below what the conferences charged. As they added ships, they became credible competitors, drawing shippers that had been wedded to the conferences. Neither company had owned a containership before 1973. By 1981, Maersk's twenty-five ships made it the world's third-largest containership operator, while Evergreen, with fifteen vessels, ranked eighth.[32]

Other independent lines proliferated, particularly in the Pacific. Taiwan's Orient Overseas, owned by shipping magnate C. Y. Tung, became the first independent carrier to run containerships between Asia and New York in 1972, charging 10 to 15 percent less than the conference. Korea Shipping Corp., another nonconference operator, laid out $88 million for eight containerships in 1973. Far Eastern Shipping Company, a Russian independent, sent two containerships a month from Yokohama to Long Beach and Oakland. Conference tariff books turned into comic books as shippers deserted the conference carriers in droves. The shift to flat per-container rates in the late 1970s revealed the severe erosion of conference bargaining power in a way not possible when each commodity was charged a different rate. The conference rate for shipping a 20-foot container from Felixstowe to Hong Kong fell from $3,645 in 1980 to $2,136 just three years later, and was lower in 1988 than at the start of the decade. The cost of shipping a 40-foot box from Europe to New York, $2,000 in the middle of 1979, was below $1,000 by the summer of 1980. By January 1981, so many nonconference ships were competing to carry trade from Manila that the Philippines–North America conference collapsed.[33]

The second important result of shippers' new power in the 1970s, along with their willingness to defy the shipping cartels, was their embrace of an idea that had been a heresy: the deregulation of transportation.

Trucking was tightly regulated almost everywhere in the early 1970s, with the notable exception of Australia. Most railroads were state-owned, damping any competitive instincts. So long as political power rested with the transportation companies and their unions, rather than their customers, the regulatory structure stood strong. If its collapse can be dated to a single event, it was the bankruptcy of the Penn Central, the largest railroad in the United States, in June 1970. The Penn Central's failure, followed in short order by half a dozen other rail bankruptcies, drew attention to the regulations that kept the railroads from adapting to truck competition. The costly and controversial government rescue program altered the political equation, and Republicans and Democrats alike began calling for reductions in regulation. In November 1975, President Gerald Ford proposed eliminating much of the Interstate Commerce Commission's authority over interstate trucking. The following year, Congress took the first steps to ease regulation of railroads.[34]

An intense national debate ensued. On one side, along with railroads wanting more flexibility to compete with truckers, were shippers and consumer advocates who argued that deregulation would reduce costs. Some trucking companies, especially those that handled smaller shipments, were eager to get rid of regulations. On the other side, many companies that handled full truckloads of freight were bitterly opposed to changes that would encourage partial truckloads, and the unions representing railroad workers and truck drivers fought changes that would weaken union power and eliminate union jobs. The regulators, who were easing regulations slowly and gradually, warned Congress against haste. "Certain shippers command substantial and sometimes overwhelmingly superior bargaining power," the ICC's chairman cautioned, asserting the need for the government to keep control in order to protect truckers and railroads from their customers.[35]

In the midst of this heated campaign, the container became a poster child for the inefficiency caused by outdated regulation.

The basic concept of the container was that cargo could move seamlessly among trains, trucks, and ships. Two decades after Malcom McLean's first containership, though, container shipping was anything but seamless. In principle, a truck line or a railroad could offer an exporter a "through rate" between St. Louis and Spain, but the through rate was simply the published truck or rail rate for that product from St. Louis to a port, plus the published ship rate for that commodity across the Atlantic. Domestically, truckers did not much like carrying containers long distances from the ports because they might well have to haul them back empty; domestic shippers preferred to use conventional trailers, which did not detach from their chassis, rather than detachable containers. Railroads did have a business carrying domestic "piggyback" truck trailers on flatcars, but the service was attractive only for relatively long trips; sending a trailer piggyback the four hundred miles from Minneapolis to Chicago took eighteen to twenty-two hours, against eight hours or so in a truck. Piggyback was often no bargain, either. The railroads set rates high in the forlorn hope that shippers would use boxcars instead, so putting a trailer on a train often cost more than taking it over the road.[36]

The railroads were no more aggressive when it came to containers, without the truck chassis and wheels. When Sea-Land and the railroads had discussed a transcontinental container service back in 1967, the railroads asked for three times the price that the ship line was willing to pay, and talks went no further. They tried again in 1972 with a service called "minibridge," in which a ship line and railroads would join forces to carry a container from, say, Tokyo to New York via the port of Oakland. The carriers would agree on a single rate for the entire trip, file it with both the Interstate Commerce Commission (the rail regulator) and the Federal Maritime Commission (the ship regulator), and decide how to split up the money. Ship lines claimed that minibridge cut costs by eliminating the long, fuel-consuming voyage through the Panama Canal. The real advantage, less publicized, was that loading and unloading ships was much cheaper in Pacific coast ports than on the East Coast: almost no one bothered to export from California to Europe by minibridge through New York. The railroads were so uninterested in the concept that they could not even be bothered to design equipment more efficient than their standard flatcars. Shippers often saw little saving. Sending televisions from Japan to New York by minibridge took several days less than by all-water service but was no less expensive. Sending synthetic rubber from Texas to Japan, a U.S. government study found in 1978, cost three times as much by minibridge through Los Angeles as when the rubber was trucked to Houston and loaded on a ship.[37]

Deregulation changed everything. In two separate laws passed in 1980, Congress freed interstate truckers to carry almost anything almost anywhere at whatever rates they could negotiate. The ICC lost its role approving rail rates, except for a few commodities such as coal and chemicals. Trucks and railcars that had often been forced to return empty were able to get cargo for backhauls. Another definitive break from the past proved critical to driving down the cost of international shipping. For the first time, railroads and their customers could negotiate long-term contracts setting rates and terms of service. The long-standing principle that all customers should pay the same price to transport the same product gave way to a system that yielded huge discounts for the biggest customers. Within five years, 41,021 contracts between railroads and shippers were filed with the ICC. Freight transportation within the United States was reshaped dramatically. Costs fell so steeply that by 1988, U.S. shippers—and, ultimately, U.S. consumers—saved nearly one-sixth of their total land freight bill.[38]

Perhaps no part of the freight industry was altered more than container shipping. The ability to sign long-term contracts gave railroads an incentive to develop a business that had languished for two decades, with assurance that their investment would not go to waste. Equipment manufacturers went back to work on low-slung railcars designed for fast loading of containers stacked two-high, the sort of cars Malcom McLean had tried—and failed—to convince railroads to use back in 1967. Deregulation meant that those double-stack cars could be used to haul international containers in one direction and containers filled with domestic freight in the other—impractical before 1980—so the international shipment did not have to bear the cost of returning an empty container to the port.

In July 1983, American President Lines sponsored the first experimental train composed only of the new double-stack cars. Within months, ship lines and railroads had negotiated ten-year contracts under which dedicated double-stack container trains would speed imports from Seattle, Oakland, and Long Beach directly to specially designed freight yards in the Midwest. Days were shaved off the delivery time. The rates, set by negotiation rather than regulation, were far lower than before and were designed to fall further as volumes rose. On average, it cost four cents to ship one ton of containerized freight one mile by rail in 1982. Adjusted for inflation, that cost dropped 40 percent over the next six years. Rail rates fell so steeply that by 1987, more than one-third of the containers headed from Asia to the U.S. East Coast crossed the United States by rail rather than making the voyage entirely by sea. A major obstacle to international trade had given way.[39]

With U.S. trucks and trains deregulated, shipper interests turned their attention to the maritime industry. Once more, they won a sweeping victory. The Shipping Act of 1984 rewrote the rules governing international shipping through U.S. ports. Shippers could now sign long-term contracts with ship lines. In return for guaranteeing a minimum amount of cargo, a shipper could negotiate a low rate and specific terms of service, such as the frequency of ships. These "service contracts" had to be made public, so other shippers with similar freight could demand the same deal. While conferences were still permitted to set rates, individual conference members were free to depart from conference rates whenever they wished, so long as they served public notice.

Shippers' newfound power put enormous downward pressure on freight rates. The official rates published by railroads and ship lines did not fall; if the improbable figures in *Lloyd's Shipping Economist* are to be believed, the conference tariff for a 20-foot container from Britain to New York doubled between 1980 and 1988. But the official rates meant nothing. A better indication of true market conditions comes from the rates bid for U.S. military freight. The military market was open only to U.S.-flag ship lines, which submitted sealed bids every six months to carry general cargo in containers at least 32 feet long. Ship lines were not obligated to bid, so whatever bids were submitted were presumably above the rates the carriers thought they could earn from commercial cargo. In October 1979, the low bidders offered $40.94 to carry 40 cubic feet of cargo either way across the Pacific. By 1986, the transpacific rates had collapsed to $2.39 westbound, $15.89 from Asia to the U.S. West Coast. Even as U.S. producer prices were rising by nearly one-third between 1979 and 1986, maritime freight rates were plummeting.[40]

After the middle of the 1970s, the growth of nonconference ship lines and the ability of shippers to negotiate rates made official tariff schedules useless as indicators of what exporters and importers were paying to ship their goods. "The rates actually charged vary widely and often deviate substantially from published tariffs," the World Bank confirmed. The *New York Times* was less diplomatic, reporting in 1986 that "the shipping world has been turned upside down by five catastrophic years of tumbling freight rates, rising costs, and sinking values of used ships." The magnitude of the saving to shippers and consumers cannot be calculated, but it was extremely large. When American President Lines studied the matter a few years later, it concluded that freight rates from Asia to North America had fallen 40 to 60 percent because of the container.[41]

## Chapter 14

# Just in Time

**Barbie was conceived as** the all-American girl. In truth, she never was: at her inception, in 1959, Mattel Corp. arranged to make her at a factory in Japan. A few years later it added a plant in Taiwan, along with a large cadre of Taiwanese women who sewed Barbie's clothes in their homes. By the middle of the 1990s, Barbie's citizenship had become even less distinct. Workers in China produced her statuesque figure, using molds from the United States and other machines from Japan and Europe. Her nylon hair was Japanese, the plastic in her body from Taiwan, the pigments American, the cotton clothing from China. Barbie, simple girl though she is, had developed her very own global supply chain.[1]

Supply chains like Barbie's are a direct result of the changes wrought by the rise of container shipping. They were unheard-of back in 1956, when Malcom McLean placed his first containers on board the *Ideal-X*, and in 1976, when high oil prices brought sky-high freight costs that stifled the flow of world trade. Until then, vertical integration was the norm in manufacturing: a company would obtain raw materials, sometimes from its own mines or oil wells; move them to its factories, sometimes with its own trucks or ships or railroad; and put them through a series of processes to turn them into finished products. As freight costs plummeted starting in the late 1970s and as the rapid exchange of cargo from one transportation carrier to another became routine, manufacturers discovered that they no longer needed to do everything themselves. They could contract with other companies for raw materials and components, locking in supplies, and then sign transportation contracts to assure that their inputs would arrive when needed. Integrated production yielded to disintegrated production. Each supplier, specializing in a narrow range of products, could take advantage of the latest technological developments in its industry and gain economies of scale in its particular product lines. Low transport costs helped make it economically sensible for a factory in China to produce Barbie dolls with Japanese hair, Taiwanese plastics, and American colorants, and ship them off to eager girls all over the world.

These possibilities first drew notice in the early 1980s, when the world discovered just-in-time manufacturing. Just-in-time, a concept originated by Toyota Motor Company in Japan, involves raising quality and efficiency by eliminating large inventories. Rather than making most of its own components, as competitors did, Toyota signed long-term contracts with outside suppliers. The suppliers were intimately involved with Toyota, helping design its products and knowing the details of its production plans. They were required to adopt strict quality standards, with very low rates of error, so that Toyota would not need to test the components before using them. The suppliers agreed to make their goods in small batches, as required for Toyota's assembly lines, and to deliver them within very narrow time windows for immediate use—hence the name, just-in-time. Keeping inventory to a minimum brought discipline to the entire manufacturing process. With few components in stock, there was little margin for error, forcing every firm in the supply chain to perform as required.[2]

The wonders of just-in-time were unmentioned outside Japan before 1981. In 1984, as Toyota agreed to assemble cars at a General Motors plant in California, U.S. business publications ran thirty-four articles on just-in-time. In 1986, there were eighty-one, and companies around the world were seeking to emulate Toyota's high-profile success. In the United States, two-fifths of the Fortune 500 manufacturers had started just-in-time programs by 1987. Overwhelmingly, these companies found that just-in-time required them to deal with transportation in a very different way. No more would manufacturers offer a load or two to some truck line's hungry salesman. Now, they wanted large-scale relationships with a much smaller number of carriers able to meet stringent requirements for on-time delivery. Customers demanded written contracts that imposed penalties for delays. Even shipments from another continent were expected to arrive on schedule. Railroads, ship lines, and truck lines with large route networks and sophisticated cargo-tracking systems had the edge.[3]

Before the 1980s, logistics was a military term. By 1985, logistics management—the task of scheduling production, storage, transportation, and delivery—had become a routine business function, and not just for manufacturers. Retailers discovered that they could manage their own supply chains, cutting out the wholesalers that had stood between manufacturers and consumers. With modern communications and container shipping, the retailer could design its own shirts and transmit the designs to a factory in Thailand, which used local labor to combine Chinese fabric made from American cotton, Malaysian buttons made from Taiwanese plastics, Japanese zippers, and decorations embroidered in Indonesia. The finished order, loaded into a 40-foot container, would be delivered in less than a month to a distribution center in Tennessee or a *hypermarché* in France. Global supply chains became so routine that in September 2001, when U.S. customs authorities stepped up border inspections following the terrorist attack that destroyed the World Trade Center in New York, auto plants in Michigan began shutting down within three days for lack of imported parts.

The improvement in logistics shows up statistically in reduced inventory levels. Inventories are a cost: whoever owns them has had to pay for them but has yet to receive money from selling them. Better, more reliable transport has permitted companies to obtain goods closer to the time they need them, instead of weeks or months in advance, tying up less money in goods sitting uselessly on warehouse shelves. In the United States, inventories began falling in the mid-1980s, as the concepts of just-in-time manufacturing took root. Manufacturers such as Dell and retailers such as Wal-Mart Stores have taken the concept to extremes, designing their entire business strategies around moving goods from factory floor to customer with minimal time in between. In 2004, nonfarm inventories in the United States were about $1 trillion lower than they would have been had they stayed at the level of the 1980s, relative to sales. Assume that the money needed to finance those inventories would have to be borrowed at 8 or 9 percent, and inventory reductions are saving U.S. businesses $80–$90 billion per year.[4]

This precision performance would have been unattainable without containerization. So long as cargo was handled one item at a time, with long delays at the docks and complicated interchanges between trucks, trains, planes, and ships, freight transportation was too unpredictable for manufacturers to take the risk that supplies from faraway places would arrive right on time. They needed to hold large stocks of components to ensure that their production lines would keep moving. The container, combined with the computer, sharply reduced that risk, opening the way to globalization. Companies can make each component, and each retail product, at the cheapest location, taking wage rates, taxes, subsidies, energy costs, and import tariffs into account, along with considerations such as transit times and security. The cost of transportation is still a factor in the cost equation, but in many cases it is no longer a large one.

Globalization, historians and economists have hastened to point out, is not a new phenomenon. The world economy became highly integrated in the nineteenth century. The decline of tariffs and other trade barriers in the years following the Napoleonic Wars led international trade to increase after decades of stagnation, and the introduction of the oceangoing steamship in the 1840s sharply reduced transport costs. Ocean freight rates fell 70 percent between 1840 and 1910, encouraging increased shipment of commodities and manufactured goods around the world, while the telegraph—the nineteenth-century counterpart of the Internet—gave people in one location current information about prices in another. Prices of grain, meat, textiles, and other commodities converged across borders, as traders found it easy to increase imports whenever domestic prices rose or domestic wages got out of hand.[5]

The globalization of the late twentieth century took on quite a different character. International trade is no longer dominated by essential raw materials or finished products. Fewer than one-third of the containers imported through southern California in 1998 contained consumer goods. Most of the rest were links in global supply chains, carrying what economists call "intermediate goods," factory inputs that have been partially processed in one place and will be processed further someplace else. The majority of the metal boxes moving around the world hold not televisions and dresses, but industrial products such as synthetic resins, engine parts, wastepaper, screws, and, yes, Barbie's hair.[6]

In international production-sharing arrangements of this sort, the manufacturer or retailer at the top of the chain will find the most economical place for each part of the process. This used to be impossible: high transportation costs acted as a trade barrier, very similar in effect to high tariffs on imports, sheltering the jobs of production workers from foreign competition but imposing higher prices on consumers. As the container made international transportation cheaper and more dependable, it lowered that barrier, decimating manufacturing employment in North America, Western Europe, and Japan, by making it much easier for manufacturers to go overseas in search of low-cost inputs. The labor-intensive assembly will be done in a low-wage country—but there are many low-wage countries. The various components and raw materials will come from whichever location can supply them most cheaply—but costs in different locations often are quite similar. Even small changes in transport costs can be decisive in determining where each stage of the process will occur.[7]

The economics of containerization have shaped these global supply chains in peculiar ways. Distance matters, but not hugely so. A doubling of the distance cargo is shipped—from Hong Kong to Los Angeles, for example, rather than Tokyo to Los Angles—raises the shipping cost only 18 percent. Places far from the end market can still be part of an international supply chain, so long as they have well-run ports and a lot of volume.[8]

Container shipping thrives on volume: the more containers moving through a port or traveling on a ship or train, the lower the cost per box. Places with lower demand or poorer infrastructure will face higher transport costs and will be far less attractive manufacturing sites for the global market. In the 1970s and 1980s, when many U.S. industrial centers were dying, Los Angeles thrived as a factory location because it was home to the nation's busiest containerport, and Los Angeles thrived as a port because it was well located to handle import volume from Asia, not just for California, but for the entire United States. The Pacific Rim became the world's workshop for consumer goods, in good part, because large ports for containers gave it some of the world's lowest shipping costs. Antwerp spent a stunning $4 billion on port expansion between 1987 and 1997, including expropriation of 4,500 acres (2,000 hectares) of land, just to keep itself in the game. Conversely, African countries with inefficient ports and little containership service are at such a transport-cost disadvantage that even rock-bottom labor costs will not attract investment in manufacturing.[9]

Shippers in places with busy ports and good land-transport infrastructure not only enjoy lower freight rates, but they also benefit from the shortest shipping times. Before the container, when breakbulk vessels like the *Warrior* carried most of the world's trade, cargo typically left the factory weeks before the ship departed, sailed at a glacial 16 knots, and spent an unproductive week in the hold each time the vessel called at an additional port. In the container age, a machine manufactured on Monday can be dropped at Port Newark on Tuesday and delivered in Stuttgart, Germany, in less time than it once would have taken to be loaded aboard a ship such as the *Warrior*. Yet time still matters. By one estimate, each day seaborne goods spend under way raises the exporter's costs by 0.8 percent, which means that a typical 13-day voyage from China to the United States has the same effect as a 10 percent tariff. The time savings represent a huge competitive advantage to shippers located near a major port. Those served by smaller ports may have to endure longer wait times between ships or shuttle links to a larger port, adding time, and hence costs, to every shipment. Air freight all but eliminates the costs of time, but it is too expensive for most goods that are made in poor countries precisely because little value is added in their production.[10]

"Any change in technology," the economist Joel Mokyr observed, "leads almost inevitably to an improvement in the welfare of some and to a deterioration in that of others." That was as true of the container as of other technologies, but on an international scale. Containerization did not create geographical disadvantage, but it has arguably made it a more serious problem.[11]

Before the container, shipping was expensive for everyone. The most expensive part of international freight transportation, loading cargo aboard ship, affected all shippers equally. Containerization has reduced international transport costs for some much more than for others. Landlocked countries, inland places in countries with poor infrastructure, and countries without enough economic activity to generate high demand for container shipping may have a tougher competitive situation now than they did in breakbulk days. Being landlocked, one study calculated, raises a country's average shipping costs by half. Another study found that it cost $2,500 to ship a container from Baltimore, on the U.S. Atlantic coast, to Durban, in South Africa—and $7,500 more to haul it by road the 215 miles from Durban to Maseru, in Lesotho. Within China, the World Bank reported in 2002, transporting a container from a central city to a port cost three times as much as shipping it from the port to America.[12]

And if high shipping costs, high port costs, and long waiting times do not leave a country at an economic disadvantage, a cargo imbalance might. Relatively few routes, it turns out, have an evenly balanced flow of maritime exports and imports. When the flow is out of balance, shippers in the more heavily trafficked direction have to pick up the cost of sending empty containers back in the other direction. In 1998, nearly three-quarters of the containers sent northbound from Caribbean islands to the United States were empty, resulting in much higher shipping costs for the southbound imports of food and consumer goods on which these island-states depend.[13]

The revolutionary days of container shipping were over by the early 1980s. Yet the aftereffects of the container revolution continued to reverberate. Over the next two decades, as container shipping began to drive international freight costs down, the volume of sea freight shipped in containers rose four times over. Hamburg, Germany's largest port, handled 11 million tons of general cargo in 1960; in 1996, more than 40 million tons of general cargo crossed the Hamburg docks, 88 percent of it in containers, and more than half of it from Asia. The prices of electronics, clothing, and other consumer goods tumbled as imports displaced domestic products from store shelves in Europe, Japan, and North America. Low-cost products that would not be viable to trade without container shipping diffused quickly around the world. Declining goods prices in the late 1990s, thanks largely to imports, helped bring three decades of inflation to an end.[14]

Container shipping, it is clear, has helped some cities and countries become part of the new global supply chains, while leaving others to the side. It has assisted the rapid economic growth of Korea while offering precious little to Paraguay. Yet the trade patterns that containerization has helped to create are not immutable. In the 1980s, ship lines' commitments assured the success of several late entrants to containerization, such as Busan, in Korea; Charleston, South Carolina; and Le Havre, in France. In the 1990s, they repeated the trick on a much larger scale in Asia.

By the end of the twentieth century, the container shipping industry was dominated by a handful of alliances of global scope. These companies' megaships may have sailed between two ports, but the cargo they carried was increasingly unlikely to have been produced in or to be destined for the end points of the voyage. By deciding where to employ their vessels, the big ship lines had the power to determine which ports succeeded and which struggled. In some cases, that choice was made for unavoidable reasons; not all ports had the depths required to handle the biggest ships. In other cases, though, ship lines joined with government officials and private port operators to change comparative advantage. The list of the world's largest containerports around the turn of the century is instructive. Of the twenty ports handling the greatest number of containers in 2003, seven had seen little or no container traffic in 1990, and three of those seven had not even existed before.

These new ports, by and large, were privately managed, and in some cases privately financed. Their creation was a deliberate response to the economics of container shipping, in which keeping the ship moving is what matters most. Only the biggest ports are worth a time-consuming stop. The ports that can load the most containers per hour consume less of a vessel's precious time. Efficient ports, with access to large flows of cargo, will receive large ships and frequent service, with direct sailings to every corner of the world. The massive ports constructed in China, Malaysia, and Thailand during the 1990s were investments in globalization. Factories whose goods use those ports will have the lowest rates and the lowest costs in lost time, saving money on imported inputs and gaining a cost advantage in export markets. Manufacturers in poorer countries, where ports are less busy or less well managed, will find that their high logistics costs make competing in foreign markets a difficult proposition.[15]

That disadvantage goes far beyond the occasional lost export sale. A country cursed with outmoded or badly run ports is a country that faces great obstacles to finding a larger role in the world economy. If Peru were as effective at port management as Australia, the World Bank estimated, that alone would increase its foreign trade by one-quarter. If it cannot be, it will receive the maritime equivalent of branchline service on a single-track railway. The big containerships that link national economies in the global supply chain, carrying nothing but stacks of metal boxes, will pass it by.[16]

Global supply chains were not in anyone's mind in the spring of 1956. Over the next half century, freight transportation developed in ways that could not have been imagined by the dignitaries watching the *Ideal-X* take on those first containers at Port Newark. Per-

Table 6
The World's Largest Containerports: Containers Handled
(Million 20–Foot Equivalents)

| *Port* | *Country* | *1990* | *2003* |
|---|---|---|---|
| Hong Kong | China | 5.1 | 20.8 |
| Singapore | Singapore | 5.2 | 18.4 |
| Shanghai | China | 0.5 | 11.4 |
| Shenzhen | China | 0.0 | 10.7 |
| Busan | Korea | 2.3 | 10.4 |
| Kaoshung | Taiwan | 3.5 | 8.8 |
| Rotterdam | Netherlands | 3.7 | 7.1 |
| Los Angeles | United States | 2.6 | 6.6 |
| Hamburg | Germany | 2.0 | 6.1 |
| Antwerp | Belgium | 1.6 | 5.4 |
| Dubai | United Arab Emirates | 1.1 | 5.1 |
| Port Klang | Malaysia | 0.5 | 4.8 |
| Long Beach | United States | 1.6 | 4.7 |
| Qingdao | China | 0.1 | 4.2 |
| New York | United States | 1.9 | 4.0 |
| Tanjung Pelepas | Malaysia | 0.0 | 3.5 |
| Tokyo | Japan | 1.5 | 3.3 |
| Bremen/Bremerhaven | Germany | 1.2 | 3.2 |
| Laem Chabang | Thailand | 0.1 | 3.2 |
| Gioia Tauro | Italy | 0.0 | 3.0 |

*Sources*: *Containerisation International Yearbook* and UN Economic and Social Commission for Asia and the Pacific.

haps the most remarkable fact about the remarkable history of the box is that time and again, even the most knowledgeable experts misjudged the course of events. The container proved to be such a dynamic force that almost nothing it touched was left unchanged, and those changes often were not as predicted.

Malcom McLean's genius was acknowledged unanimously: almost everyone save the dockworkers' unions thought that putting freight into containers was a brilliant concept. The idea that the container would cause a revolution in shipping, though, seemed more than a little far-fetched. At best, the container was expected to help ships recover a tiny share of the domestic freight business and to benefit Hawaii and Puerto Rico. Truckers ignored it. Railroads shunned it. Even as ship lines talked it up, most of them treated the container as an adjunct to the business they knew, just another one of the many shapes and sizes of cargo that they were accustomed to storing in their holds. Labor was no better informed. When West Coast longshore union leader Harry Bridges negotiated the 1960 contract that allowed unlimited automation of the docks, he drastically underestimated the speed with which containers would alter work on the waterfront, and demanded far too little for his members as a result. When New York longshore leader Teddy Gleason warned in 1959 that the container would eliminate 30 percent of his union's jobs in New York, he was simply wrong: between 1963 and 1976, longshore hours worked in New York City fell by three-quarters.

The economics of container shipping were equally treacherous for ship operators themselves. Many ship lines sacrificed the potential advantages of containerization by ordering vessels that carried containers along with other types of cargo or even passengers. Others guessed wrong about how big their ships or their containers should be. McLean himself went badly astray several times: he ordered fuel-guzzling SL-7s just ahead of the 1973 oil shock, built the sluggish but fuel-efficient Econships just as fuel prices plummeted, and sailed the Econships on a round-the-world route that left some legs heavily booked but others operating well below capacity. The "experts" who deemed container shipping uncompetitive on long routes, such as those across the Pacific, were proven to be wildly off course, and Asia's containerports, filled with boxes destined for North America and Europe, soon became the largest in the world.

Haste, contrary to what many in the shipping industry had assumed, was not a prerequisite for survival in the container era. Matson, previously active only in U.S. domestic trades, raced to become the first line to carry containers across the Pacific in the belief that an early start would assure it loyal customers; as it learned when other companies rudely barged into the business, customer loyalty counted for little. Moore-McCormack may have been the first line to carry containers across the Atlantic, but it could not turn that head start into a viable business. Nor did Grace Line's role as the first container carrier to South America make it a survivor.

The companies that emerged as the world's largest containership operators in the early twenty-first century were relative latecomers to the game. A. P. Møller's Maersk Line built its first containership only in 1973, seventeen years after the *Ideal-X* and seven years after container shipping came to the North Atlantic. Mediterranean Shipping Company, based in Switzerland, did not even exist until 1970, and Evergreen Marine was founded only in 1968. These companies arrived with financial and managerial skills foreign to many of the carriers they replaced, skills appropriate to an industry in which raising capital and managing information systems were far more important than maritime knowledge. They operated without the legacy of government subsidies and directives that had crippled many of their predecessors by forcing them to buy ships built in their home countries or to sail routes determined by regulators. In an industry that almost everywhere wrapped itself in nationalist pride, the long-term survivors were profoundly international. Maersk's headquarters were in Denmark, but by 2005 it had gained control of more than five hundred containerships and one-sixth of the world market by absorbing companies as diverse as Britain's Overseas Containers Ltd., South African Marine, the Dutch shipping giant Nedlloyd, and Malcom McLean's old company, Sea-Land Service.

If the market repeatedly misjudged the container, so did the state. Governments in New York City and San Francisco ignored the consequences of containerization as they wasted hundreds of millions of dollars reconstructing ports that were outmoded before the concrete was dry. The British government's planning efforts led to the costly creation of new ports; officials never dreamed that a privately owned dock in an out-of-the-way town would turn itself into the country's largest container terminal overnight. Transportation regulators did little better. Japan's Ministry of Transport thought that it could avert overcapacity and keep Japanese ship lines profitable by forcing them to work together, only to be surprised as ship rates in the Pacific tumbled. Regulators and politicians in America, desperate to preserve a system that sought to protect shipbuilders, ship operators, truckers, and railroads, delayed reforms that could have allowed the container to reduce international shipping costs much earlier. By holding on to policies that supposedly strengthened U.S. shipping with a panoply of subsidies and restrictions meant to favor one interest group or another, they ultimately destroyed the competitiveness of the U.S.-flag fleet.[17]

The huge increase in long-distance trade that came in the container's wake was foreseen by no one. When he studied the role of freight in the New York region in the late 1950s, Harvard economist Benjamin Chinitz predicted that containerization would favor metropolitan New York's industrial base by letting the region's factories ship to the South more cheaply than could plants in New England or the Midwest. Apparel, the region's biggest manufacturing sector, would not be affected by changes in transport costs, because it was not "transport-sensitive." The possibility that falling transport costs could decimate much of the U.S. manufacturing base by making it practical to ship almost everything long distances simply did not occur to him. Chinitz was hardly alone in failing to recognize the extent to which lower shipping costs would stimulate trade. Through the 1960s, study after study projected the growth of containerization by assuming that existing import and export trends would continue, with the cargo gradually being shifted into containers. The prospect that the container would permit a worldwide economic restructuring that would vastly increase the flow of trade was not taken seriously.[18]

"The market" got many things wrong when it came to the container, and so did "the state." Both private-sector and public-sector misjudgments slowed the growth of containerization and delayed the economic benefits it would bring. Yet in the end, the logic of shipping freight in containers was so compelling, the cost savings so enormous, that the container took the world by storm. Half a century after the *Ideal-X*, the equivalent of 300 million 20-foot containers were making their way across the world's oceans each year, with 26 percent of them originating in China alone. Countless more were being shipped cross-border by truck or train.[19]

Containers had become ubiquitous—and in addition to cheap goods, they were bringing a new set of social problems. Stacks of abandoned containers, too beaten up to use, too expensive to repair, or simply unneeded, littered landscapes around the world. The exhaust of containerships and the trucks and trains serving them had become a massive environmental problem, and the endless growth of traffic in and out of expanding ports was subjecting nearby communities to congestion, noise, and high rates of cancer attributed to diesel emissions; the price tag for a cleanup in Los Angeles and Long Beach alone was estimated to be $11 billion. The flood of containers had become a major headache for security officials concerned that a single box, loaded with a radioactive "dirty" bomb timed to explode upon arrival in a major port, could contaminate an entire city and throw international commerce into chaos; radiation detectors went up at the gates to many terminals in an effort to keep terrorist containers from being loaded aboard ships. The use of containers outfitted with mattresses and toilets to smuggle immigrants had become routine, with immigration inspectors unable to detect more than a tiny share of containers with human cargo among the hundreds of thousands of boxes filled with legitimate goods.[20]

None of these problems, serious as they were, posed the most remote threat to the growth of container shipping. Containers themselves kept getting larger, with 48-foot and even 53-foot boxes allowing trucks to haul more freight on each trip. The world's fleet expanded steadily, with the capacity of pure containerships rising 10 percent per year from 2001 through 2005. And ships themselves reached unprecedented size. Dozens of vessels able to carry 4,000 40-foot containers had joined the world's fleet by 2006, and even larger ones were on order.

Where vessel size had once been limited by the locks in the Panama Canal, containerships had grown so large that twenty-first-century naval architects were constrained by the Straits of Malacca, the busy shipping lane between Malaysia and Indonesia. If a containership ever reaches Malacca-Max, the maximum size for a vessel able to pass through the straits, it will be a quarter mile long and 190 feet wide, with its bottom some 65 feet below the waterline. If it should sink, it will take nearly $1 billion of cargo with it. Its capacity will be 18,000 TEUs, or 9,000 standard 40-foot containers, enough to fill a 68-mile line of trucks each time it arrives in port. Where it will call is a serious question, because few ports anywhere are deep enough to accommodate it. The answer may well be brand-new ports built in deep water offshore, with Malacca-Max ships linking offshore platforms and smaller vessels shuttling containers to land. If they ever come about, these enormously costly ships and ports will create yet more economies of scale, making it still cheaper and easier to move goods around the globe.[21]

# Abbreviations

The following abbreviations are used in the endnotes.

| | |
|---|---|
| COHP | Containerization Oral History Project, National Museum of American History, Smithsonian Institution, Washington, DC |
| ICC | United States Interstate Commerce Commission |
| ILA | International Longshoremen's Association |
| ILWU | International Longshoremen's and Warehousemen's Union |
| *JOC* | *Journal of Commerce* |
| Marad | United States Maritime Administration |
| NACP | National Archives at College Park, MD |
| NBER | National Bureau of Economic Research |
| NYMA | New York Municipal Archives |
| *NYT* | *New York Times* |
| OAB/NHC | Operational Archives Branch, Naval Historical Center, Washington, DC |
| OECD | Organisation for Economic Co-operation and Development |
| PANYNJ | Port Authority of New York and New Jersey |
| PNYA | Port of New York Authority |
| ROHP | Regional Oral History Program, Bancroft Library, University of California at Berkeley, Berkeley, CA |
| UNCTAD | United Nations Conference on Trade and Development |
| VVA | Virtual Vietnam Archive, Texas Tech University, Lubbock, TX, on-line at http://www.vietnam.ttu.edu/virtualarchive/ |

# Notes

## Chapter 1
## The World the Box Made

1. Steven P. Erie, *Globalizing L.A.: Trade, Infrastructure, and Regional Development* (Stanford, 2004).

2. Christian Broda and David E. Weinstein, "Globalization and the Gains from Variety," Working Paper 10314, NBER, February 2004.

3. As Jefferson Cowie shows in a definitive case study, the relocation of capital in search of lower production costs is not a new phenomenon; see *Capital Moves: RCA's Seventy-Year Quest for Cheap Labor* (New York, 1999). The argument of this book is not that containerization initiated the geographic shift of industrial production, but rather that it greatly increased the range of goods that can be manufactured economically at a distance from where they are consumed, the distances across which those products can feasibly be shipped, the punctuality with which that movement occurs, and the ability of manufacturers to combine inputs from widely dispersed sources to make finished products.

4. For a description of life aboard a modern containership, see Richard Pollak, *The Colombo Bay* (New York, 2004).

5. Former U.S. Coast Guard commander Stephen E. Flynn estimated in 2004 that it takes 5 agents 3 hours to completely inspect a loaded 40-foot container, so physically inspecting every box imported through Los Angeles and Long Beach on the average day would require 270,000 man-hours. This equates to approximately 35,000 customs inspectors for those two ports alone. See the thorough discussion of ways to improve the security of container shipping in his *America the Vulnerable: How the U.S. Has Failed to Secure the Homeland and Protect Its People from Terror* (New York, 2004), chap. 5.

6. Several factors make freight-cost data particularly treacherous. Average costs are greatly affected by the mix of cargo; the now defunct ICC

used to report the average cost per ton-mile of rail freight, but year-to-year changes in the average depended mainly upon demand for coal, which traveled at much lower rates per ton than manufactured goods. Second, most historical cost information concerns a single aspect of the process—the ocean voyage between two ports—rather than the total door-to-door cost of a shipment. Third, a proper measure of freight costs over time would have to account for changes in service quality, such as faster ocean transit and reduced cargo theft, and no freight cost index does this. Fourth, a large number of freight shipments occur either within a large company or at prices privately negotiated between the shipper and transportation carriers, so the information required to measure costs economywide often is not publicly available. Edward L. Glaeser and Janet E. Kohlhase, "Cities, Regions, and the Decline of Transport Costs," Working Paper 9886, NBER, July 2003, p. 4.

7. U.S. Congress, Joint Economic Committee, *Discriminatory Ocean Freight Rates and the Balance of Payments*, November 19, 1963 (Washington, DC, 1964), p. 333; John L. Eyre, "Shipping Containers in the Americas," in Pan American Union, "Recent Developments in the Use and Handling of Unitized Cargoes" (Washington, DC, 1964), pp. 38–42. Eyre's data were developed by the American Association of Port Authorities.

8. Estimate of freight rates reaching 25 percent of value is in Douglas C. MacMillan and T. B. Westfall, "Competitive General Cargo Ships," *Transactions of the Society of Naval Architects and Marine Engineers* 68 (1970): 843. Ocean freight rates for pipe and refrigerators are in Joint Economic Committee, *Discriminatory Ocean Freight Rates*, p. 342. Trade shares are taken from U.S. Bureau of the Census, *Historical Statistics of the United States* (Washington, DC, 1975), p. 887.

9. Eyre,"Shipping Containers in the Americas," p. 40.

10. Paul Krugman, "Growing World Trade: Causes and Consequences," *Brookings Papers in Economic Activity* 1995, no. 1 (1995): 341; World Trade Organization, *World Trade Report 2004* (Geneva, 2005), pp. 114–129.

11. Robert Greenhalgh Albion, *The Rise of New York Port* (New York, 1939; reprint, 1971), pp. 145–146; Peter L. Bernstein, *Wedding of the Waters: The Erie Canal and the Making of a Great Nation* (New York, 2005); Douglass North, "Ocean Freight Rates and Economic Development 1750–1913," *Journal of Economic History* 18 (1958): 537–555. W. W. Rostow, among many others, argues that railroads were essential to the "take off" of U.S. growth in the 1840s and 1850s; see his *Stages of Economic Growth* (Cambridge, UK, 1960), pp. 38–55. Alfred D. Chandler, Jr., *The Visible Hand: The Managerial Revolution in American Business* (Cambridge, MA, 1977), also assigns a critical role to railroads, although for very different

reasons. Robert William Fogel, *Railroads and American Economic Growth* (Baltimore, 1964), rejects the Rostow view, asserting that "the railroad did not make an overwhelming contribution to the productive potential of the economy," p. 235. Albert Fishlow also rejects Rostow's claim that railroad construction was essential in stimulating American manufacturing, but contends that cheaper freight transportation had important effects on agriculture and led to a reorientation of regional economic relationships; see *American Railroads and the Transformation of the Ante-Bellum Economy* (Cambridge, MA, 1965) as well as "Antebellum Regional Trade Reconsidered," *American Economic Review* (1965 supplement): 352–364. On the role of railroads in Chicago's rise, see William Cronon, *Nature's Metropolis: Chicago and the Great West* (New York, 1991), and Mary Yeager Kujovich, "The Refrigerator Car and the Growth of the American Dressed Beef Industry," *Business History Review* 44 (1970): 460–482. For an example from Britain, see Wray Vamplew, "Railways and the Transformation of the Scottish Economy," *Economic History Review* 24 (1971): 54. On transportation and urban development, see James Heilbrun, *Urban Economics and Public Policy* (New York, 1974), p. 32, and Edwin S. Mills and Luan Sendé, "Inner Cities," *Journal of Economic Literature* 35 (1997): 731. On aviation, see Caroline Isard and Walter Isard, "Economic Implications of Aircraft," *Quarterly Journal of Economics* 59 (1945): 145–169.

12. The seminal article along this line was Robert Solow, "Technical Change and the Aggregate Production Function," *Review of Economics and Statistics* 39, no. 2 (1957): 65–94. On the problems of innovation, see Joel Mokyr, "Technological Inertia in Economic History," *Journal of Economic History* 52 (1992): 325–338; Nathan Rosenberg, "On Technological Expectations," *Economic Journal* 86, no. 343 (1976): 528; and Erik Brynjolfsson and Lorin M. Hitt, "Beyond Computation: Information Technology, Organizational Transformation, and Business Performance," *Journal of Economic Perspectives* 14, no. 4 (2000): 24. Electricity was first used in manufacturing in 1883; for discussion of its relatively slow acceptance in manufacturing, see Warren D. Devine, Jr., "From Shafts to Wires: Historical Perspective on Electrification," *Journal of Economic History* 43 (1983): 347–372. Examples of the debate over computers include Paul A. David, "The Dynamo and the Computer: An Historical Perspective on the Modern Productivity Paradox," *American Economic Review* 80 (1990): 355–361; Stephen D. Oliner and Daniel E. Sichel, "The Resurgence of Growth in the Late 1990s: Is Information Technology the Story?" *Journal of Economic Perspectives* 14, no. 4 (2000): 3–22; and Dale W. Jorgenson and Kevin J. Stiroh, "Information Technology and Growth," *American Economic Review* 89, no. 2 (1999): 109–115.

13. Paul M. Romer, "Why, Indeed, in America? Theory, History, and the Origins of Modern Economic Growth," Working Paper 5443, NBER, January 1996.

14. David Ricardo, *The Principles of Political Economy and Taxation* (London, 1821; reprint, New York, 1965), pp. 77–97. Richard E. Caves and Ronald W. Jones point out that the widely taught Heckscher-Ohlin model, which shows that a country has a comparative advantage in producing goods that make more intensive uses of its more abundant factor of production, assumes that transport costs will not affect trade; see their *World Trade and Payments: An Introduction*, 2nd ed. (New York, 1977). More typically, Miltiades Chacholiades, *Principles of International Economics* (New York, 1981), p. 333, describes international market equilibrium under the unstated assumption that trade is costless.

15. The seminal article in this field was Paul Krugman, "Increasing Returns and Economic Geography," *Journal of Political Economy* 99, no. 3 (1991): 483–499. The impact of changing transportation costs is further developed in Krugman and Anthony J. Venables, "Globalization and the Inequality of Nations," *Quarterly Journal of Economics* 110, no. 4 (1995): 857–880, and in Masahisa Fujita, Paul Krugman, and Anthony J. Venables, *The Spatial Economy: Cities, Regions, and International Trade* (Cambridge, MA, 1999).

16. David Hummels, "Have International Transportation Costs Declined?" Working Paper, University of Chicago Graduate School of Business, 1999, and the International Monetary Fund, *World Economic Outlook*, September 2002, p. 116, contend that the cost of sea freight has not fallen significantly in recent decades. James E. Anderson and Eric van Wincoop, "Trade Costs," *Journal of Economic Literature* 42 (September 2004): 691–751, and Céline Carrere and Maurice Schiff, "On the Geography of Trade: Distance Is Alive and Well," World Bank Policy Research Working Paper 3206, February 2004, are among those arguing the continued significance of transport costs in determining trade flows. David Coe and three coauthors offer a technical critique of those arguments and conclude that long-distance international trade has in fact increased, implying that lower transport costs may have encouraged globalization; see "The Missing Globalization Puzzle," International Monetary Fund Working Paper WP/02/171, October 2002.

17. The closest approximation to a general history of the container is Theodore O. Wallin, "The Development, Economics, and Impact of Technological Change in Transportation: The Case of Containerization" (Ph.D. diss., Cornell University, 1974).

## Chapter 2
## Gridlock on the Docks

1. Dramatic photos of cargo-handling operations on the West Coast, which were similar to those on the New York docks, can be found in Otto Hagel and Louis Goldblatt, *Men and Machines: A Story about Longshoring on the West Coast Waterfront* (San Francisco, 1963). Description of coffee handling is from Debra Bernhardt interview with Brooklyn longshoreman Peter Bell, August 29, 1981, New Yorkers at Work Oral History Project, Robert F. Wagner Labor Archive, New York University, Tape 10A. See also recollection of former longshoreman Jock McDougal in Ian McDougall, *Voices of Leith Dockers* (Edinburgh, 2001), p. 28; of former San Francisco longshoreman Bill Ward in the ILWU oral history collection, viewed July 5, 2004, at http://www.ilwu.org/history/oral-histories/bill-ward.cfm? renderforprint=1. Grace Line anecdote from interview with Andrew Gibson, Box AC NMAH 639, COHP.

2. Alfred Pacini and Dominique Pons, *Docker à Marseille* (Paris, 1996), p. 174; T. S. Simey, ed., *The Dock Worker: An Analysis of Conditions of Employment in the Port of Manchester* (Liverpool, 1956), p. 199; New York Shipping Association, "Annual Accident Report Port of Greater New York and Vicinity," January 15, 1951, in Jensen Papers, Collection 4067, Box 13, Folder "Accidents-Longshore Ind."

3. Charles R. Cushing, "The Development of Cargo Ships in the United States and Canada in the Last Fifty Years" (manuscript, January 8, 1992); Peter Elphick, *Liberty: The Ships That Won the War* (London, 2001), p. 403.

4. Ward interview, ILWU; interview with former longshoreman George Baxter in McDougall, *Voices of Leith Dockers*, p. 44.

5. See the colorful descriptions of unloading in Pacini and Pons, *Docker à Marseille*, p. 137.

6. U.S. Department of Commerce, Bureau of Economic Analysis, "Estimates of Non-Residential Fixed Assets, Detailed Industry by Detailed Cost," available at http://www.bea.gov/bea/dn/faweb/Details/Index.html; Andrew Gibson interview; Paul Richardson interview, July 1, 1997, COHP, Box AC NMAH 639. Cost estimate of merchant ships appears in testimony of Geoffrey V. Azoy, Chemical Bank, in U.S. House of Representatives, Committee on Merchant Marine and Fisheries, *Hearings on HR 8637, To Facilitate Private Financing of New Ship Construction*, April 27, 1954, p. 54. MacMillan and Westfall estimated that cargo handling and port expenses accounted for 51.8 percent of the total cost of a short voyage on a C2 freighter in 1958 and 35.9 percent of the total cost of a long voyage. "Competitive General Cargo Ships," p. 837.

7. For examinations of dockworkers' conditions in many countries, see Sam Davies et al., eds., *Dock Workers: International Explorations in Comparative Labour History, 1790–1970* (Aldershot, UK, 2000).

8. U.S. Bureau of the Census and Bureau of Old-Age and Survivors Insurance, *County Business Patterns*, First Quarter, 1951 (Washington, DC, 1953), p. 56; George Baxter interview in McDougall, *Voices of Leith Dockers*, p. 44; unnamed longshoreman quoted in William W. Pilcher, *The Portland Longshoremen: A Dispersed Urban Community* (New York, 1972), p. 41); Pacini and Pons, *Docker à Marseille*, p. 46; Paul T. Hartman, *Collective Bargaining and Productivity* (Berkeley, 1969), p. 26; David F. Wilson, *Dockers: The Impact of Industrial Change* (London, 1972), p. 23.

9. Many of the problems on the docks were eloquently discussed in the 1951 report of a New York State Board of Inquiry into waterfront conditions; for a summary, see "Employment Conditions in the Longshore Industry," New York State Department of Labor *Industrial Bulletin* 31, no. 2 (1952): 7. ILA president Joseph P. Ryan, who eventually lost his post owing to charges of corruption, proposed in 1951 that employers should offer loans to his men to give them an alternative to loan sharks; see "Ryan Message to Members 1951" in Jensen Papers, Collection 4067, Box 13, Folder "Bibliography—Longshoremen Study Outlines," and Waterfront Commission of New York Harbor, *Annual Report*, various years. Mullman's testimony is reported in "Newark Kickback Inquiry," *NYT*, December 16, 1954. On mandatory betting, see Paul Trilling, "Memorandum and Recommendations on the New York Waterfront," December 14, 1951, in Jensen Papers, Collection 4067, Box 12, Folder "Appendix Materials." Information on New Orleans taken from Eric Arnesen, *Waterfront Workers of New Orleans: Race, Class, and Politics 1863–1923* (New York, 1991), p. 254.

10. Some of these schemes are reviewed in Peter Turnbull, "Contesting Globalization on the Waterfront," *Politics and Society* 28, no. 3 (2000): 367–391, and in Vernon H. Jensen, *Hiring of Dock Workers and Employment Practices in the Ports of New York, Liverpool, London, Rotterdam, and Marseilles* (Cambridge, MA, 1964), pp. 153, 200, and 227. On Rotterdam, see also Erik Nijhof, "Des journaliers respectables: les dockers de Rotterdam et leurs syndicates 1880–1965," in *Dockers de la Méditerranée à la Mer du Nord* (Avignon, 1999), p. 121.

11. Wilson, *Dockers*, p. 34.

12. In Amsterdam and Rotterdam, most dockworkers were in the direct employ of stevedoring firms, and most dockers who were not in full-time employment received guarantees of 80 percent of regular pay if they reported to the hiring center twice daily; on average, they received 39 hours' wages and 9 hours of guarantee per 48-hour workweek. See untitled typescript from Scheefvaart Vereeniging Noord dated May 1, 1953, in Jensen

Papers, Collection 4067, Box 13, Folder "Reports on Foreign Dock Workers." On the UK pension scheme, see Wilson, *Dockers*, p. 118. On Hamburg, see Klaus Weinhauer, "Dock Labour in Hamburg: The Labour Movement and Industrial Relations, 1880s–1960s," in Davies et al., *Dock Workers*, 2:501.

13. Raymond Charles Miller, "The Dockworker Subculture and Some Problems in Cross-Cultural and Cross-Time Generalizations," *Comparative Studies in Society and History* 11, no. 3 (1969): 302–314. For belief that it did not pay to work well and quickly, see Horst Jürgen Helle, "Der Hafenarbeiter zwischen Segelschiff und Vollbeschäftigung," *Economisch en Sociaal Tijdschrift* 19, no. 4 (1965): 270. Oregon comments in Pilcher, *The Portland Longshoremen*, p. 22. Marseilles dockers went on strike in 1955 to demand regular shifts. See Pacini and Pons, *Docker à Marseille*, p. 118. According to data from the British Ministry of Labour, the base weekly pay of a full-time docker before World War II was 30–40 percent above the corresponding pay in construction and heavy manufacturing; dockers' average weekly earnings, however, were only 10 percent higher than in those other sectors, because dockers' work was more sporadic. See Wilson, *Dockers*, p. 19.

14. Richard Sasuly, "Why They Stick to the ILA," *Monthly Review*, January 1956, 370; Simey, *The Dock Worker*, pp. 44–45; Malcolm Tull, "Waterfront Labour at Fremantle, 1890–1990," in Davies et al., *Dock Workers*, 2:482; U.S. Bureau of the Census, *U.S. Census of Population and Housing 1960* (Washington, DC, 1962), Report 104, Part I.

15. The proportion of African American dockworkers is from census data reported in Lester Rubin, *The Negro in the Longshore Industry* (Philadelphia, 1974), pp. 34–44. For a detailed analysis of racial preferences and discrimination among dockworkers in New York, New Orleans, and California, see Bruce Nelson, *Divided We Stand: American Workers and the Struggle for Black Equality* (Princeton, 2001), chaps. 1–3. On the New Orleans dockers, see Daniel Rosenberg, *New Orleans Dockworkers: Race, Labor, and Unionism, 1892–1923* (Albany, 1988), and Arnesen, *Waterfront Workers of New Orleans*; odd details, carefully omitting any mention of race, are in William Z. Ripley, "A Peculiar Eight Hour Problem," *Quarterly Journal of Economics* 33, no. 3 (1919): 555–559. On racial discrimination, see Robin D. G. Kelley, " 'We Are Not What We Seem': Rethinking Black Working-Class Opposition in the Jim Crow South," *Journal of American History* 80, no. 1 (1993): 96; Seaton Wesley Manning, "Negro Trade Unionists in Boston," *Social Forces* 17, no. 2 (1938): 259; Roderick N. Ryon, "An Ambiguous Legacy: Baltimore Blacks and the CIO, 1936–1941," *Journal of Negro History* 65, no. 1 (1980): 27; Clyde W. Summers, "Admission Policies of Labor Unions," *Quarterly Journal of Economics* 61, no. 1 (1946):

98; Wilson, *Dockers*, p. 29. The Portland grain workers' case is mentioned in Charles P. Larrowe, *Harry Bridges: The Rise and Fall of Radical Labor in the United States* (New York, 1972), p. 368.

16. On Portland, see Pilcher, *The Portland Longshoremen*, p. 17; on Antwerp, Helle, "Der Hafenarbeiter," p. 273; for Edinburgh, see interviews with dockers Eddie Trotter and Tom Ferguson in McDougall, *Voices of Leith Dockers*, pp. 132 and 177; for Manchester, see Simey, *The Dock Worker*, p. 48. Macmillan quotation appears in Wilson, *Dockers*, p. 160.

17. On the docker culture, see Pilcher, *The Portland Longshoremen*, pp. 12 and 25–26; Wilson, *Dockers*, p. 53; and Miller, "The Dockworker Subculture," passim. Rankings are reported in John Hall and D. Caradog Jones, "Social Grading of Occupations," *British Journal of Sociology* 1 (1950): 31–55.

18. Wilson, *Dockers*, pp. 101–102; Clark Kerr and Abraham Siegel, "The Interindustry Propensity to Strike—an International Comparison," in *Industrial Conflict*, ed. Arthur Kornhauser, Robert Dublin, and Arthur M. Ross (New York, 1954), p. 191; Miller, "The Dockworker Subculture," p. 310. The most notable exception to labor militancy was in New York, where, as Nelson shows, a combination of corrupt union leadership and appeals to Irish Catholic solidarity against other ethnic groups undermined labor radicalism and allowed the port to operate without a strike between 1916 and 1945; see Nelson, *Divided We Stand*, pp. 64–71.

19. Rupert Lockwood, *Ship to Shore: A History of Melbourne's Waterfront and Its Union Struggles* (Sydney, 1990), pp. 223–225; Arnesen, *Waterfront Workers of New Orleans*, p. 254; David F. Selvin, *A Terrible Anger* (Detroit, 1996), pp. 41 and 48–52; Pacini and Pons, *Docker à Marseille*, pp. 46 and 174; interview with former longshoreman Tommy Morton in McDougall, *Voices of Leith Dockers*, p. 112.

20. Thievery as a response to reductions in pay is discussed in Selvin, *A Terrible Anger*, p. 54. The docker joke is one of several in Wilson, *Dockers*, p. 53. Theft is discussed, among many other places, in the interview with longshoreman Tommy Morton in McDougall, *Voices of Leith Dockers*, p. 115; in Pilcher, *The Portland Longshoremen*, p. 100; and in Andrew Gibson interview in COHP.

21. The Welt had originated as a way to give longshoremen a break when they were working in refrigerated holds, but it spread to general cargo in Liverpool and in Glasgow, where it was known as "spelling." See Wilson, *Dockers*, pp. 215 and 221. For productivity, see Miller, "The Dockworker Subculture," p. 311; MacMillan and Westfall, "Competitive General Cargo Ships," p. 842; Wilson, *Dockers*, p. 308; and William Finlay, *Work on the Waterfront: Worker Power and Technological Change in a West Coast Port* (Philadelphia, 1988), p. 53.

22. See the two interesting excerpts of articles on containers from 1920 and 1921 in "Uniform Containerization of Freight: Early Steps in the Evolution of an Idea," *Business History Review* 43, no. 1 (1969): 84.

23. For early efforts to promote containers in America, see G. C. Woodruff, "The Container Car as the Solution of the Less Than Carload Lot Problem," speech to Associated Industries of Massachusetts, October 23, 1929, and "Freight Container Service," speech to Traffic Club of New York, March 25, 1930. A prescient summary of the possibilities of containerization, including the potential economic benefits to the public, is in Robert C. King, George M. Adams, and G. LLoyd Wilson, "The Freight Container as a Contribution to Efficiency in Transportation," *Annals of the American Academy of Political and Social Science* 187 (1936): 27–36.

24. The ICC ruling requiring commodity-based rates can be found at 173 ICC 448. The North Shore Line's rates are discussed in ICC Docket 21723, June 6, 1931. On the implications of the ICC case, see Donald Fitzgerald, "A History of Containerization in the California Maritime Industry: The Case of San Francisco" (Ph.D. diss., University of California at Santa Barbara, 1986), pp. 15–20.

25. On Australia, see photo in Lockwood, *Ship to Shore*, p. 379. On early containerization in Europe, see Wilson, *Dockers*, p. 137, and René Borruey, *Le port de Marseille: du dock au conteneur, 1844–1974* (Marseilles, 1994), pp. 296–306. Examples of North American ship lines carrying containers are in H. E. Stocker, "Cargo Handling and Stowage," Society of Naval Architects and Marine Engineers, November 1933. Information about the Central of Georgia comes from George W. Jordan, personal correspondence, November 15, 1997. See also "Steel Containers," *Via—Port of New York*, July 1954, pp. 1–5.

26. *Containers: Bulletin of the International Container Bureau*, no. 5 (June 1951): 12 and 68; Fitzgerald, "A History of Containerization," p. 35; Padraic Burke, *A History of the Port of Seattle* (Seattle, 1976), p. 115; Lucille McDonald, "Alaska Steam: A Pictorial History of the Alaska Steamship Company," *Alaska Geographic* 11, no. 4 (1984).

27. Pierre-Edouard Cangardel, "The Present Development of the Maritime Container," *Containers*, no. 35 (June 1966): 13 (author's translation). Container census data appear in *Containers*, no. 13 (June 1955): 9, and no. 2 (December 1949): 65. Belgian example appears in *Containers*, no. 19 (December 1957): 18 and 39.

28. Peter Bell interview discussed handling of early containers. The "hindrance" comment by Waldemar Isbrandtsen of Isbrandtsen Company is in International Cargo Handling Coordination Association, "Containerization Symposium Proceedings, New York City, June 15, 1955," p. 11, and the comment about forklifts by Frank McCarthy of Bull-Insular Line

is on p. 19. See also presentation by A. Vicenti, president, Union of Cargo Handlers in the Ports of France, *Containers*, no. 12 (December 1954): 20. Levy address appears in *Containers*, no. 1 (April 1949): 48 (author's translation). Customs duties posed an obstacle as well: until an agreement in 1956, receiving countries frequently levied duties on the value of an arriving container as well as its contents. *Containers*, no. 33 (June 1965): 18. The military study is reported in National Research Council, Maritime Cargo Transportation Conference, *Transportation of Subsistence to NEAC* (Washington, DC, 1956), p. 5.

29. National Research Council, Maritime Cargo Transportation Conference, *The SS Warrior* (Washington, DC, 1954), p. 21.

30. National Research Council, Maritime Cargo Transportation Conference, *Cargo Ship Loading* (Washington, DC, 1957), p. 28.

## Chapter 3
## The Trucker

1. Fitzgerald, "A History of Containerization," pp. 30–31.

2. *North Carolina: A Guide to the Old North State* (Chapel Hill, 1939), p. 537; *Robesonian*, February 26, 1951.

3. Malcolm P. McLean, "Opportunity Begins at Home," *American Magazine* 149 (May 1950): 21; *News and Observer* (Raleigh), February 16, 1942, p. 7; *Robesonian*, February 26, 1951.

4. McLean, "Opportunity," p. 122.

5. For detail on McLean Trucking's early history, see "Malcolm P. McLean, Jr., Common Carrier Application," ICC *Motor Carrier Cases* (hereafter MCC) at 30 MCC 565 (1941). McLean's attempt to block his competitors' merger was decided in *McLean Trucking Co. v. U.S.*, 321 U.S. 67, January 14, 1944. McLean's new service was approved in September 1944; 43 MCC 820. McLean's first purchase, of McLeod's Transfer Inc., occurred in 1942 and was approved over the objections of three protestants; 38 MCC 807. He acquired another trucking company, American Trucking, late in the war; 40 MCC 841 (1946). Revenue figure for 1946 appears at 48 MCC 43 (1948).

6. Intercity truck lines handled 30.45 billion ton-miles of freight in 1946. By 1950, they were carrying 65.65 billion. See ICC, *Transport Economics*, December 1957, p. 9. Total railroad ton-miles were unchanged over that period. Railroads' revenues per ton-mile between 1942 and 1956 varied between 23 percent and 26.8 percent those of truckers. *Transport Economics*, November 1957, p. 8.

7. Information about managing under ICC oversight from author's interview with Paul Richardson, Holmdel, NJ, January 14, 1992.

8. *M.P. McLean, Jr.—Control; McLean Trucking Co.—Lease—Atlantic States Motor Lines Incorporated*, ICC No. MC-F-3300, 45 MCC 417; *M.P. McLean, Jr.—Control; McLean Trucking Company, Inc.—Purchase (Portion)—Garford Trucking, Inc.*, ICC No. MC-F-3698, 50 MCC 415.

9. The cigarette case is *Cigarettes and Tobacco from North Carolina Points to Atlanta*, 48 MCC 39 (1948).

10. Author's interviews with Paul Richardson, Holmdel, NJ, July 20, 1992, and Walter Wriston, New York, June 30, 1992. McLean's success in using his management techniques to turn around Carolina Motor Express, a troubled company of which he had assumed temporary control in 1952, is detailed in *M.P. McLean, Jr.—Control; McLean Trucking Company—Control—Carolina Motor Express Lines, Inc. (Earl R. Cox, Receiver)*, 70 MCC 279 (1956).

11. *M.P. McLean, Jr.—Control; McLean Trucking Co.—Lease—Atlantic States Motor Lines Incorporated*, 45 MCC 417; *M.P. McLean, Jr.—Control; McLean Trucking Company, Inc.—Purchase (Portion)—Garford Trucking, Inc.*, 50 MCC 415; ICC, *Transport Statistics in the United States 1954*, Part 7, Table 30; Wriston interview.

12. Author's telephone interview with William B. Hubbard, July 1, 1993.

13. Author's telephone interview with Earl Hall, May 12, 1993; author's telephone interview with Robert N. Campbell, June 25, 1993.

14. The first public notice of the container scheme appeared in A. H. Raskin, "Union Head Backs 'Sea-Land' Trucks," *NYT*, February 17, 1954.

15. PANYNJ, *Foreign Trade 1976* (New York, 1977), p. 23; author's interview with Paul Richardson, Holmdel, NJ, July 20, 1992; PNYA, Weekly Report to Commissioners, March 13, 1954, 16, in Doig Files; PNYA, Minutes of Committee on Port Planning, April 8, 1954, 2, in Meyner Papers, Box 43.

16. Pan-Atlantic Steamship Corporation, "Summary of Post–World War II Coastwise Operations," mimeo, n.d.; Wriston interview; Phillip L. Zweig, *Wriston: Walter Wriston, Citibank, and the Rise and Fall of American Financial Supremacy* (New York, 1995), p. 78.

17. The details of this convoluted transaction are reviewed in ICC, Case No. MC-F-5976, *McLean Trucking Company and Pan-Atlantic Steamship Corporation—Investigation of Control*, July 8, 1957.

18. McLean's net worth as of September 1955 appears in "I.C.C. Aide Urges Waterman Sale," *NYT*, November 28, 1956. McLean quotation is from author's interview with Gerald Toomey, New York, May 5, 1993.

19. Wriston interview. The McLean quotation is from Zweig, who interviewed McLean for *Wriston*, p. 79.

20. Wriston interview; Zweig, *Wriston*, p. 81; Janet Berte Neale, "America's Maritime Innovator," program for AOTOS Award 1984. Many of the relevant financial details were not included in McLean Industries' financial reports.

21. McLean Industries, *Annual Report* for the year ending December 31, 1955.

22. The program of which McLean took advantage was intended to help traditional ship lines, not upstart challengers. As Andrew Gibson and Arthur Donovan point out, "It took an innovator from another sector of the transportation industry to see how the trade-in program, designed to renew the subsidized fleet, could be used to help launch a revolution that eventually transformed the entire industry." See their *The Abandoned Ocean: A History of United States Maritime Policy* (Columbia, SC, 2000), p. 176.

23. "Railroads Assail Sea-Trailer Plan," *NYT*, February 11, 1955; ICC, *McLean Trucking Company and Pan-Atlantic Steamship Corporation—Investigation of Control*; McLean Industries, *Annual Report*, 1955, pp. 5 and 11; U.S. Department of Commerce, *Annual Report of the Federal Maritime Board and Maritime Administration, 1955* (Washington, DC, 1955), p. 14, and 1956 (Washington, DC, 1956), p. 7; K. W. Tantlinger, "U.S. Containerization: From the Beginning through Standardization" (paper presented to World Port Conference, Rotterdam, 1982); "T-2's Will 'Piggy Back' Truck Trailers," *Marine Engineering/Log* (1956), p. 83. Cost analysis from author's interview with Guy F. Tozzoli, New York, January 13, 2004.

24. Pan-Atlantic gave up the idea of building roll on–roll off ships by late 1956, supposedly to save money on construction costs and to gain greater flexibility. See "Pan Atlantic Changes Plans for Roll-On Ships," *Marine Engineering/Log* (December 1956), p. 112.

25. Much of this section is drawn from Tantlinger, "U.S. Containerization"; author's telephone interview with Keith Tantlinger, December 1, 1992; and author's interview with Keith Tantlinger, San Diego, January 3, 1993. These containers were designed for Ocean Van Lines and carried, 36 to a barge, by Alaska Freight Lines between Seattle, Anchorage, and Seward. They are distinct from the much smaller steel "Cargo Guard" boxes first used by Alaska Steamship Company in 1953 and the 12-foot wooden "crib boxes" that Alaska Steamship carried aboard the *Susitna*, which some identify as the first containership. See Tippetts-Abbett-McCarthy-Stratton, *Shoreside Facilities for Trailership, Trainship, and Containership Services* (Washington, DC, 1956), p. 45; McDonald, "Alaska Steam," p. 112; and Burke, *A History of the Port of Seattle*, p. 115.

26. Tantlinger, "U.S. Containerization"; author's interview with Keith Tantlinger, January 3, 1993; author's telephone interview with Earl Hall, May 14, 1993.

27. The spreader bar is covered by U.S. Patent 2,946,617, issued July 26, 1960.

28. Information about delays taken from Tantlinger interview, and the announcement of the start date is in "Tank Vessels Begin Trailer Runs in April," *JOC*, February 19, 1956. Houston comment is cited in Marc Felice, "The Pioneer," article appearing in program for the AOTOS Award 1984. For cost figures, see Pierre Bonnot, "Prospective Study of Unit Loads," *Containers*, no. 36 (December 1956): 25–29.

29. Pan-Atlantic Steamship Corporation, "Summary of Operations."

30. "ICC Aide Urges Waterman Sale," *NYT*, November 28, 1956, p. 70; ICC, *McLean Trucking Company and Pan-Atlantic Steamship Corporation—Investigation of Control.*

31. Borruey, *Le port de Marseille*, p. 296. Fitzgerald, "A History of Containerization," p. 2. For photos of Seatrain's vessels on trial in 1928, see *Fairplay*, June 17, 1976, p. 15.

32. Cangardel, "The Present Development of the Maritime Container."

## Chapter 4
## The System

1. Author's telephone interview with Robert N. Campbell, June 25, 1993.

2. Tantlinger, "U.S. Containerization"; Cushing, "The Development of Cargo Ships."

3. The containers, chassis, refrigerated units, and twist locks all are covered by patent 3,085,707, issued after much delay on April 16, 1963.

4. Campbell interview; Tantlinger, "U.S. Containerization." Skagit Steel and Iron was closed in the early 1990s, and most of the company's records were destroyed.

5. *Marine Engineering/Log* (November 1955), p. 104; Tantlinger, "U.S. Containerization"; PNYA, Minutes of Committee on Operations, February 2, 1956, Meyner Papers, Box 44; Paul F. Van Wicklen, "New York—The Port That Gave Containerization Its Oomph" in Containerization and Intermodal Institute, "Containerization: The First 25 Years" (New York, 1981); "Tanker to Carry 2-Way Loads," *NYT*, April 27, 1956. The conversion of the C-2s is discussed in "Full-Scale Container Ship Proves Itself," *Marine Engineering/Log* (December 1957), p. 67, and in author's telephone interview with Robert N. Campbell, June 25, 1993. Bonner quotation appears in McLean Industries, *Annual Report*, 1957, p. 8.

6. McLean Industries, *Annual Report*, 1957 and 1958.

7. McLean Industries, *Annual Report*, 1958; Campbell interview.

8. Author's telephone interview with Earl Hall, October 2, 1992; author's telephone interview with William Hubbard, July 1, 1993; author's interview with Charles Cushing, New York, April 7, 1993.

9. William L. Worden, *Cargoes: Matson's First Century in the Pacific* (Honolulu, 1955), p. 120.

10. Ibid., pp. 114–120; Fitzgerald, "A History of Containerization," pp. 39–41.

11. Matson's caution was described in author's telephone interview with Leslie A. Harlander, November 2, 2004. Observation about hiding pedigrees is from Cushing interview. On Weldon's background, see statement of Matson president Stanley Powell, Jr., U.S. House of Representatives, Committee on Merchant Marine and Fisheries, *Cargo Container Dimensions*, November 1, 1967, pp. 48–49. Weldon comment appears in his "Cargo Containerization in the West Coast–Hawaiian Trade," *Operations Research* 6 (September–October 1958): 650.

12. Weldon,"Cargo Containerization," p. 652–655.

13. Ibid., p. 661–663.

14. Les Harlander, interview by Arthur Donovan and Andrew Gibson, June 19, 1997, COHP.

15. Harlander interview, COHP; letter, Keith Tantlinger to George D. Saunders, December 3, 1992 (copy in possession of author). In the letter, Tantlinger states, "I caught Les Harlander prowling the vessel to apparently see what he could learn, and I asked him to leave the ship." In a telephone interview with the author, November 2, 2004, Harlander recalled that he had visited the ship as a guest of Pan-Atlantic.

16. Harlander interview, COHP; American Society of Mechanical Engineers, *The PACECO Container Crane*, brochure prepared for dedication of national historic mechanical engineering landmark, Alameda, California, May 5, 1983. Details of the antiswing device are in L. A. Harlander, "Engineering Development of a Container System for the West Coast–Hawaiian Trade," *Transactions of the Society of Naval Architects and Marine Engineers* 68 (1960): 1079.

17. Harlander interview, COHP; Harlander, "Engineering Development," p. 1053. The containers apparently were well made; in 1981, 23 years after they were built, 85 percent of the original production run of 600 containers were still in service. Harlander interview, COHP.

18. Negotiations with PACECO are recounted in Harlander interview, COHP; the lashing system is described in Harlander, "Engineering Development," p. 1084.

19. Foster Weldon, "Operational Simulation of a Freighter Fleet," in National Research Council, *Research Techniques in Marine Transportation*, Publication 720 (Washington, DC, 1959), pp. 21–27.

20. Fitzgerald, "A History of Containerization," p. 47; American Society of Mechanical Engineers, *The PACECO Container Crane.*

21. Leslie A. Harlander, "Further Developments of a Container System for the West Coast–Hawaiian Trade," *Transactions of the Society of Naval Architects and Marine Engineers* 69 (1961): 7–14; Fitzgerald, "A History of Containerization," pp. 57–59; Worden, *Cargoes*, pp. 143–144.

22. Benjamin Chinitz, for example, devoted only a couple of mentions to containerization, predicting in 1960 that "in the next few decades" few places would have piggyback (container on railcar) service and even fewer would have maritime service with containers; see *Freight and the Metropolis: The Impact of America's Transport Revolution on the New York Region* (Cambridge, MA, 1960), pp. 83, 86, and 161. Jerome L. Goldman, "Designed to Cut Cargo-Handling Costs," *Marine Engineering/Log* (1958), p. 43. McLean Industries, *Annual Reports*, 1957–60; Campbell interview; John Niven, *American President Lines and Its Forebears, 1848–1984* (Newark, DE, 1987), p. 211; Grace's plans were described in U.S. Department of Commerce, *Annual Report of the Federal Maritime Board and Maritime Administration, 1958*, p. 4; Edward A. Morrow, "All-Container Ship Welcomed by Port on Her Debut," *NYT*, January 13, 1960; John P. Callahan, "Container Vessel on First Run," *NYT*, January 30, 1960; "Grace Initiates Seatainer Service," *Marine Engineering/Log* (1960), p. 55; Harold B. Meyers, "The Maritime Industry's Expensive New Box," *Fortune*, November 1967. The ILA may have been behind Venezuelan dockers' refusal to handle Grace's containers; see George Panitz, "NY Dockers Map Annual Wage Drive," *JOC*, December 20, 1961.

23. PNYA, *Annual Report*, various years; "Puerto Rico Trailer Service," *NYT*, April 22, 1960; "Bull Line Gets Container Ships," *NYT*, May 5, 1961; "Transport News: Sea-Land Service," *NYT*, December 17, 1959. Financial information for Pan-Atlantic and Sea-Land Service is from ICC, *Transport Statistics*, Part 5, Table 4, various years. For the parent company's losses, see McLean Industries, *Annual Report*, 1960. Gerald Toomey, then with Consolidated Freightways, a large truck line, recalled that Consolidated's chairman predicted in 1962 that Sea-Land would not last two years; author's interview, New York, May 5, 1993.

24. Edward A. Morrow, "Seatrain Spurns Shipping Merger," *NYT*, August 12, 1959. Campbell interview; McLean Industries, *Annual Report*, 1958.

25. "Just recruiting" comment from author's interview with Gerald P. Toomey, May 5, 1993. On use of intelligence and personality tests, see Arthur Donovan and Andrew Gibson interview with Scott Morrison, July 8, 1998, COHP. Comment on pitching pennies from Cushing interview, April 7, 1993.

26. Author's interview with Paul Richardson, Holmdel, NJ, January 14, 1992; author's telephone interview with Kenneth Younger, December 16, 1991; author's telephone interview with William Hubbard, July 1, 1993.

27. Container tonnage from PNYA *Annual Reports*. Quotation is from author's interview with naval architect Charles Cushing, who joined Sea-Land in 1960.

28. Sea-Land Service, presentation to Sea-Land management meeting, Hotel Astor, New York, December 12–14, 1963, mimeo.

29. Werner Baer, "Puerto Rico: An Evaluation of a Successful Development Program," *Quarterly Journal of Economics* 73, no. 4 (1959): 645–671; A. W. Maldonado, *Teodoro Moscoso and Puerto Rico's Operation Bootstrap* (Gainesville, 1997).

30. Author's interview with Gerald Toomey, May 5, 1993; author's interview with William B. Hubbard, July 1, 1993; Edward A. Morrow, "U.S. Antitrust Inquiry Begun into Proposed Sale of Bull Lines," *NYT*, March 29, 1961.

31. Sea-Land's practice was to write off its ships over six years, an unusually short period for long-lived assets. Very high write-offs made the short-term profit picture look bleak, but it meant that Sea-Land could report very high profits a few years later, once the ships had been fully depreciated. This accounting, deliberately designed to depress short-term profitability, was not widely appreciated by analysts who examined the company's financial reports. In the mid-1960s, the Internal Revenue Service forced Sea-Land to depreciate its ships over fifteen years instead of six, and its financial reporting became less obscure. Author's telephone interview with Earl Hall, May 21, 1993, and McLean Industries *Annual Report*, 1965. Concerning the bid for Bull Line, see George Horne, "Bull Steamship Company Sold to Manuel Kulukundis Interests," *NYT*, April 22, 1961; Edward A. Morrow, "Decision Put Off in Bull Line Case," *NYT*, August 4, 1961. The attempt to block the sale of the ships to Bull was one of the more embarrassing episodes of McLean's career. He told a hastily called congressional hearing on the issue that the government program to sell old vessels to nonsubsidized ship lines was a "give-away program," and was then forced to admit that Waterman had applied for ships under the same program; the Waterman application, he said, was "a mistake," although one that he had not tried to correct. "M'Lean Attacks Ship Exchanges," *NYT*, August 17, 1961.

32. "Bull Line Stops Puerto Rico Runs," *NYT*, June 25, 1962; "Sea-Land to Add to Trailer Runs," *NYT*, June 26, 1962; author's interview with Gerald Toomey, May 5, 1993; author's interview with William B. Hubbard, July 1, 1993; author's telephone interview with Amadeo Francis, April 28, 2005.

33. Toomey interview; U.S. Census Bureau, *Statistical Abstract*, various issues.

34. Sea-Land Service, "The Importance of Containerized Ocean Transportation Service to Puerto Rico," mimeo, n.d. (1969).

35. McLean Industries, *Annual Reports*, 1962 and 1965; Cushing interview; McLean Industries, *Annual Report*, 1962; Toomey interview.

36. Employment figures from ICC, *Transport Statistics*, 1963, Part 5, Table 4. Author's interview with Richard Healey, January 19, 1994; Toomey interview, Richardson interview, July 12, 1992; Hubbard interview.

37. "It wasn't unusual" from Healey interview; Campbell interview; Hubbard interview; Richardson interview, January 14, 1992; George Panitz, "Sea-Land Plans Alaska Service," *JOC*, April 1, 1964.

38. Hall interview; presentations to Sea-Land management meeting, Hotel Astor, New York, December 12–14, 1963; ICC, *Transport Statistics*, various issues.

## Chapter 5
## The Battle for New York's Port

1. Chinitz, *Freight and the Metropolis*, pp. 21, 50. The number of piers is given in a letter from Edward F. Cavanagh, Jr., New York City commissioner of marine and aviation, to Board of Inquiry on Longshore Work Stoppage, January 14, 1952, in Jensen Papers, Collection 4067, Box 16. For a description of the New Jersey freight yards, see Carl W. Condit, *The Port of New York*, vol. 2, *The History of the Rail and Terminal System from the Grand Central Electrification to the Present* (Chicago, 1981), pp. 103–107. Attempts by New Jersey interests to eliminate the single rate led to the formation of the PNYA in 1921. See Jameson W. Doig, *Empire on the Hudson: Entrepreneurial Vision and Political Power at the Port of New York Authority* (New York, 2001).

2. Estimates of truck share of total cargo are based on unpublished PNYA data cited in Chinitz, *Freight and the Metropolis*, p. 41. Average waiting time appears in PNYA, "Proposal for Development of the Municipally Owned Waterfront and Piers of New York City," February 10, 1948, p. 64; *NYT*, May 17, 1952.

3. Waterfront Commission of New York Harbor, *Annual Report for the Year Ended June 30, 1954*, p. 33, and *Annual Report for the Year Ended June 30, 1955*, p. 13. Interesting light on the union's view of public loaders can be found in a July 28, 1952, letter from Waldman & Waldman, the ILA's counsel, to ILA president Joseph P. Ryan recommending changes in the operation of Local 1757, in Vertical File, "International Longshoremen's Association," Tamiment Library, New York University. A formal list of

"authorized" public loader charges appears in Truck Loading Authority, "Official Loading Charges in the Port of New York," in Jensen Papers, Collection 4067, Box 13. As late as 1963, the trucking industry complained that truckers spent $1 million a year on bribes to gain precedence in waiting lines at piers. See New York City Council on Port Development and Promotion, minutes of November 18, 1963, Wagner Papers, Reel 40532, Frame 728.

4. *County Business Patterns*, 1951, p. 56.

5. *County Business Patterns*, 1951, pp. 2, 56; Chinitz, *Freight and the Metropolis*, pp. 31, 96. Detail on plant locations in selected industries in the early part of the century is in Robert Murray Haig, *Major Economic Factors in Metropolitan Growth and Arrangement* (New York, 1927; reprint, New York, 1974), esp. pp. 64–65 and 96–97. Haig's maps make clear that other industries, notably apparel, were not at all reliant on waterfront access.

6. *County Business Patterns*, 1951. Brooklyn estimate from New York City marine and aviation commissioner Vincent A. G. O'Connor, Address to Brooklyn Rotary Club, October 17, 1956, Wagner Papers, Reel 40531, Frame 1585.

7. PNYA, *Outlook for Waterborne Commerce through the Port of New York*, November 1948, Table VIII; Census Bureau, *Historical Statistics*, p. 761; Thomas Kessner, *Fiorello H. LaGuardia and the Making of Modern New York* (New York, 1989), p. 559.

8. Chinitz, *Freight and the Metropolis*, pp. 77–78.

9. Ibid., p. 202. For additional trucking charges, see PNYA, "Proposal for Development," p. 65. Shippers filed eighty-nine "informal complaints" about bills for waiting time, wharf demurrage, and terminal charges during the year ending June 30, 1955, "the greater portion of which were against rates for truck loading and unloading waterborne freight in the Port of New York area." U.S. Department of Commerce, *Annual Report of the Federal Maritime Board and Maritime Administration, 1955*, p. 33.

10. Nelson, *Divided We Stand*, pp. 71–73; Vernon Jensen, *Strife on the Waterfront* (Ithaca, NY, 1974), pp. 105–110 and chap. 6; Philip Taft, "The Responses of the Bakers, Longshoremen and Teamsters to Public Exposure," *Quarterly Journal of Economics* 74, no. 3 (1960): 399. The executive director of the Waterfront Commission, Samuel M. Lane, charged in January 1955 that the New York Shipping Association, the organization of port employers, was "hopeless" when it came to cleaning up waterfront corruption. See Waterfront Commission Press Release 1040, January 27, 1955, in Jensen Papers, Collection 4067, Box 16.

11. The original proposals for the Waterfront Commission came from the Dewey-appointed New York State Crime Commission and from the Port Authority. See State of New York, *Record of the Public Hearing Held by*

*Governor Thomas E. Dewey on the Recommendations of the New York State Crime Commission for Remedying Conditions on the Waterfront of the Port of New York*, June 8–9, 1953, and PNYA, "Comparison of Plans for Improvement of Waterfront Labor Conditions in the Port of New York," January 29, 1953; A. H. Raskin, "C-Men on the Waterfront," *NYT Magazine*, October 9, 1955, p. 15; letters from Lee K. Jaffe, director of public relations, PNYA, to Steve Allen, NBC Television, November 1, 1957, and from Daniel P. Noonan, director of Public Relations, Department of Marine and Aviation, to Steve Allen, October 31, 1957, Wagner Papers, Reel 40531, Frames 1920 and 1922. The film, a musical about a political campaign between rebels and old-line union leaders on the waterfront, was eventually shot on a privately owned pier.

12. A pier inventory, compiled after some of the oldest had been demolished, can be found in New York City Planning Commission, *The Waterfront* (New York, 1971), p. 89; Cavanagh letter to Board of Inquiry; George Horne, "City Action Seen on Port Program," *NYT*, August 7, 1952; Austin J. Tobin, "Transportation in the New York Metropolitan Region during the Next Twenty-five Years" (New York, 1954), p. 7.

13. A massive truck terminal in lower Manhattan, opened in 1932, was the main exception. See Doig, *Empire on the Hudson*, pp. 84–104 and 118–119.

14. Wallace S. Sayre and Herbert Kaufman, *Governing New York City: Politics in the Metropolis* (New York, 1960), p. 341; cover letter in PNYA, *Marine Terminal Survey of the New Jersey Waterfront* (New York, 1949); Doig, *Empire on the Hudson*, pp. 259–260. A prominent article by Cullman published within nine months of the war's end discussed the urgent need for improved port facilities and airports and noted the Port Authority's success at carrying out large capital projects; the subheadline—written at a time when the agency had no responsibility whatsoever for ports or airports—was: "Now the Port Authority, with 25 years behind it, prepares for a new era of sea, land, and air traffic." See "Our Port of Many Ports," *NYT Magazine*, May 5, 1946, p. 12.

15. John I. Griffin, *The Port of New York* (New York, 1959), p. 91; PNYA, "Proposal for Development"; Austin J. Tobin, statement to New York City Board of Estimate, July 19, 1948, Doig Files; PNYA, *Annual Report 1949*, p. 7; PNYA, *Marine Terminal Survey*, 5; Doig, *Empire on the Hudson*, pp. 353–354 and 538. As early as 1946, the city's commissioner of marine and aviation was rejecting suggestions that the Port Authority should organize a port improvement campaign, commenting that "the Port Authority has nothing to do with the Port of New York, and has no authority in it." See "Rejuvenated Port to Rise in Future," *NYT*, November 23, 1946. The

ILA's role in opposition is noted in Joshua Freeman, *Working-Class New York* (New York, 2000), p. 161.

16. PNYA, Weekly Report to Commissioners, April 5, 1952; "Betterments Set for Port Newark," *NYT,* April 9, 1952; Charles Zerner, "Big Port Terminal Near Completion," *NYT,* January 31, 1954; Edward P. Tastrom, "Newark Port to Start Operating New $6 Million Terminal Soon," *JOC,* March 9, 1954; "Awaits Bid for Piers," *Newark Evening News,* December 8, 1952; "Modernizing the Docks," *New York World-Telegram,* December 9, 1952; "City's Port Costs Show Blunder in Rejecting Authority's Aid," *Brooklyn Eagle,* December 17, 1952.

17. McLean's plans developed quickly enough that they became public within two or three months; see Raskin, "Union Head Backs 'Sea-Land' Trucks." Tobin, "Transportation in the New York Metropolitan Area during the Next Twenty-five Years," pp. 10–12.

18. PNYA, Minutes of Committee on Port Planning, September 2, 1954, Meyner Papers, Box 43; PNYA, Minutes of the Commissioners, December 9, 1954, 232, Meyner Papers, Box 43; June 29, 1955, 216; October 26, 1955, 316 and 322, all in Meyner Papers, Box 44; PNYA, *Thirty-fifth Annual Report,* 1956, pp. 1–4.

19. Press release, Office of the Governor, December 2, 1955; PNYA, Minutes of Committee on Port Planning, January 5, 1956, Meyner Papers, Box 44. The Port Authority's previous view of Elizabeth's potential was expressed at *Marine Terminal Survey,* p. 26, which discussed the potential for port development in Newark, Jersey City, Hoboken, Weehawken, and North Bergen, but emphasized with italic type that the Elizabeth waterfront was best suited for *industrial* use.

20. Newark share derived from data in PNYA, *Annual Report 1955,* p. 9, and PANYNJ, *Foreign Trade 1976.*

21. Chris McNickle, *To Be Mayor of New York: Ethnic Politics in the City* (New York, 1993), pp. 97–107; proposed 1954 capital budget, Wagner Papers, Reel 7709, Frame 1372; John J. Bennett, chairman, City Planning Commission, to Henry L. Epstein, deputy mayor, March 11, 1954, Wagner Papers, Reel 7709, Frame 1179; New York Department of Marine and Aviation, press release, August 24, 1955, Wagner Papers, Reel 40531, Frame 1220; Jensen, *Strife on the Waterfront,* p. 147; Wagner letter to City Planning Commission in Wagner Papers, Reel 40507, Frame 843.

22. Cullman to Lukens, December 9, 1955; Lukens to file, December 12, 1955, in Doig Files.

23. O'Connor address to New York Symposium on Increasing Port Efficiency, November 28, 1956, Wagner Papers, Reel 40531, Frame 1554; Department of Marine and Aviation, "Rebuilding New York City's Waterfront," September 5, 1956, Wagner Papers, Reel 40531, Frames 1603–1639.

24. The New York Council on Port Promotion and Development, established by the city, estimated in 1963 that handling general cargo cost $10 per ton in New York versus $5 per ton in Baltimore. Wagner Papers, Reel 40532, Frame 866; "Statement of Vincent A. G. O'Connor, Commissioner of Marine & Aviation, regarding Operation of Grace Line Terminal at Marine & Aviation Piers 57 and 58, North River," Wagner Papers, Reel 40531, Frame 1268; O'Connor address to convention of ILA, July 11, 1955, Wagner Papers, Reel 40531, Frame 1314. Individual longshore gangs had established priority hiring rights on individual piers as "regular" and "regular extra" gangs, and the shift of a carrier from one pier to another could lead to violent disputes over priority status at the new location.

25. Department of City Planning, Newsletter, November 1956, Wagner Papers, Reel 40507, Frame 1596; oral history interviews with Robert F. Wagner, May 21, 1988, Julius C. C. Edelstein, April 5, 1991, and Thomas Russell Jones, June 10, 1993, in LaGuardia and Wagner Archive, LaGuardia Community College, Queens, NY; McNickle, *To Be Mayor of New York*, p. 121; Downtown–Lower Manhattan Association, "Lower Manhattan" (1958), 6.

26. Press release, September 4, 1957, Wagner Papers, Reel 40531, Frame 1945; press release, September 11, 1957, Wagner Papers, Reel 40531, Frame 1957; O'Connor statement at Board of Estimate capital budget hearing, November 18, 1958, Wagner Papers, Reel 40532, Frame 1149; interview with Guy F. Tozzoli, New York, January 13, 2004; letter from Howard S. Cullman and Donald V. Lowe to Mayor Wagner and the Board of Estimate, September 18, 1957, Wagner Papers, Reel 40531, Frame 1448; "Statement by Vincent A. G. O'Connor, Commissioner of Marine and Aviation, regarding Port of New York Authority's Attack on Lease with Holland-America Line for $18,723,000 Terminal, New Pier 40, to Be Built at the Foot of West Houston Street, Manhattan," September 19, 1957, Wagner Papers, Reel 40531, Frame 1936.

27. James Felt, chairman, City Planning Commission, to O'Connor, September 23, 1959, Wagner Papers, Reel 40508, Frame 691; City of New York Department of City Planning, "Redevelopment of Lower Manhattan East River Piers," September 1959, Wagner Papers, Reel 4058, Frame 693; Moses to Felt, September 29, 1959, Wagner Papers, Reel 40508, Frame 688; O'Connor to Board of Estimate, November 25, 1959, Wagner Papers, Reel 40531, Frame 2179. Moses, still a powerful figure in the city and the region during this period, appears to have had no interest in freight transportation. The port, maritime affairs, and freight transportation in general receive no mention in Robert A. Caro's authoritative biography, *The Power Broker: Robert Moses and the Fall of New York* (New York, 1974), and Moses's own memoir says nothing about shipping beyond the observa-

tion that, in the late 1940s, "[o]ur magnificent port was literally dying." See Moses's *Public Works: A Dangerous Trade* (New York, 1970), p. 894. According to Guy Tozzoli, who knew Moses for many years, Moses was very interested in autos and passenger transportation, but had no interest in port-related matters or in the freight-handling problems of New York businesses. Author's interview, New York, January 13, 2004.

28. Condit, *The Port of New York*, 2:346.

29. U.S. Department of Commerce, *Annual Report of the Federal Maritime Board and Maritime Administration*, 1957 (Washington, DC, 1957), p. 12; PNYA, Minutes of the Commissioners, February 14, 1957, p. 98, Meyner Papers, Box 44; PNYA, Weekly Report to the Commissioners, November 15, 1965, Doig Files; "Full-Scale Container Ship Proves Itself," 6; U.S. National Academy of Sciences, *Roll-On, Roll-Off Sea Transportation* (Washington, DC, 1957), p. 9; "Propeller Club Annual Convention," *Marine Engineering/Log* (November 1958), pp. 64–65.

30. PNYA, "Report on Port Authority Operation of Port Newark & Newark Airport, January 1, 1960–December 31, 1960"; Chinitz, *Freight and the Metropolis*, p. 156.

31. Elizabeth officials protested that the Port Authority was violating a 1951 agreement that it would not condemn land in Elizabeth without the city's consent. See PNYA, Weekly Report to the Commissioners, March 31, 1956; letter, Austin J. Tobin to Elizabeth mayor Nicholas LaCorte, May 21, 1956; New Jersey governor Robert B. Meyner to Elizabeth city attorney Jacob Pfeferstein, June 4, 1956; Memo, Francis A. Mulhearn, PNYA legal department to Tobin, June 29, 1956, all in Doig Files. On the differing reactions to the container, see PNYA, Minutes of Committee on Construction, March 26, 1958, Meyner Papers, Box 44; O'Connor address on Marine and Aviation Day, May 23, 1961, Wagner Papers, Reel 40532, Frame 325; "Creation of a Container Port," *Via—Port of New York, Special Issue: Transatlantic Transport Preview* (1965), p. 31; Anthony J. Tozzoli and John S. Wilson, "The Elizabeth, N.J. Port Authority Marine Terminal," *Civil Engineering*, January 1969, pp. 34–39.

32. New York Department of Marine and Aviation, press release, January 23, 1961, Wagner Papers, Reel 40532, Frame 357; Remarks by Mayor Robert F. Wagner, August 30, 1962, Wagner Papers, Reel 40532, Frame 457; Walter Hamshar, "Face-Lift for the Waterfront," *New York Herald Tribune*, November 2, 1963; "NY Port Development Scored," *JOC*, December 23, 1963; New York City Planning Commission, "The Port of New York: Proposals for Development" (1964), pp. 8, 13, and plate 2; Minutes of New York City Council on Port Development and Promotion, November 18, 1963, Wagner Papers, Reel 40532, Frame 728; "Report on Recommendations by the Steering Committee to the Committee for Alleviating

Truck Congestion and Delay at the Waterfront of the City of New York," October 7, 1965, Wagner Papers, Reel 40532, Frame 978.

33. King to Tobin, November 8, 1965; PNYA, Minutes of the Commissioners, November 10, 1965; PNYA, press release, November 15, 1965; PNYA, Minutes of the Commissioners, September 8, 1966; PNYA, transcript of "New Jersey Observations," WNDT-TV, November 15, 1965, all in Doig files; "One Dispute at a Time," *NYT*, July 12, 1966.

34. PNYA, *Annual Report*, 1996, p. 14; First National City Bank, "The Port of New York: Challenge and Opportunity," June 1967, pp. 27, 30; *Longshore News*, October–November 1966, p. 4,

35. Edward C. Burks, "Jersey Facilities Set Port Agency Pace," *NYT*, May 11, 1975; Edith Evans Asbury, "Port Agency Scored on Jersey Project," *NYT*, July 17, 1966; PANYNJ, *Foreign Trade 1976*, p. 12.

36. Brown to Lindsay, May 12, 1966, in Mayor John V. Lindsay Papers, NYMA, Reel 45087, Frame 1560; PNYA, "The 1970 Outlook for Deep Sea Container Services (New York, 1967)," p. 2; PNYA, *Container Shipping: Full Ahead* (New York, 1967); "Containers Widen Their World," *Business Week*, January 7, 1967; George Horne, "Container Revolution, Hailed by Many, Feared," *NYT*, September 22, 1968; memo, Halberg to Brown, May 11, 1966, Lindsay Papers, Reel 45087, Frame 1561.

37. Halberg to Deputy Mayor Robert W. Sweet, September 29, 1967, in Lindsay Papers, Department of Marine and Aviation, Reel 45087, Frame 1653; *Longshore News*, April 1967, p. 4, November 1967, p. 4, October 1968, p. 1, and October 1969, p. 1; Werner Bamberger, "A 90-Second Depot for Containerships Studied," *NYT*, December 1, 1966; Paul F. Van Wicklen, "Elizabeth: The Port of New York's Prototype for the Container Era" (manuscript prepared for *Ports and Terminals*, April 28, 1969); memo, Patrick F. Crossman, commissioner of economic development, to Lindsay, April 2, 1970, in Lindsay Papers, Confidential Subject Files, Reel 45208, Frame 707; Lindsay to Tobin, June 29, 1970, in Lindsay Papers, Confidential Subject Files, Reel 45208, Frame 668. The proposed vertical terminal for two thousand containers was developed by a New York company called Speed-Park Inc.; see R. D. Fielder, "Container Storage and Handling," *Fairplay*, January 5, 1967, p. 31.

38. Joseph P. Goldberg, "U.S. Longshoremen and Port Development," in *Port Planning and Development as Related to Problems of U.S. Ports and the U.S. Coastal Environment*, ed. Eric Schenker and Harry C. Brockel (Cambridge, MD, 1974), pp. 76–78; *Containerisation International Yearbook 1974* (London, 1974), p. 76; Waterfront Commission of New York Harbor, *Annual Report*, various years; *County Business Patterns*, 1964, 34–91, and *County Business Patterns*, 1973, pp. 34–111.

39. Condit, *The Port of New York*, 1:346; Bill D. Ross, "The New Port Newark Is Prospering," *NYT*, December 12, 1973; Goldberg, "U.S. Longshoremen and Port Development," p. 78; David F. White, "New York Harbor Tries a Comeback," *New York*, October 16, 1978, p. 75; Richard Phalon, "Port Jersey Development Could Cut Brooklyn Jobs," *NYT*, January 14, 1972; New York City Planning Commission, *The Waterfront*, p. 35; William DiFazio, *Longshoremen: Community and Resistance on the Brooklyn Waterfront* (South Hadley, MA: Bergin & Garvey, 1985), pp. 34–35.

40. Bureau of the Census, *U.S. Census of Population and Housing 1960* (Washington, DC, 1962), Report 104, Part I, and *1970 Census of Population and Housing* (Washington, DC., 1972), New York SMSA, Part I. Tract boundaries in 1970 were not identical to those in 1960, so definitive conclusions about economic change in small geographic areas are possible only in scattered instances. Housing data from New York City Planning Commission, "New Dwelling Units Completed in 1975," Mayor Abraham Beame Papers, NYMA, Departmental Correspondence, City Planning Commission, Reel 61002, Frame 167.

41. *County Business Patterns*, 1964, 1967, and 1976, Part 34.

42. By the late 1970s, according to one estimate, trucking a container from the waterfront to the railroad yard cost $85 to $120 in Brooklyn, but only $21 in New Jersey; see White, "New York Harbor Tries a Comeback," p. 78. After adjusting for differences in the industrial mix, Edgar M. Hoover and Raymond Vernon found that plants in the New York region built between 1945 and 1956 occupied 4,550 square feet of land per worker, compared to 1,040 square feet in plants built prior to 1922; they also calculated that taxes on industries in big-city locations were much higher than in other parts of the New York region. See *Anatomy of a Metropolis* (Cambridge, MA, 1959), pp. 31, 57–58. Factory relocation data are taken from Marilyn Rubin, Ilene Wagner, and Pearl Kamer, "Industrial Migration: A Case Study of Destination by City-Suburban Origin within the New York Metropolitan Area," *Journal of the American Real Estate and Urban Economics Association* 6 (1978): 417–437.

43. Ellen M. Snyder-Grenier, *Brooklyn! An Illustrated History* (Philadelphia, 1996), pp. 152–163; "Red Hook," in *The Columbia Gazeteer of North America*, 2000 on-line edition; Finlay, *Work on the Waterfront*, p. 61; Richard Harris, "The Geography of Employment and Residence in New York since 1950," in *Dual City: Restructuring New York*, ed. John Mollenkopf and Manual Castells (New York, 1992), p. 133; New York State Department of Labor, *Population and Income Statistics*; Brian J. Godfrey, "Restructuring and Decentralization in a World City," *Geographical Review* (thematic issue, *American Urban Geography*) 85 (1995): 452.

## Chapter 6
## Union Disunion

1. New York Shipping Association, "Proposed Revision of General Cargo Agreement for the Period October 1, 1954 to September 30, 1956," October 20, 1954, and "Proposed Revision of the General Cargo Agreement for the Period October 1, 1954 to September 30, 1956," December 28, 1954, both in ILA files, Robert F. Wagner Labor Archive, New York University, Collection 55, Box 1.

2. Information on the ILA's relations with McLean comes from author's interviews with Thomas W. Gleason, New York, September 29, 1992, and with Guy F. Tozzoli, New York, January 14, 2004. For background on ILA concerns during this period, see Jensen, *Strife on the Waterfront*, pp. 173–83; Philip Ross, "Waterfront Labor Response to Technological Change: A Tale of Two Unions," *Labor Law Journal* 21, no. 7 (1970): 400; and "General Cargo Agreement Negotiated by the New York Shipping Association Inc. with the International Longshoremen's Association (IND) for the Port of Greater New York and Vicinity, October 1, 1956–September 30, 1959," in Jensen Papers, Collection 4096, Box 5.

The Waterfront Commission was seeking to change hiring procedures in the port to eliminate the corruption that came from shape-up. In general, employers hired twenty-one-man gangs rather than individual workers. Each pier (or each employer in Port Newark, where there were no traditional piers) had one or more "regular" gangs that had first call on work. If additional men were needed on a given day, the employer would call for "extra" gangs, with the rules for determining assignment of extra gangs varying greatly in different sections of the port. Pan-Atlantic, for example, had eight "regular" gangs, four Negro and four white. As there was not enough work for all regular gangs every day, a "regular" gang on one pier might also be a "regular extra" gang at another pier if work was available. Employers wanted the ability to choose among available "extra" gangs, but the ILA objected that employers would favor younger workers, leaving gangs with older longshoremen without work. The issue was an extremely difficult one for the union. Newark and parts of Brooklyn had arrangements to equalize earnings among gangs, and union leaders in those areas objected strongly to any attempts to standardize hiring throughout the port, as the Waterfront Commission sought. Manhattan, Jersey City, and Hoboken locals appear to have been much more willing to reach agreement with the commission. Despite the intensity of concern, the ILA does not appear to have had much success in equalizing earnings; of the six gangs that worked for Pan-Atlantic between October 1956 and September 1957, one had average earnings of more than $6,000, two had average earnings of $4,500 to $4,999, and one had average earnings of less than $3,500. See transcript of Waterfront Commission of New York Harbor union-management conferences on seniority issues in ILA District 1 Papers, Kheel Center, Catherwood Library, Cornell University, Collection 5261, Box 1. Wage data are in New York Shipping Association, "Port-Wide Survey of Gang Earnings," September 12, 1958, in Jensen Papers, Collection 4067, Box 13.

The ILA was not formally segregated in the Port of New York, but there were two identifiably "Negro" locals, Local 968 in Brooklyn and Local 1233 in Newark. The Brooklyn local never succeeded in controlling its own pier, and its leaders complained that employers discriminated against Negroes in hiring extra gangs; see the testimony of Thomas Fauntleroy, business agent of Local 968, "In the Matter of the Arbitration between ILA-Independent, and Its Affiliated Locals, and New York Shipping Association," September 29, 1958, in Jensen Papers, Collection 4096, Box 5. In 1959, Local 968 merged into the large Local 1814. The Newark local fared better, because, unlike the situation in New York City, Newark custom did not give priority to any local or gangs. Individual gangs were identified in Waterfront Commission records by codes such as "I" (Italian), "N" (Negro), and "S" (Spanish). See P. A. Miller, Jr., "Current Hiring Customs and Practices in All Areas in the Port of New York," Waterfront Commission, December 20, 1955, in Jensen Papers, Collection 4067, Box 14. On race relations on the New York docks, see Rubin, *The Negro in the Longshore Industry*, pp. 59–69, and Nelson, *Divided We Stand*, pp. 79–86.

3. New York Shipping Association, "Proposals for Renewal of the General Cargo Agreement Submitted by the New York Shipping Association, Inc., to the I.L.A. (Ind.)," October 29, 1956; ILA Locals 1418 and 1419 proposal, September 5, 1956; New Orleans Steamship Association counterproposal, October 1, 1956; Board of Inquiry Created by Executive Order No. 10689, "Report to the President on the Labor Dispute Involving Longshoremen and Associated Occupations in the Maritime Industry on the Atlantic and Gulf Coast," November 24, 1956, all in ILA files, Collection 55, Box 1, Folder "Agreement, Negotiations, & Strikes, June–Dec. 1956, 1 of 2."

4. McLean Industries, *Annual Report*, 1958, p. 4; Pacific Maritime Association, *Monthly Research Bulletin*, January 1959; "Hopes Dim for Accord between Dock Union, New York Shippers, Pacts Expire Tonight," *Wall Street Journal*, September 26, 1959. Field comment in Jensen, *Strife on the Waterfront*, p. 228.

5. *NYT*, November 18, 1958; and November 27, 1958; Port of New York Labor Relations Committee press release, December 17, 1958, in Jensen Papers, Collection 4067, Box 13.

6. Jacques Nevard, "I.L.A. Demands Six-Hour Day and Curbs on Automation," *NYT*, August 11, 1959; Ross, "Waterfront Labor Response," p. 401.

7. Jack Turcott, "Pier Strike Ties Up E. Coast, Spurs Revolt," *New York Daily News*, October 2, 1959; Jensen, *Strife on the Waterfront*, pp. 235–247.

8. Jensen, *Strife on the Waterfront*, pp. 247–250; "Dock Union, Shippers Sign Agreement on Labor Contract," *Wall Street Journal*, December 4, 1959. Barnett's comment appears in New York Shipping Association, "Progress Report 1959," p. 5, and his views were echoed in Walter Hamshar, "I.L.A. Container Pact Gives N.Y. Cargo Lead," *Herald Tribune*, January 3, 1960; Jacques Nevard, "Port Gains Noted in New Pier Pact," *NYT*, January 3, 1960.

9. Jensen, *Strife on the Waterfront*, pp. 250–253. Industry concern about the long-term cost is reflected in the statement by New York Shipping Association chairman Alexander Chopin in New York Shipping Association, "Progress Report 1959," p. 8.

10. For background on the ILWU, see Bruce Nelson, *Workers on the Waterfront: Seamen, Longshoremen, and Unionism in the 1930s* (Champaign, 1990); Selvin, *A Terrible Anger*; Larrowe, *Harry Bridges*; Howard Kimeldorf, *Reds or Rackets? The Making of Radical and Conservative Unions on the Waterfront* (Berkeley, 1988); Stephen Schwartz, *Brotherhood of the Sea: A History of the Sailors' Union of the Pacific, 1885–1985* (Piscataway, NJ, 1986); Henry Schmidt, "Secondary Leadership in the ILWU, 1933–1966," interviews by Miriam F. Stein and Estolv Ethan Ward (Berkeley, 1983); and ILWU, *The ILWU Story: Two Decades of Militant Unionism* (San Francisco, 1955). The number of stoppages is from Charles P. Larrowe, *Shape Up and Hiring Hall* (Berkeley, 1955), p. 126. Andrew Herod, *Labor Geographies: Workers and the Landscapes of Capitalism* (New York, 2001), emphasizes the importance of spatial location in the maintenance of longshore unions' power; although he specifically discusses the ILA, his discussion is equally applicable to the ILWU. A list of forty-eight "hip-pocket rules" in the port of Los Angeles, presented by the Pacific Maritime Association, appears in U.S. House of Representatives, Committee on Merchant Marine and Fisheries, *Study of Harbor Conditions in Los Angeles and Long Beach Harbor*, July 16, 1956, p. 14. The special importance of work rules to unions in an industry that relies on casual labor is emphasized by Hartman, *Collective Bargaining*, p. 41. For examples of the many rules in West Coast ports, see Hartman, *Collective Bargaining*, pp. 46–72, and Lincoln Fairley, *Facing Mechanization: The West Coast Longshore Plan* (Los Angeles, 1979), pp. 16–17.

11. "Working Class Leader in the ILWU, 1935–1977," interview with Estolv Ethan Ward, 1978 (Berkeley, 1980), p. 803.

12. J. Paul St. Sure, "Some Comments on Employer Organizations and Collective Bargaining in Northern California since 1934" (Berkeley, 1957), pp. 598–609.

13. Louis Goldblatt, "Working Class Leader in the ILWU, 1935–1977," interviews by Estolv Ethan Ward (Berkeley, 1977), p. 784; Clark Kerr and Lloyd Fisher, "Conflict on the Waterfront," *Atlantic* 183, no. 3, (1949): 17.

14. St. Sure, "Some Comments," p. 643, claimed that Bridges carefully avoided calling a strike or letting the ILWU contract expire to avoid jurisdictional challenges. See also Larrowe, *Harry Bridges*, p. 352. Bridges's testimony and much other information about the state of the Los Angeles port appears in the record of the Merchant Marine and Fisheries Committee hearings, *Study of Harbor Conditions in Los Angeles and Long Beach Harbor*, October 19–21, 1955 and July 16, 1956.

15. Larrowe, *Harry Bridges*, p. 352.

16. The formal committee statement is in "Report of the Coast Labor Relations Committee to the Longshore, Ship Clerks and Walking Bosses Caucus," March 13–15, 1956, in ILA District 1 Files, Collection 5261, Box 1, Folder "Pacific Coast Experience."

17. Herb Mills, "The San Francisco Waterfront—Labor/Management Relations: On the Ships and Docks. Part One: 'The Good Old Days' " (Berkeley, 1978), p. 21; Fairley, *Facing Mechanization*, p. 48; Hartman, *Collective Bargaining*, pp. 73–83.

18. Jennifer Marie Winter, "Thirty Years of Collective Bargaining: Joseph Paul St. Sure, Management Labor Negotiator 1902–1966" (M.A. thesis, California State University at Sacramento, 1991), chap. 4. In one well-known incident, the permanent labor arbitrator in the port of San Francisco was called to a ship to deal with a safety grievance and found only four workers on the job, sitting in the hold, drinking coffee. The rest of their gang, he was informed, had gone to a ball game and would come to work at midnight. See Larrowe, *Harry Bridges*, p. 352. Hartman, *Collective Bargaining*, pp. 84–88; ILWU, "Coast Labor Relations Committee Report," October 15, 1957.

19. Hartman, *Collective Bargaining*, pp. 87–89; Sidney Roger, "A Liberal Journalist on the Air and on the Waterfront," interview by Julie Shearer (Berkeley, 1998), p. 616.

20. Fairley, *Facing Mechanization*, p. 64, discussed how the six-hour day was "perverted" by subsequent practices. For details on the vote, see Hartman, *Collective Bargaining*, p. 91. The leadership's difficulty convincing members on this issue is evident from a cartoon appearing in the *Dispatcher*, the ILA newspaper, showing a tombstone with the epitaph "Here Lies YOUNG Mr. Overtimer—Survived by a Loving Family Who Wishes He

Had Worked Less and Lived Longer." See ILWU, "Report of the Officers to the Thirteenth Biennial Convention," Part I, April 6, 1959, p. 11.

21. Longshoreman comment from interview with Bill Ward, then a member of ILWU Local 13 in Wilmington, CA, in ILWU–University of California at Berkeley Oral History Project. Warning about automation appears in ILWU, "Report . . . to the Thirteenth Biennial Convention," p. 10. Comment about Bridges is from Roger, "A Liberal Journalist," p. 187.

22. For more on these talks, see Fairley, *Facing Mechanization*, pp. 103–104; Hartman, *Collective Bargaining*, pp. 90–94; Larrowe, *Harry Bridges*, pp. 352–353. The full text of the ILA proposal is reprinted in Fairley, *Facing Mechanization*, p. 80.

23. Fairley, *Facing Mechanization*, pp. 122–129; Hartman, *Collective Bargaining*, pp. 96–97; Winter, "Thirty Years of Collective Bargaining," chap. 5. Savings per hour calculated from data in Hartman, *Collective Bargaining*, p. 123.

24. Fairley, *Facing Mechanization*, pp. 132–133, and Germain Bulcke, "Longshore Leader and ILWU–Pacific Maritime Association Arbitrator," interview by Estolv Ethan Ward, (Berkeley, 1984), p. 66.

25. Pacific Maritime Association and ILWU, "Memorandum of Agreement on Mechanization and Modernization," October 18, 1960; Ross, "Waterfront Labor Response," p. 413.

26. The split among PMA members is detailed in Fairley, *Facing Mechanization*, p. 125, and Winter, "Thirty Years of Collective Bargaining," chap. 5. Hartman, *Collective Bargaining*, pp. 99–100, discusses opposition within the ILWU. Nearly one-third of San Francisco longshoremen were over age fifty-four, and only 11 percent were younger than thirty-five. See Robert W. Cherny, "Longshoremen of San Francisco Bay, 1849–1960," in Davies et al., *Dock Workers*, 1:137.

27. Hartman, *Collective Bargaining*, pp. 164–66.

28. Ibid., pp. 124–44 and 272–279; Finlay, *Work on the Waterfront*, p. 65.

29. Hartman quotation, *Collective Bargaining*, p. 150; on "Bridges loads," see Larrowe, *Harry Bridges*, p. 356.

30. Bridges statement in ILWU/PMA joint meeting, August 7, 1963, quoted in Hartman, *Collective Bargaining*, p. 147; arbitration award ibid., p. 148.

31. Cost savings from ibid., p. 178; container statistics derived from ibid., pp. 160 and 270. Hartman estimates that the container accounted for about 4 percent of the total productivity increase between 1960 and 1963 and perhaps 7 or 8 percent by 1964 (p. 162).

32. The Canadian version of the Mechanization and Modernization Agreement was signed November 21, 1960, just a month after the U.S.

pact, and can be found in Jensen Papers, Accession 4067, Box 15. Kempton column is cited in Jensen, *Strife on the Waterfront*, p. 261.

33. Goldberg, "U.S. Longshoremen and Port Development," 68–81; New York Shipping Association, "Progress Report 1959." Gleason discusses his background in a July 31, 1981, interview with Debra Bernhardt, New Yorkers at Work Oral History Collection, Robert Wagner Labor Archive, New York University, Tape 44, although much of the information provided in this interview is unreliable. Peter Bell, interview by Debra Bernhardt, August 29, 1981, New Yorkers at Work Oral History Collection, Robert Wagner Labor Archive, New York University, Tape 10A, discusses dissidents.

34. Waterfront Commission of New York Harbor, *Annual Report 1961–62*, p. 16.

35. Werner Bamberger, "Container Users Study Royalties," *NYT*, November 24, 1960; "Container Board Set Up," *NYT*, April 11, 1961; Panitz, "NY Dockers." Longshore work hours were compiled by the New York Shipping Association, and a partial set can be found in the Vernon Jensen files; unfortunately, missing records make it impossible to reconstruct the full series of hours.

36. Field comment appears in an undated memo to all ILA members in the Port of New York in ILA Files, Collection 55, Box 1, while demand for portwide seniority is in "Local No. 856, ILA, Proposals to 1962 Atlantic Coast District Wage Scale Committee and New York District Council," n.d., Collection 55, Box 1.

37. ILA, "Changes to Be Made in General Cargo Master Agreement," June 13, 1962; New York Shipping Association, "Monetary Offer to International Longshoremen's Association," August 1, 1962; Memo from Walter L. Eisenberg, Ph.D., Economic Consultant, to Thomas W. Gleason, Chairman, ILA Negotiating Committee, Re Employer Proposals of August 1, 1962, n.d.; all in ILA Files, Collection 55, Box 1, Folder "Agreements, Negotiations, & Strikes 1961–63"; John P. Callahan, "Anastasia Balks at I.L.A. Demands," *NYT*, July 17, 1962. Gleason speech to World Trade Club, September 10, 1962, quoted in Jensen, *Strife on the Waterfront*, p. 269.

38. Jensen, *Strife on the Waterfront*, pp. 271–279.

39. "Statement by the Mediators," "Mediators' Proposal," and "Memorandum of Settlement," mimeographed, January 20, 1963; *Congressional Record*, January 22, 1963, p. 700; *Herald Tribune*, September 12, 1963, p. 27.

40. New York Department of Marine and Aviation, press release, January 23, 1961, Wagner Papers, Reel 40532, Frame 357; Remarks by Mayor Robert F. Wagner, August 30 1962, Wagner Papers, Reel 40532, Frame 457; Walter Hamshar, "Face-Lift for the Waterfront," *Herald Tribune*, No-

vember 2, 1963; Minutes of New York City Council on Port Development and Promotion, November 18, 1963, Wagner Papers, Reel 40532, Frame 728; John P. Callahan, "Automation Fear Haunts Dockers," *NYT,* June 9, 1964. Gleason comment in Jensen, *Strife on the Waterfront*, p. 301. The comments of Philip Ross also are relevant; by 1964, he argues, ILA leaders believed that the government would not tolerate further strikes, especially where featherbedding was involved. See Ross, "Waterfront Labor Response," p. 404.

41. James J. Reynolds, chairman, Theodore W. Kheel, and James J. Healy, "Recommendation on Manpower Utilization, Job Security and Other Disputed Issues for the Port of New York," September 25, 1964. The ILA's reaction appeared in the *Brooklyn Longshoreman*, September 1964. Jensen argues, probably correctly, that Gleason wanted to avoid a strike in his first negotiation as union president, but that he lacked the power to deliver; see *Strife on the Waterfront*, p. 307.

42. On the Johnson administration's concerns about labor settlements increasing inflation, see Edwin L. Dale Jr., "Johnson Voices Inflation Fear," *NYT,* May 10, 1964. Quotation is from ILA Local 1814, "Shop Stewards Information Bulletin," December 17, 1964, ILA Files, Collection 55, Box 1.

43. George Panitz, "New York Pier Talks Hit Surprising Snag," *JOC*, January 5, 1965; Gleason interview by Debra Bernhardt. The local-by-local tally on the vote was published in the *Congressional Record,* January 12, 1965, p. 582. The settlement in the South Atlantic and Gulf ports reduced minimum gang sizes there to eighteen; see George Horne, "2 Southern Lines in Dockers' Pact," *NYT,* February 17, 1965. Despite the ILA agreement, local disputes over manning in Boston led Sea-Land to cancel plans to open service there; see Alan F. Schoedel, "Boston Talks in Deadlock," *JOC,* June 29, 1966, "Boston Containership Handling Dispute Ends," *JOC*, August 4, 1966, and "No Progress Reported in Boston Port Dispute," *JOC*, November 22, 1966.

44. The U.S. Department of Labor's broad concerns are laid out in Norman G. Pauling, "Some Neglected Areas of Research on the Effects of Automation and Technological Change on Workers," *Journal of Business* 37, no. 3 (1964): 261–273. Following an international conference in London in December 1962, the American Foundation on Automation and Employment published "A Report to the President of the United States," April 30, 1963. The labor movement's official view is in Arnold Beichman, "Facing Up to Automation's Problems," *AFL-CIO Free Trade Union News* 18, no. 2 (1963). On the UAW, see Reuben E. Slesinger, "The Pace of Automation: An American View," *Journal of Industrial Economics* 6, no. 3 (1958): 254, esp. Kennedy's comments were made at a press conference on February 14, 1962. For an interesting discussion of automation issues in the con-

text of the printing industry, which presents many similar issues, see Michael Wallace and Arne L. Kalleberg, "Industrial Transformation and the Decline of Craft: The Decomposition of Skill in the Printing Industry, 1931–1978," *American Sociological Review* 47, no. 3 (1982): 307–324.

45. Ben B. Seligman, *Most Notorious Victory: Man in an Age of Automation* (New York, 1966), pp. 227 and 231; Juanita M. Kreps, *Automation and Employment* (New York, 1964), p. 20.

46. Seligman, *Most Notorious Victory*, pp. 238–241; Benjamin S. Kirsh, *Automation and Collective Bargaining* (New York, 1964), pp. 175–176.

47. Goldblatt, "Working Class Leader," p. 860. Herod, *Labor Geographies*, offers a sophisticated discussion of these disputes revolving around the nature and location of longshore work. Concern about jobs lost to barge carriers, known as LASH (lighter aboard ship) vessels, appears in *Longshore News*, December 1969, p. 3. Critics of the ILWU and ILA agreements have made much of the routinization and "de-skilling" of longshore work due to containerization. See, for example, Herb Mills, "The Men along the Shore," *California Living*, September 1980. Containerization undoubtedly eliminated the need for some skills but greatly increased the need for others. Sea-Land, as one example, employed almost twice as many mechanics at Port Elizabeth in 1980 as were employed in the entire Port of New York two decades earlier. David J. Tolan, interview by Debra Bernhardt, August 1, 1980, New Yorkers at Work Oral History Collection, Robert F. Wagner Labor Archives, New York University, Tape 123. See also Finlay, *Work on the Waterfront*, pp. 20, 121.

48. Bell interview; Finlay, *Work on the Waterfront*, pp. 174–176; Roger, "A Liberal Journalist," p. 569. Stanley Aronowitz, *From the Ashes of the Old: American Labor and America's Future* (Boston, 1998), p. 31, blames the ILWU and the ILA for creating a situation in which longshoremen's sons "are obliged to seek work in low-wage, nonunion retail and service jobs which typically pay half of what factory and transportation jobs pay," under the rather romantic assumption that greater union resistance would have kept the docks as they were.

## Chapter 7
## Setting the Standard

Many of the source materials for chapter 7 were obtained from private sources and may not be available in public archives.

1. European container census of 1955 reported in *Containers* 7, no. 13 (1955): 9; "Grace Initiates Seatainer Service," *Marine Engineering/Log* (February 1960), p. 56. Marine Steel advertisement is in International Cargo Handling Coordination Association, "Containerization Symposium Pro-

ceedings, New York City, June 15, 1955," p. 3. Figures on the U.S. container fleet are from a Reynolds Metals Co. study cited in John G. Shott, *Progress in Piggyback and Containerization* (Washington, DC, 1961), p. 11.

2. Douglas J. Puffert, "The Standardization of Track Gauge on North American Railways, 1830–1890," *Journal of Economic History* 60, no. 4 (2000): 933–960, and "Path Dependence in Spatial Networks: The Standardization of Railway Track Gauge," *Explorations in Economic History* 39 (2002): 282–314.

3. Puffert, "Path Dependence," p. 286; A. T. Kearney & Co., "An Evaluation of the 35′ Container Size as a Major Factor in Sea-Land's Growth," typescript, 1967; Weldon, "Cargo Containerization"; "Grace Initiates Seatainer Service," *Marine Engineering/Log* (February 1960), p. 56.

4. On "lock-in," see W. Brian Arthur, *Increasing Returns and Path Dependence in the Economy* (Ann Arbor, 1994), chap. 2. There is an extensive literature exploring the economic costs of technological incompatibilities; see especially Joseph Farrell and Garth Saloner, "Installed Base and Compatibility: Innovation, Product Preannouncements, and Predation," *American Economic Review* 76, no. 5 (1986): 940–955; Michael L. Katz and Carl Shapiro, "Systems Competition and Network Effects," *Journal of Economic Perspectives* 8, no. 2 (1994): 93–115; and S. J. Liebowitz and Stephen E. Margolis, "Network Externality: An Uncommon Tragedy," *Journal of Economic Perspectives* 8, no. 2 (1994): 133–150.

5. Minutes of November 18, 1958, meeting of Committee on Standardization of Van Container Dimensions (hereafter Marad Dimensions Committee).

6. Minutes of November 19, 1958, meeting of Committee on Construction and Fittings (hereafter Marad Construction Committee); author's telephone interview with Vincent Grey, May 1, 2005.

7. Minutes of MH-5 Van Container Subcommittee, February 25, 1959.

8. Marad Dimensions Committee, December 9, 1958; Minutes of MH-5 Van Container Subcommittee, February 25, 1959.

9. On railroads' capacities, see Tippetts-Abbett-McCarthy-Stratton, *Shoreside Facilities*, p. 8, while railroad standardization is treated in John G. Shott, *Piggyback and the Future of Freight Transportation* (Washington, DC, 1960), p. 33, and *Progress in Piggyback*, p. 19. Concerning Bull Line, see F. M. McCarthy, "Aspects on Containers," presented to Marad Construction Committee, December 10, 1958. Bull Line's choice of sizes is justified in International Cargo Handling Coordination Association, "Containerization Symposium Proceedings," p. 19.

10. Minutes of Marad Dimensions Committee, April 16, 1959; letter, Ralph B. Dewey, Pacific American Steamship Association, to L. C. Hoffman, Marad, May 25, 1959; memorandum to various steamship company

officials from George Wauchope, Committee of American Steamship Lines, June 16, 1959; minutes of Marad Dimensions Committee, June 24, 1959. Matson's position on height is laid out in a "Report on why the standard container height and regional supplementary standard van container lengths, as proposed by the ASA Sectional Committee MH5, should not be approved," submitted to Pacific American Steamship Association, February 15, 1960; Edward A. Morrow, "Line Chides I.C.C. on Rate Policies," *NYT*, April 17, 1960.

11. Letter, W. H. Reich, chairman, Marad/Industry Container Standardization Committee on Construction and Fittings, to L. C. Hoffman, Marad, June, 25, 1959.

12. Morris Forgash, "Transport Revolution at the Last Frontier—The Thought Barrier," in *Revolution in Transportation*, ed. Karl M. Ruppenthal (Stanford, 1960), p. 59; "Uniformity Urged in Big Containers," *NYT*, September 12, 1959.

13. Minutes of MH-5 Size Task Force, September 16, 1959. See testimony of Les Harlander to the House Merchant Marine and Fisheries Committee, November, 1967. For comments on Hall, see Vince Grey, "Setting Standards: A Phenomenal Success Story," in Jack Latimer, *Friendship among Equals* (Geneva, 1997), p. 40. Pan-Atlantic had not been a participant in the standardization process until that point; Matson had been, but was not notified of the September 16 meeting until the previous day and did not attend; letter of Robert Tate, Matson, to J. M. Gilbreth, Van Container Subcommittee, September 15, 1959. On Hall's interest in preferred numbers, see MH-5 Executive Committee, minutes, May 4, 1961.

14. Ralph B. Dewey, Pacific American Steamship Association, to Herbert H. Hall, November 12, 1959; Dewey to L. C. Hoffman, Marad, November 12, 1959; Hoffman to Dewey, n.d.; Marad Dimensions Committee, January 14, 1960; Pacific American Steamship Association, minutes of special containerization committee, February 8, 1960; Dewey letter and statement to MH-5 committee, February 25, 1960. The vote is given in a letter from Hall to Dewey, June 20, 1961. Grace Line and American President Lines were so concerned by the government's threat not to fund nonstandard containerships that they amended pending application for construction subsidies so that their proposed ships would handle 20-foot containers rather than 17-footers, which Grace was already using.

15. Letter from George C. Finster, standards manager, American Society of Mechanical Engineers, to members of MH-5 committee, June 29, 1960; letter, George Wauchope to Committee of American Steamship Lines members, July 26, 1960; Pacific American Steamship Association, minutes of containerization committee, August 4, 1960; "U.S. Body Enters Container Field," *NYT*, April 28, 1961. For Hall's view on "modular" sizes, see MH-

5 committee minutes, June 6, 1961. On the procedures by which the standards were deemed to have been approved, see the testimony of Fred Muller Jr., U.S. House of Representatives, Committee on Merchant Marine and Fisheries, *Cargo Container Dimensions*, November 16, 1967. The standards were codified as ASA MH5.1–1961. Federal Maritime Board and Maritime Administration press release NR 61–35, April 28, 1961.

16. MH-5 minutes, June 6, 1961.

17. Tineke M. Egyedi, "The Standardized Container: Gateway Technologies in Cargo Transportation," Working Paper, Delft University of Technology, 2000.

18. *Containers*, no. 30 (December 1963): 26; Egyedi, "The Standardized Container"; "Is Container Standardization Here?" *Via—Port of New York, Special Issue: Transatlantic Transport Preview* (1965), p. 28.

19. Cost estimate appears in "Memorandum of Comment" by John J. Clutz, Association of American Railroads, to MH-5 Van Container Subcommittee #3, December 13, 1961.

20. Minutes, MH-5 Van Container Subcommittee #3, December 14, 1961; Tantlinger, "U.S. Containerization."

21. Tantlinger, "U.S. Containerization"; letter, M. R. McEvoy, president, Sea-Land Service, to Vincent G. Grey, American Standards Association, January 29, 1963.

22. Letter, James T. Enzensperger, Pacific American Steamship Association, to Eugene Spector, American Merchant Marine Institute, November 5, 1964; Tantlinger, "U.S. Containerization."

23. American Merchant Marine Institute, "Van Containers in Service," n.d. (circulated January 1965); Pacific American Steamship Association, minutes of containerization committee, January 21, 1965; telegram, K. L. Selby, president, National Castings Co., to R. K. James, executive director, Committee of American Steamship Lines, January 7, 1965.

24. Pacific American Steamship Association, "SAAM Proposed Cargo Container Standards," January 20, 1965; Herbert H. Hall, "Facts Concerning the ASA-MH5 Sectional Committee Proposed Van Container Corner Fitting," June 14, 1965; Memorandum, Tantlinger to W. E. Grace, Fruehauf Corporation, August 12, 1965.

25. Murray Harding, "Final World Standards Set for Van Freight Containers," *JOC*, October 5, 1965; Harlander interview, COHP.

26. "Is Container Standardization Here?" p. 30.

27. Various countries' findings are detailed in letter, Harlander to Martin Rowbotham, chairman, second ad hoc panel on corner fittings, January 13, 1967, and letter, Robotham to panel members, February 1, 1967. Other sources include Grey, "Setting Standards," p. 41; ISO, "Report of Ad Hoc Panel Convened at London Meeting," January 1967; and author's tele-

phone interview with Les Harlander, November 2, 2004. Ship lines' opposition is reported in the minutes of a meeting of "some members" of the MH-5 Securing and Handling Subcommittee, February 16, 1967. The ISO container and fitting specifications are in *Jane's Freight Containers*, 1st ed. (New York 1968), p. 4–11.

28. Minutes of MH-5 Demountable Container Subcommittee, July 20, 1967; Edward A. Morrow, "Rail Aide Scores Sea Containers," *NYT*, September 17, 1967.

29. ASA-MH-5 committee, cited in L. A. Harlander, "Container System Design Developments over Two Decades," *Marine Technology* 19 (1982): 366; Meyers, "The Maritime Industry's Expensive New Box."

30. The possibility of such additional restrictions on nonstandard operators was much discussed at the 1967 hearings of the U.S. House of Representatives Committee on Merchant Marine and Fisheries, reprinted in *Cargo Container Dimensions* (Washington, DC, 1968).

31. Minutes of MH-5 Demountable Container Subcommittee, November 9, 1965; memo, L. A. Harlander to S. Powell and others, Matson Navigation Company, November 12, 1965.

32. Minutes of ASA Group 1 Demountable Container Subcommittee, February 2, 1966; minutes of MH-5 Sectional Committee, June 23, 1966; letter, Hall to Tantlinger, November 1, 1966; Harlander interview, COHP; L. A. Harlander, "The Role of the 24-Foot Container in Intermodal Transportation," submitted to ASA MH-5 committee, June 1966; Statement of Michael R. McEvoy, president, Sea-Land Service, in House Merchant Marine and Fisheries Committee, *Cargo Container Dimensions*, p. 130; MH-5 Executive Committee, minutes, June 1, 1967.

33. *Congressional Record*, November 6, 1967, pp. 31144–31151; House Merchant Marine and Fisheries Committee, *Cargo Container Dimensions*, Gulick testimony, October 31, 1967, p. 28; Ralph B. Dewey testimony, November 16, 1967, pp. 162–169.

34. House Merchant Marine and Fisheries Committee, *Cargo Container Dimensions*, Powell testimony November 1, 1967, p. 50, and McLean comment November 16, 1967, p. 121.

35. Ibid., Powell testimony November 1, 1967, pp. 70–71; Harlander interview, COHP.

36. Minutes, combined meeting of MH-5 Load and Testing and Handling and Securing Subcommittees, November 30, 1966; Leslie A. Harlander, "Intermodal Compatibility Requires Flexibility of Standards," *Container News*, January 1970, p. 20; Minutes of MH-5 committee, January 29 and May 20–21, 1970; L. A. Harlander, "Container System Design Developments," p. 368.

37. Marad, "Intermodal Container Services Offered by U.S. Flag Operators," January 1973 (unpaginated).

## Chapter 8
## Takeoff

1. New York figure estimated from PNYA data; West Coast figure taken from Hartman, *Collective Bargaining*, p. 160.

2. Ernest W. Williams, Jr., *The Regulation of Rail-Motor Rate Competition* (New York, 1958), p. 208; Werner Bamberger, "Containers Cited as Shipping 'Must,' " *NYT*, January 21, 1959, and "Industry Is Exhibiting Caution on Containerization of Fleet," *NYT*, December 4, 1960. Military freight accounted for one-fifth of the revenues of U.S.-flag international ship lines in 1964; see Werner Bamberger, "Lines Ask Rule on Cargo Bidding," *NYT*, July 14, 1966.

3. McLean Industries, *Annual Reports*, 1957–60; Werner Bamberger, "Lukenbach Buys 3 of 5 Vessels Needed for Containership Fleet," *NYT*, November 26, 1960; George Horne, "Luckenbach Ends Domestic Service," *NYT*, February 21, 1961; "Ship Line Drops Florida Service," *NYT*, March 2, 1961; "Grace Initiates Seatainer Service," *Marine Engineering/Log* (1960), p. 55; Niven, *American President Lines*, p. 211.

4. "Coast Carriers Win Rate Ruling," *NYT*, January 5, 1961.

5. United Cargo Corporation, a freight forwarder, offered container service from the United States to Europe as early as 1959, but the service involved boxes only 10½ feet long, which were carried in ships' holds along with other freight. Jacques Nevard, "Container Line Plans Extension," *NYT*, June 6, 1959.

6. Census Bureau, *Historical Statistics*, pp. 711 and 732; Beverly Duncan and Stanley Lieberson, *Metropolis and Region in Transition* (Beverly Hills, 1970), pp. 229–245.

7. Census Bureau, *Historical Statistics*, pp. 732–733; ICC, *Transport Economics*, July 1956, p. 10.

8. For information on piggyback operations prior to 1950, see Kenneth Johnson Holcomb, "History, Description and Economic Analysis of Trailer-on-Flatcar (Piggyback) Transportation" (Ph.D. diss., University of Arkansas, 1962), pp. 9–13.

9. *Movement of Highway Trailers by Rail*, 293 ICC 93 (1954).

10. U.S. Census Bureau, *Statistical Abstract 1957*, Table 705, p. 564; Wallin, "The Development, Economics, and Impact," p. 220; ICC Bureau of Economics, "Piggyback Traffic Characteristics," December 1966, p. 6. On Teamster opposition, see Irving Kovarsky, "State Piggyback Statutes and Federalism," *Industrial and Labor Relations Review* 18, no. 1 (1964): 45.

11. Curtis D. Buford, *Trailer Train Company: A Unique Force in the Railroad Industry* (New York, 1982); Comments of Roy L. Hayes, "Panel Presentations: Railroad Commercial Panel," *Transportation Law Journal* 28, no. 2 (2001): 516; Walter W. Patchell, "Research and Development," in

*Management for Tomorrow*, ed. Nicholas A. Glaskowsky, Jr. (Stanford, 1958), pp. 31–34; Shott, *Piggyback and the Future of Freight Transportation*, p. 7.

12. Comments of Richard Steiner, "Panel Presentations: Railroad Commercial Panel"; Holcomb, "History, Description and Economic Analysis," pp. 43–44; Eric Rath, *Container Systems* (New York, 1973), p. 33.

13. Holcomb, "History, Description and Economic Analysis," pp. 54–67; Rath, *Container Systems*, p. 33.

14. Details here are taken from the ensuing U.S. District Court decision, *New York, New Haven and Hartford v. ICC*, 199 F. Supp 635.

15. The relevant sentence in the Transportation Act of 1958 reads, "Rates of a carrier shall not be held up to a particular level to protect the traffic of any other mode of transportation, giving due consideration to the objectives of the national transportation policy declared in this Act." "Coast Carriers Win Rate Ruling," *NYT*, January 5, 1961; Robert W. Harbeson, "Recent Trends in the Regulation of Intermodal Rate Competition in Transportation," *Land Economics* 42, no. 3 (1966). The case was finally decided in the railroads' favor by a unanimous Supreme Court, *ICC v. New York, New Haven & Hartford*, 372 U.S. 744, April 22, 1963. The dubious economics of determining a railroad's "fully-distributed cost" of carrying a particular load are, fortunately, beyond the scope of this book.

16. Holcomb, "History, Description and Economic Analysis," p. 220; Bernard J. McCarney, "Oligopoly Theory and Intermodal Transport Price Competition: Some Empirical Findings," *Land Economics* 46, no. 4 (1970): 476.

17. Five of the ten leading users of the New York Central's Flexi-Van service were freight forwarders, but four leading manufacturers and the Montgomery Ward department-store chain also were on the list; see memo, R. L. Milbourne, New York Central, to managers, July 10, 1964, in Penn Central Archives, Hagley Museum and Library, Wilmington, Delaware, Accession 1810/Box B-1872/Folder 15. Alexander Lyall Morton, "Intermodal Competition for the Intercity Transport of Manufactures," *Land Economics* 48, no. 4 (1972): 360.

18. ICC, "Piggyback Traffic Characteristics," pp. 6 and 58–60; Forgash, "Transport Revolution at the Last Frontier," p. 63; Robert E. Bedingfield, "Personality: Champion of the Iron Horse," *NYT*, February 22, 1959; "Trains and Trucks Take to the Ocean," *Via—Port of New York, Special Issue: Transatlantic Transport Preview* (1965), p. 26; ICC, *Transport Statistics in the United States, Part 9: Private Car Lines*, Table 5, various years.

19. ICC, "Piggyback Traffic Characteristics," p. 28. Canada's piggyback carloadings from 1959 through 1961 were about one-third those of the United States, despite a much smaller economy. *Containers*, no. 35 (June 1966): 33.

20. Edward A. Morrow, "3-Way Piggyback Introduced Here," *NYT*, August 10, 1960; Robert E. Bedingfield, "PiggyBack Vans Span Ocean Now," *NYT*, March 12, 1961; *Containers*, no. 31 (June 1964): 25.

21. Author's interview with Bernard Czachowski, New York, January 24, 1992.

22. PNYA, *Annual Reports*, various years; Hartman, *Collective Bargaining*, p. 270; McLean Industries, *Annual Report*, 1965.

23. U.S. Department of Commerce, Marad, "United States Flag Containerships," April 25, 1969.

24. "Operators Uneasy on New Ships; Fear of Rapid Obsolescence Cited," *NYT*, May 24, 1959. On 1964 discussions about entering the transatlantic trade, see Scott Morrison interview, COHP.

25. Hall interview; George Horne, "Intercoastal Trade," *NYT*, January 29, 1961, "Line Will Renew U.S. Coastal Run," *NYT*, February 23, 1961, and "U.S. Aid Is Denied for Coastal Runs," *NYT*, May 13, 1961. Some of the details here are from Jerry Shields, *The Invisible Billionaire: Daniel Ludwig* (Boston, 1986), p. 224.

26. Earl Hall interview, October 2, 1992; Sea-Land, *Annual Report*, 1965.

27. Morrison interview, COHP.

28. Ibid.; Werner Bamberger, "Rules on Cargo Boxes Revised to Spur Use and Ease Shipping," *NYT*, March 17, 1966; Edward Cowan, "Container Service on Atlantic Begins," *NYT*, April 24, 1966.

29. Cowan, "Container Service"; Edward A. Morrow, "New Stage Nears in Container Race," *NYT*, March 28, 1966; A. D. Little, "Containerisation on the North Atlantic" (London, 1967), p. 14.

30. On whiskey, see Morrison interview, COHP. The estimate of Sea-Land's military cargo comes from memorandum, B. P. O'Connor, director of international freight sales, to J. R. Sullivan, Weehawken division superintendent, New York Central Railroad, April 27, 1966, in Penn Central Archives, 1810/B-1675/8. On competitive bidding, see OAB/NHC, Post 1946 Command Files, MSTS, Box 889, Folder 1/1966; U.S. Department of Defense, news release No. 750-66, August 31, 1966; "US Is Firm on Its Plan for Bidding," *JOC*, June 29, 1966.

31. PNYA, *Annual Reports*; "The 1970 Outlook for Deep Sea Container Services," p. 2; Edward Cowan, "Container Service on Atlantic Begins," *NYT*, April 24, 1966.

32. Wallin, "The Development, Economics, and Impact," p. 16; PNYA, *Container Shipping: Full Ahead*; "Countdown on for Container Ships," *Via—Port of New York, Special Issue: Transatlantic Transport Preview* (1965), p. 8; Werner Bamberger, "A Danger Is Seen in Container Rise," *NYT*,

September 9, 1967; "Containerization Comes of Age," *Distribution Manager*, October 1968.

33. "Containers Widen Their World," *Business Week*, January 7, 1967; Frank Broeze, *The Globalisation of the Oceans: Containerisation from the 1950s to the Present* (St. Johns, NF, 2002), p. 41.

34. Statement of Lester K. Kloss, A. T. Kearney & Co., in U.S. House of Representatives, Merchant Marine and Fisheries Committee, *Container Cargo Dimensions*, November 16, 1967, p. 183; "Containerization Comes of Age"; comment by U.S. Navy Capt. D. G. Bryce, "MSTS Area Commanders' Conference," March 4–7, 1969, OAB/NHC, Command Histories, Box 193, Folder 2/1989, p. 137.

35. Press release, German Federal Railroad, July 26, 1967, in Penn Central Archives, 1810/B-1675/6.

36. Letter, J. R. Sullivan, New York Central, to H. W. Large, Vice President–Traffic, Pennsylvania Railroad, April 11, 1966, in Penn Central Archives, 1810/B-1675/8.

37. Aaron Cohen, "Report on Containerization in Export-Import Trade," Traffic Executive Association–Eastern Railroads, April 20, 1966, in Penn Central Archives, 1810/B-1675/9. Charges for empty containers are discussed in statement of James A. Hoyt, Grace Line, to Traffic Executive Association Eastern Railroads, January 30, 1967, in Penn Central Archives, 1810/B-1675/10. On the Whirlpool proposal, see letter, Harold E. Bentsen, manager, international distribution, Whirlpool Corp., to B. P. O'Connor, director, international freight sales, New York Central Railroad, June 28, 1967 and letter, O'Connor to Bentsen, July 6, 1967, Penn Central Archives, 1810/B-1675/8; on Matson, see memo, D. L. Werby to W. R. Brooks, New York Central, July 20, 1967, Penn Central Archives, 1810/B-1675/10.

38. Letter, John A. Daily to J. R. Sullivan, New York Central, February 6, 1967, in Penn Central Archives, 1810/B-1675/10.

39. Kenneth Younger interview, December 16, 1991.

40. "A Railroader on Containerization," *Distribution Manager*, October 1968; ICC, *Transport Statistics*.

## Chapter 9
## Vietnam

1. The formal decision to expand the war was communicated in National Security Action Memorandum No. 328, April 6, 1965.

2. Command History 1964, Military Assistance Command Vietnam (MACV), Record Group (RG) 472, NACP; Edward J. Marolda and Oscar P. Fitzgerald, *The United States Navy and the Vietnam Conflict*, vol. 2, *From*

*Military Assistance to Combat, 1959–1965* (Washington, DC, 1986), pp. 357–358; *Sealift* 15, no. 6 (1965): 5.

3. Memorandum for the Commander in Chief, Pacific. Terms of reference for Honolulu conference, April 8, 1965, Historians Background Material Files, 1965, MACV, RG 472, NACP. Information on backups is in MACV Fact Sheet, June 19, 1965, Mission Council Action Memorandums, Historians Background Material Files 1965, MACV, RG 472, NACP.

4. Command History 1966, MACV, pp. 709–715, RG 472, NACP; "No Congestion at Saigon Port," Vietnam Feature Service, Record 154933, VVA, Texas Tech University; Memorandum from W. S. Post Jr., Acting Commander, MSTS, to Secretary of Navy, Monthly Background Reports 1964–65, MSTS Command File, Box 895, OAB/NHC, Washington, DC; William D. Irvin, "Reminiscences of Rear Admiral William D. Irvin" (Annapolis, 1980), p. 634.

5. On the push system, see interview with Lt. Col. Dolan, transportation officer, 1st Logistics Command, by Maj. John F. Hummer, March 30, 1966, in Classified Organizational History Files, 1966, 1st Logistics Command, U.S. Army Pacific, RG 550, NACP. Quotation is from Joseph M. Heiser, Jr., *A Soldier Supporting Soldiers* (Washington, DC, 1991), p. 104.

6. Edwin B. Hooper, *Mobility Support Endurance: A Story of Naval Operational Logistics in the Vietnam War, 1965–1968* (Washington, DC, 1972), p. 62; General Frank S. Besson Jr., speech to Council on World Affairs, Dallas, TX, May 7, 1968, in Oral History Program Former Commanders—Frank S. Besson, Jr., Historical Office, Headquarters, U.S. Army Materiel Command, 1986. James F. Warnock, Jr., "Recorded Recollections of Lt. Col. James F. Warnock Jr., Executive Officer, 29th Quartermaster Group, 1st Logistics Command, 9 April 1966," Port Study, April 29, 1966, Classified Organizational History Files, 1st Logistics Command, Records of US Army Pacific, RG 550, NACP; Logistics Summary for the week ending July 30, 1965, General Records, Assistant Chief of Staff for Logistics, MACV, RG 472, NACP.

7. Westmoreland and Killen memorandum to the ambassador, March 12, 1965, Historians Background Material Files, MACV, RG 472, NACP; Joint Chiefs of Staff, Historical Division, *The Joint Chiefs of Staff and the War in Vietnam, 1960–1968, Part II*, pp. 21-23 and 21-28, Historical Division, Joint Secretariat, Joint Chiefs of Staff, Record 33179, VVA; Command History 1965, MACV, pp. 107–108 and 409.

8. Quarterly Command Report, Second Quarter, FY 1966, Classified Organizational History Files, 1st Logistical Command, Records of U.S. Army Pacific, RG 550, NACP; "MACV Fact Sheet," June 19, 1965; MACV, Historians Background Material Files, Minutes of Mission Council Meetings of June 28, 1965, July 6, 1965, and July 13, 1965, NACP

472/270/75/33/03, Box 20; *The Joint Chiefs of Staff and the War in Vietnam, 1960–1968, Part II*, p. 21-25; telegram, Secretary of State Dean Rusk to Vietnam Coordinating Committee, August 8, 1965, in Mission Council Action Memorandums, 1965, Historians Background Material Files, MACV, RG 472, NACP; Talking Paper–End of Year Press Conference—Engineer Effort in Vietnam, December 21, 1965; Miscellaneous Memoranda, Historians Background Material Files 1965, MACV, RG 472, NACP.

9. Briefing for Secretary McNamara, Ambassador Lodge, General Wheeler, November 28, 1965, Historians Background Material Files, MACV, RG 472, NACP. The Military Sea Transportation Service had declared Vietnam to be a "danger area" on May 11, which entitled seamen to double pay for each day there plus additional bonuses if their ship was attacked or if a harbor was attacked while their ship was in port; see memorandum, Glynn Donaho, Commander, MSTS, to Secretary of Navy, May 11, 1965, in Monthly Reports, MSTS Command File, 1964–65, OAB/NHC. Ship diversions to Philippines from author's telephone interview with Milton Stickles, June 1, 2004. Quotation about second-class ports from MACV Command History 1965, p. 118; congressional visit in MACV, Historians Background Material Files, 1965, NACP 472/270/75/33/1–2, box 8.

10. *Sealift*, March 1966, p. 14; Command History, 1965, p. 121, MACV, RG 472, NACP; "AB&T Employees Perform Critical Tasks in Vietnam," *Sealift*, August–September 1969, p. 6; Lawson P. Ramage, "Reminiscences of Vice Admiral Lawson P. Ramage" (Annapolis, 1970), p. 535.

11. Command History 1965, p. 119, MACV, RG 472, NACP; Testimony of General Frank S. Besson Jr. to U.S. House of Representatives, Committee on Government Operations, Military Operations Subcommittee, August 4, 1970, p. 53.

12. On palletization and other changes, see Highlights, U.S. Naval Operations Vietnam, January 1966, OAB/NHC. Quotation is from author's interview with Robert N. Campbell, June 25, 1993.

13. Author's interview with William Hubbard, August 10, 1993; and Ron Katims interview, COHP; *Baltimore Sun*, January 22, 1966.

14. *The Joint Chiefs of Staff and the War in Vietnam, 1960–1968, Part II*, pp. 37-6 to 37-8; VVA, Record 33179; HQ MACV, Command History, 1965, NARA 472/270/75/32/6–7 Box 1, p. 231-2. Sea-Land had not been involved in the last major MSTS exercise before the Vietnam buildup, which was conducted entirely with breakbulk vessels; see *Sealift*, December 1964, p. 4, January 1965, p. 5, and March 1965, p. 13. A plan for containerships to carry Conex containers is discussed in Alan F. Schoedel, "Viet Containership Plan Eyed," *JOC*, January 26, 1966. The Okinawa contact

is reported in Werner Bamberger, "Container Ships Sought for War," *NYT*, May 26, 1966.

15. On Equipment Rental Inc., see Katims interview, COHP, and Operational Report-Lessons Learned for quarter ended July 31, 1966, Command Histories, 1st Logistics Command, USARV, RG 472, NACP. On Okinawa and Vietnam contracts, see Besson speech to National Defense Transportation Association Annual Transportation and Logistics Forum, Washington, DC, October 14, 1968, Historical Office, Headquarters, U.S. Army Materiel Command; author's telephone interview with Frank Hayden, former deputy head of contracts, MSTS, June 29, 2004; Department of Defense news release 458-66, May 25, 1966, Military Sea Transportation Service, Command History 1966, OAB/NHC, Washington, DC. "Ship Run Bid Refused," *Baltimore Sun*, June 24, 1966.

16. Operational Report-Lessons Learned for quarter ended July 31, 1966, 1st Logistical Command, p. 16. Port delays are reported in Briefing Data Prepared in Conjunction with Secretary of Defense McNamara's Visit to RVN, October 1966, General Records, Assistant Chief of Staff for Logistics, MACV, RG 472, NACP. Data on man-hour requirements are from Besson presentation to Association of the United States Army, October 10, 1966, Historical Office, Headquarters, U.S. Army Materiel Command.

17. Memorandum from Donaho on inspection trip to Asia, August 2–20, 1966, in Command Files, MSTS, OAB/NHC; Financial and Statistical Report, MSTS, various issues, OAB/NHC; Logistics Summary for 5–20 August, 1966, General Records, Assistant Chief of Staff for Logistics, MACV, RG 472, NACP; Operational Report-Lessons Learned for period ending January 31, 1967, 1st Logistics Command, RG 472, NACP Logistics Summary, 15 December 1966, 1st Logistical Command, RG 472, NACP; *Pacific Stars & Stripes*, October 14, 1966.

18. Ramage, "Reminiscences," p. 532; Werner Bamberger, "Navy Augments Shipping for War," *NYT*, March 30, 1967; *Sealift*, May 1967, pp. 9–10.

19. Katims interview, COHP; Campbell interview; Logistical Summaries, June and September 1967, USRVN, RG 472, NACP; *Sealift*, October 1967, p. 20; "New Supply Concept Comes to Vietnam," *1st Logistical Command Vietnam Review* 1, no. 1 (1967).

20. Command History 1967, p. 772, MACV, RG 472, NACP; Vice Admiral Lawson P. Ramage, Remarks to Propeller Club of the United States, St. Louis, October 11, 1968, Command History, MSTS, OAB/NHC. On the feeder ships, see John Boylston, interview with Arthur Donovan and Andrew Gibson, December 7, 1998, COHP, Box 639.

21. U.S. Army Materiel Command, "Sharpe Army Depot," November 1966; MSTS Area Commanders' Conference, March 5–8, 1968, p. 92, Command Histories, MSTS, AOB/NHC; MSTS Area Commanders' Conference, March 5–8, 1968, p. 102.

22. MSTS Area Commanders' Conference, March 5–8, 1968, p. 47; Logistical Summaries, 1968, USRVN, RG 472, NACP; Operational Report-Lessons Learned, October 31, 1968, Classified Organizational History Files, 1st Logistical Command, U.S. Army Pacific, RG 550, NACP; Memorandum from COMSERVPAC to COMNAVSUPSTSCOMME, June 30, 1968, Classified Organizational History Files, Assistant Chief of Staff for Logistics, RG 472, NACP; Memorandum from Commander, MSTS, September 26, 1968, Organizational History Files, Assistant Chief of Staff for Logistics, RG 472, NACP; Memorandum for Record, Expanded Containership Service to RVN, December 31, 1968, Classified Organizational History Files, Assistant Chief of Staff for Logistics, RG 472, NACP; Joseph M. Heiser, Jr., *Vietnam Studies: Logistic Support* (Washington, DC, 1974), p. 199.

23. "Remarks of Malcom P. McLean" in MSTS, "MSTS/Industry Conference on Military Sealift, 12–23 December 1967," Command History, MSTS, OAB/NHC; Classified Organizational History Files for the Quarter Ending 30 April 1968, 1st Logistical Command, RG 472, NACP; Besson testimony, August 4, 1970, p. 46.

24. Vice Admiral Lawson P. Ramage, Speech to National Defense Transportation Agency 22nd National Transportation and Logistics Forum, October 6, 1967, Command History, MSTS, OAB/NHC; "New Supply Concept Comes to Vietnam"; Besson remarks to National Defense Transportation Association, October 14, 1968, p. 13, and congressional testimony, August 4, 1970, pp. 73–75. The Joint Logistics Review Board's recommendations were controversial and were carried out only in part; see, for example, the objections to merging the MSTS with the army's port and trucking operations, in Edwin B. Hooper, "The Reminiscences of Vice Admiral Edwin B. Hooper" (Annapolis, 1978), pp. 472–474.

25. Frank B. Case, "Contingencies, Container Ships, and Lighterage," *Army Logistician* 2, no. 2 (1970): 16–22. On containerization of ammunition, see "Operation TOCSA: A Containerization First!" *Army Logistician* 2, no. 5 (1970): 14, and *Sealift*, April 1970, pp. 14–16; Besson testimony, August 4, 1970, p. 47.

26. Military Prime Contract Files, July 1, 1965–June 30, 1973, Records of the Office of the Secretary of Defense, RG 330, NACP. On competitive bidding, see Ramage, "Reminiscences," pp. 540–542. Sea-Land's revenues are reported in ICC, *Transport Statistics*, Part 5: Carriers by Water, Table 4.

27. Katims interview, COHP; author's interview with William P. Hubbard, July 1, 1993.

28. MSTS Area Commanders Conference, March 1968, pp. 63, 92, 96; Review and Analysis, March 1968, Command History, 1st Logistical Command, RG 472, NACP.

29. Memorandum from C. F. Pfeifer, Inspector General, on Asia trip October 8–18, 1967, Command Histories, MSTS, OAB/NHC; Classified Organizational History Files for the Quarter Ending April 30, 1968, 1st Logistical Command, Records of U.S. Army Pacific, RG 550, NACP.

30. *Jane's Freight Containers*, p. 309.

31. *Jane's Freight Containers, 1969–70* (New York, 1969), pp. 179–180; Mark Rosenstein, "The Rise of Maritime Containerization in the Port of Oakland, 1950 to 1970" (M.A. thesis, New York University, 2000), p. 95; memo, H. E. Anderson, Traffic Manager, Pacific Command, October 30, 1968, General Records, Assistant Chief of Staff for Logistics, MACV, RG 472, NACP.

32. Worden, *Cargoes*, pp. 150–153; Harlander interview, COHP.

33. Scott Morrison interview, COHP; "Sea-Land Keeps Port Schedule," *Baltimore Sun*, March 18, 1968; Boylston interview, COHP; Rosenstein, "The Rise of Maritime Containerization," p. 96.

34. Marad, Office of Maritime Promotion, "Cargo Data," March 11, 1969.

## Chapter 10
## Ports in a Storm

1. Thomas B. Crowley, "Crowley Maritime Corporation: San Francisco Bay Tugboats to International Transportation Fleet," interview by Miriam Feingold Stein (Berkeley, 1983), p. 33.

2. Census Bureau, *Historical Statistics*, Q495–496, p. 757; Roger H. Gilman, "The Port, a Focal Point," *Transactions of the American Society of Civil Engineers*, 1958, p. 365.

3. Gilman, "The Port, a Focal Point" originally presented in 1956, should be seen as a call for just such involvement by government agencies; Gilman was director of port planning for PNYA.

4. U.S. Census Bureau, *Statistical Abstract 1957*, pp. 590–591.

5. Seattle Port Commission, *Shipping Statistics Handbook* (1963); Erie, *Globalizing L.A.*, p. 80.

6. Fitzgerald, "A History of Containerization," pp. 48, 91–93.

7. Booz-Allen & Hamilton, "General Administrative Survey, Port of Seattle," January 20, 1958, pp. VI-1–VI-12; Seattle Port Commission, "Report of the Marine Terminal Task Force to the Citizens' Port Commis-

sion," October 1, 1959, pp. 7, 12, 34; Burke, *A History of the Port of Seattle*, pp. 114–117; Foster and Marshall Inc., "Port of Seattle, Washington, $7,500,000 General Obligation Bonds," May 4, 1961.

8. Erie, *Globalizing L.A.*, pp. 80–88.

9. Woodruff Minor, *Pacific Gateway: An Illustrated History of the Port of Oakland* (Oakland, 2000), p. 45; Port of Oakland, "Port of Oakland," 1957; Ben E. Nutter, "The Port of Oakland: Modernization and Expansion of Shipping, Airport, and Real Estate Operations, 1957–1977," interview by Ann Lage, 1991 (Berkeley, 1994), pp. 51, 84, 139; Rosenstein, "The Rise of Maritime Containerization," p. 45.

10. George Horne, "Intercoastal Trade," *NYT*, January 29, 1961; Nutter, "The Port of Oakland," pp. 78–79. American-Hawaiian never received the government subsidies it sought to finance its ships.

11. Rosenstein, "The Rise of Maritime Containerization," pp. 47, 69; Nutter, "The Port of Oakland," pp. 79–80; Port of Oakland, "60 Years: A Chronicle of Progress," 1987, pp. 17–18.

12. Erie, *Globalizing L.A.*, p. 89; Walter Hamshar, "Must U.S. Approve All Pier Leases," *Herald Tribune*, April 5, 1964.

13. Nutter, "The Port of Oakland," p. 82; Rosenstein, "The Rise of Maritime Containerization," pp. 98–104.

14. Ting-Li Cho, "A Conceptual Framework for the Physical Development of the Port of Seattle," Port of Seattle Planning and Research Department, April 1966, p. 15; Arthur D. Little Inc., *Community Renewal Programming: A San Francisco Case Study* (New York, 1966), p. 34.

15. Rosenstein, "The Rise of Maritime Containerization," pp. 65 and 85–86; Worden, *Cargoes*, 148; Nutter, "The Port of Oakland," pp. 112, 120; Port of Oakland, "1957 Revenue Bonds, Series P, $20,000,000," October 17, 1978, p. 15; Erie, *Globalizing L.A.*, p. 90; Seattle Port Commission, "Container Terminals 1970–1975: A Development Strategy," November 1969, pp. 1, 10.

16. Burke, *A History of the Port of Seattle*, pp. 116, 122; Erie, *Globalizing L.A.*, pp. 85–89; Minor, *Pacific Gateway*, p. 53; Fitzgerald, "A History of Containerization," pp. 91–93; Niven, *American President Lines*, pp. 250–251; Nutter, "The Port of Oakland," p. 84.

17. U.S. Department of Commerce, Marad, "Review of United States Oceanborne Trade 1966" (Washington, DC, 1967), p. 11.

18. Executive Office of the President, Economic Stabilization Program, Pay Board, "East and Gulf Coast Longshore Contract," May 2, 1972.

19. Alan F. Schoedel, "Boston Talks in Deadlock," *JOC*, June 29, 1966, and "No Progress Reported in Boston Port Dispute," *JOC*, November 22, 1966.

20. John R. Immer, *Container Services of the Atlantic*, 2nd ed. (Washington, DC, 1970), chaps. 14 and 15; Philadelphia Maritime Museum, "Delaware Riever Longshoremen Oral History Project: Background Paper," Vertical File, ILA Local 1291, Tamiment Labor Archive, New York University; *Longshore News*, December 1969; Charles F. Davis, "Ports of Philadelphia Posts Impressive Record," *JOC*, February 5, 1970; Bremer Ausschuß für Wirtschaftsforschung, *Container Facilities and Traffic in 71 Ports of the World Midyear 1970* (Bremen, 1971).

21. Matson Research Corporation, "The Impact of Containerization on the U.S. Economy" (Washington, DC, 1970), 1:88–98.

22. Robert J. McCalla, "From 'Anyport' to 'Superterminal,' " in *Shipping and Ports in the Twenty-first Century*, ed. David Pinder and Brian Slack (London, 2004), pp. 130–134; U.S. Department of Commerce, Marad, "Containerized Cargo Statistics Calendar Year 1974" (Washington, DC, 1974), p. 7; Austin J. Tobin, "Political and Economic Implications of Changing Port Concepts," in Schenker and Brockel, *Port Planning and Development*, p. 269. On Richmond's brief experience as a containerport, see John Parr Cox, "Parr Terminal: Fifty Years of Industry on the Richmond Waterfront," interview by Judith K. Dunning (Berkeley, 1992), pp. 181–183.

23. PNYA, *Via—Port of New York, Special Issue: Transatlantic Transport Preview*, (1965), pp. 12–16.

24. Anthony G. Hoare, "British Ports and Their Export Hinterlands: A Rapidly Changing Geography," *Geografiska Annaler, Series B. Human Geography* 68, no. 1 (1986): 30–32; *Fairplay*, September 14, 1967, p. 5.

25. Wilson, *Dockers*, pp. 137, 309.

26. Ibid., pp. 181–191; Anthony J. Tozzoli, "Containerization and Its Impact on Port Development," *Journal of the Waterways, Harbors and Coastal Engineering Division, Proceedings of the American Society of Civil Engineers* 98, no. WW3 (1972): 335; *Fairplay*, May 16, 1968, p. 51.

27. McKinsey & Company, "Containerization: The Key to Low-Cost Transport," June 1967; A. D. Little, "Containerisation on the North Atlantic," p. 61; Turnbull, "Contesting Globalization," pp. 367–391.

28. "Developments in London," *Fairplay*, November 17, 1966, p. 29.

29. Wilson, *Dockers*, p. 239; J. R. Whittaker, *Containerization* (Washington, DC, 1975), pp. 35–42.

30. Wilson, *Dockers*, p. 152; *Fairplay*, July 18, 1968, p. 9.

31. Morrison interview, COHP; "UK Dockers Accept Pay Offer," *JOC*, March 23, 1970; Edward A. Morrow, " 'Intermodal' Fee Stirs a Dispute," *NYT*, April 8, 1968; "Shipping Events: Inquiry Barred," *NYT*, July 26, 1968.

32. Hoare, "British Ports," pp. 35–39; D. J. Connolly, "Social Repercussions of New Cargo Handling Methods in the Port of London," *International Labour History* 105 (1972): 555. Connolly charges "the application of

cargo handling technology" with "the decline of the traditional dockland communities, and consequently, the debasement of social life among the dockworkers concerned," p. 566.

33. Turnbull, "Contesting Globalization," pp. 387–388; Wilson, *Dockers*, pp. 243–244; *Fortune*, November 1967, p. 152.

34. Bremer Ausschuß für Wirtschaftsforschung, *Container Facilities*, pp. 48–51.

35. National Ports Council, *Container and Roll-On Port Statistics, Great Britain, 1971: Part 1* (London, 1971), p. 31; National Ports Council, *Annual Digest of Port Statistics 1974*, Vol. 1 (London, 1975), Table 41; Henry G. Overman and L. Alan Winters, "The Geography of UK International Trade," Working Paper CEPDP0606, Centre for Economic Performance, London, January 2004. Overman and Winters's figures have been recalculated to exclude airborne trade.

36. *Fairplay*, April 3, 1975, p. 15, and April 17, 1975, p. 56; National Ports Council, *Annual Digest*. Overman and Winters attribute the shift in port performance to the changed pattern of British trade after 1973, and neglect the impact of containerization on the growth or decline of individual ports. See also Whittaker, *Containerization*, p. 33, and UK Department for Transport, "Recent Developments and Prospects at UK Container Ports" (London, 2000), Table 4. Department for Transport, *Transport Statistics Report: Maritime Statistics 2002* (London, 2003), Table 4.3, provides 1965 tonnage figures for sixty-eight British ports, but data for Felixstowe are not available.

37. Katims interview, COHP.

38. *Jane's Freight Containers*, p. 324; A. G. Hopper, P. H. Judd, and G. Williams, "Cargo Handling and Its Effect on Dry Cargo Ship Design," *Quarterly Transactions of the Royal Institution of Naval Architects* 106, no. 2 (1964).

39. Bremer Ausschuß für Wirtschaftsforschung, *Container Facilities*; *Fairplay*, October 5, 1967.

40. *Jane's Freight Containers*, pp. 303–309; *Jane's Freight Containers 1969–70*, pp. 175–194; Daniel Todd, "The Interplay of Trade, Regional and Technical Factors in the Evolution of a Port System: The Case of Taiwan," *Geografiska Annaler, Series B. Human Geography* 75, no. 1 (1993): 3–18.

41. Port of Singapore Authority, *Reports and Accounts*, 1964 and 1966.

42. Port of Singapore Authority, *A Review of the Past and a Look into the Future* (Singapore, 1971), p. 8.

43. Port of Singapore Authority, *Reports and Accounts*, 1968, p. 22.

44. *Fairplay*, November 7, 1974, p. 15; *Containerisation International Yearbook*; Gerald H. Krausse, "The Urban Coast in Singapore: Uses and Management," *Asian Journal of Public Administration* 5, no. 1 (1983):44–46.

45. *Containerisation International Yearbook*; Krausse, "The Urban Coast in Singapore," pp. 44–46; Port of Singapore Authority, *A Review*, p. 19; United Nations Economic and Social Commission for Asia and the Pacific, *Commercial Development of Regional Ports as Logistics Centres* (New York, 2002), p. 45.

## Chapter 11
## Boom and Bust

1. Comment by James A. Farrell Jr., chairman of Farrell Lines, to New York World Trade Club, *NYT*, June 7, 1966.

2. Matson Research Corp., *The Impact of Containerization*, 1:151; McLean Industries, *Annual Report*, 1968.

3. Tozzoli, "Containerization and Its Impact on Port Development,", pp. 336–337; Marad, "United States Flag Containerships," April 25, 1969. Grace Line's four biggest container-carrying ships, built in 1963–64, had room for 117 first-class passengers; see *Jane's Freight Containers 1969–70*, p. 389. On the complexities of moving containers on breakbulk ships, see Broeze, *The Globalisation of the Oceans*, pp. 29 and 41.

4. The first newly built vessel designed solely to carry containers in cells was the *Kooringa*, constructed in Australia in 1964 for Associated Steamships. *Kooringa* carried containers of 14.5 tons or less—smaller than standard 20-foot containers—on a domestic route between specially built terminals in Melbourne and Fremantle. The ship had two gantry cranes for loading and unloading. *Kooringa* proved to be a dead end in the development of containerization, and lost any competitive advantage after the arrival of standard-size containers. The service was discontinued in 1975 after heavy losses. See Broeze, *The Globalisation of the Oceans*, p. 34, and *The Australian Naval Architect* 2, no. 3 (1998): 6. Roy Pearson and John Fossey, *World Deep-Sea Container Shipping* (Liverpool, 1983), pp. 247–253.

5. McKinsey & Co., "Containerization: A 5-Year Balance Sheet" (1972), p. 1-1. McKinsey's estimate of the outlays was £4 billion, which was $9.6 billion at the 1970 exchange rate; I have inflated this to current value using the U.S. producer price index for capital equipment. For British carriers' earnings, see *Fairplay*, January 12, 1967, p. 92, and January 11, 1968, p. 92A.

6. ICC, *Transport Statistics*, 1965–67; John J. Abele, "Smooth Sailing or Rough Seas?" *NYT*, January 19, 1969; John J. Abele, "Investors in Conglomerates Are Seeing the Other Side of the Coin," *NYT*, April 13, 1969.

7. Toomey interview; John Boylston interview, COHP; Frank V. Tursi, Susan E. White, and Steve McQuilkin, *Lost Empire: The Fall of R. J. Reynolds Tobacco Company* (Winston-Salem, 2000), p. 174; John J. Abele, "Stock Exchange Ends Day Mixed," *NYT*, January 4, 1969.

8. Immer, *Container Services of the Atlantic*, pp. 194 and 198–200; Peter Stanford, "The SL-7: Sea-Land's Clipper Ship," *Sea History*, Fall 1978; Sea-Land advertisement, "SL-7," n.d.

9. *Lloyd's Shipping Economist*, August 1982, p. 36; "Sea-Land Line Orders 5 New Containerships," *NYT*, August 14, 1969; Tursi, White, and McQuilkin, *Lost Empire*, p. 176.

10. United Nations Economic and Social Commission for Asia and the Pacific, *Statistical Yearbook 1975* (Bangkok, 1977), pp. 205–208; Marad, *Foreign Oceanborne Trade of the United States*, 1970.

11. *Fairplay*, June 15, 1972.

12. United Nations, *Statistical Yearbook 1975*, p. 208.

13. Reuters, August 9, 1969; Marad, "Maritime Subsidies" (Washington, DC, 1971), p. 85.

14. Broeze, *The Globalisation of the Oceans*, p. 50; *Fairplay*, October 7, 1971, p. 41.

15. United Nations, *Statistical Yearbook 1975*, pp. 41–43, 127–129, 230–232, and 390; International Monetary Fund, *Direction of Trade Annual 1969–75* (Washington, DC, 1977), pp. 2–3; Matson Research Corp., *The Impact of Containerization*, 1:114–122; "Matson, Sea-Land to Expand Containership Services," *JOC*, March 18, 1970; *Fairplay*, February 16, 1967 and July 15, 1971, p. 11; OECD, *OECD Economic Surveys: Australia*, esp. 1979.

16. McKinsey & Co., "Containerization: A 5-Year Balance Sheet," p. 1-4.

17. Marad, "United States Flag Containerships," April 25, 1969; Pearson and Fossey, *World Deep-Sea Container Shipping*, p. 220.

18. Marad, "A Statistical Analysis of the World's Merchant Fleet," 1968 and 1974.

19. Pearson and Fossey, *World Deep-Sea Container Shipping*, p. 30; *Fairplay*, February 10, 1972, p. 40.

20. Matson Research Corp., *The Impact of Containerization*, 1:24.

21. P. Backx and C. Earle, "Handling Problems Reviewed," *Fairplay*, February 9, 1967, p. 36; McKinsey & Co., "Containerization: The Key to Low-Cost Transport," p. 57; *Fairplay*, November 24, 1966; Matson Research Corp., *The Impact of Containerization*, 2:4; Litton Systems Inc., "Oceanborne Shipping: Demand and Technology Forecast," June 1968, p. 6-2.

22. *Fairplay*, April 20, 1967, p. 42.

23. There is a long-standing debate on the extent to which conferences have succeeded in restricting competition and raising prices. For recent summaries, see Alan W. Cafruny, *Ruling the Waves* (Berkeley, 1987), and William Sjostrom, "Ocean Shipping Cartels: A Survey," *Review of Network Economics* 3, no. 2 (2004).

24. *Fairplay*, August 24, 1967, p. 8; J. McNaughton Sidey, "Trans-Atlantic Container Services," *Fairplay*, October 5, 1967.

25. *Fairplay*, February 9, 1967, p. 41; "U.S. Panel Weight a Boxship Accord," *NYT*, August 28, 1969.

26. Hans Stueck, "2 Big German Shipping Lines Plan Merger, *NYT*, July 4, 1969; George Horne, "U.S. Lines Plans 16-Ship Charter," *NYT*, October 4, 1969; Werner Bamberger, "Line Sets Its Course on Time Charters," *NYT*, January 11, 1970.

27. George Horne, "Grace Line Is Tentatively Sold," *NYT*, February 7, 1969; Broeze, *The Globalisation of the Oceans*, p. 48; Farnsworth Fowle, "4 Freighters Sold for $38.4 Million," *NYT*, August 6, 1970; "Cooling the Rate War on the North Atlantic," *Business Week*, April 29, 1972; "U.S. to Challenge R. J. Reynolds Bid," *NYT*, December 15, 1970; *Fairplay*, July 15, 1971, p. 62, and December 9, 1971, p. 45; ICC, *Transport Statistics*, Part 5, Table 4, 1970 and 1971.

28. Broeze, *The Globalisation of the Oceans*, pp. 42 and 57–59; UNCTAD, *Review of Maritime Transport 1972–73*, p. 97; Gilbert Massac, "Le transport maritime par conteneurs: concentrations et globalisation," *Techniques avancées*, no. 43 (April 1998); Gunnar K. Sletmo and Ernest W. Williams Jr., *Liner Conferences in the Container Age: U.S. Policy at Sea* (New York, 1981), p. 308; "Cooling the Rate War."

29. Pearson and Fossey, *World Deep-Sea Container Shipping*, p. 25; Wallin,, "The Development, Economics, and Impact," p. 883.

30. U.S. Council of Economic Advisers, *Economic Report of the President* (Washington, DC, 1982), p. 356; UNCTAD, *Review of Maritime Transport 1974*, p. 40.

31. UNCTAD, *Review of Maritime Transport 1972–73*, p. 96; Pearson and Fossey, *World Deep-Sea Container Shipping*, pp. 25, 220; Clare M. Reckert, "R. J. Reynolds Profit Up 3% in Quarter," *NYT*, February 13, 1975; "Their Ship's Finally Come In," *NYT*, September 8, 1974.

32. UNCTAD, *Handbook of International Trade and Development Statistics 1981 Supplement* (New York, 1982), p. 45; UNCTAD, *Review of Maritime Transport 1975*, p. 36, and *1976*, p. 32; Robert Lindsey, "Pacific Shipping Rate War Flares, Mostly on Soviet Vessel Build-Up," *NYT*, July 4, 1975.

33. On costs, see Sletmo and Williams, *Liner Conferences*, pp. 147 and 156. Peninsula & Oriental, a major British ship line, announced in 1968 that its planning was based on the assumption that the Suez Canal was

closed permanently, and other carriers appear to have made the same choice; see *Fairplay*, July 4, 1968, p. 79, and Pearson and Fossey, *World Deep-Sea Container Shipping*, p. 248.

34. Relative fuel costs appear in Sletmo and Williams, *Liner Conferences*, p. 162. Opposition by Sea-Land's board to the SL-7 purchase is discussed in John Boylston interview, COHP.

35. On relations between Sea-Land and R. J. Reynolds, see Tursi, White, and McQuilkin, *Lost Empire*, chaps. 15–16 and 23; R. J. Reynolds Industries, *Annual Reports* from 1975 through 1980; transcript of R. J. Reynolds Industries Analyst Meeting, September 19–21, 1976; and comment from R. J. Reynolds Industries' chief financial officer Gwain H. Gillespie at analyst presentation, November 1, 1984, p. 78. These and other relevant R. J. Reynolds documents are available on a Web site created in conjunction with antitobacco litigation, tobaccodocuments.org.

36. Colin Jones, "Heading for a Period of Consolidation," *Financial Times*, January 15, 1976.

## Chapter 12
## The Bigness Complex

1. Author's telephone interview with Earl Hall, May 21, 1993; "Malcom McLean's $750 Million Gamble," *Business Week*, April 16, 1979.

2. "Pinehurst Club Is Sold for $9-Million," *NYT*, January 1, 1971; author's telephone interview with Dena Van Dyk, May 2, 1994; William Robbins, "Vast Plantation Is Carved Out of North Carolina Wilderness," *NYT*, May 8, 1974; *Business Week*, April 1, 1979.

3. Sletmo and Williams, *Liner Conferences*, p. 39.

4. *Lloyd's Shipping Economist*, September 1982, p. 9; Pearson and Fossey, *World Deep-Sea Container Shipping*, p. 220; UNCTAD, *Review of Maritime Transport*, various issues.

5. Michael Kuby and Neil Reid, "Technological Change and the Concentration of the U.S. General Cargo Port System: 1970–88," *Economic Geography* 68, no. 3 (1993): 279.

6. American Association of Port Authorities; Marad, "Containerized Cargo Statistics," various years; Pearson and Fossey, *World Deep-Sea Container Shipping*, p. 29; *Containerisation International Yearbook*, various years. The figures for this period must be interpreted cautiously, because the statistical definition of "container" had not yet been standardized in terms of 20-foot units, and individual ports' statistics did not always distinguish between loaded and empty containers.

7. Hugh Turner, Robert Windle, and Martin Dresner, "North American Containerport Productivity: 1984–1997," *Transportation Research Part E* (2003): 354.

8. Yehuda Hayut, “Containerization and the Load Center Concept,” *Economic Geography* 57, no. 2 (1981): 170.

9. Brian Slack, “Pawns in the Game: Ports in a Global Transportation System,” *Growth and Change* 24, no. 4 (1993): 579–588; Kuby and Reid, “Technological Change,” p. 280; *Containerisation International Yearbook*, 1988.

10. Port of Seattle, Marine Planning and Development Department, “Container Terminal Development Plan,” October 1991; Eileen Rhea Rabach, “By Sea: The Port Nexus in the Global Commodity Network (The Case of the West Coast Ports)” (Ph.D. diss., University of Southern California, 2002), p. 86. Rabach’s assertion that port competition is a zero-sum game is not correct; as this study argues, declining costs throughout the transportation system have stimulated the flow of international trade.

11. UNCTAD, *Review of Maritime Transport 1979*, p. 29; Marad, “United States Port Development Expenditure Report,” 1991; Herman L. Boschken, *Strategic Design and Organizational Change: Pacific Rim Seaports in Transition* (Tuscaloosa, 1988), pp. 61–65. On the Oakland dredging saga, see Christopher B. Busch, David L. Kirp, and Daniel F. Schoenholz, “Taming Adversarial Legalism: The Port of Oakland’s Dredging Saga Revisited,” *Legislation and Public Policy* 2, no. 2 (1999): 179–216; Ronald E. Magden, *The Working Longshoreman* (Tacoma, 1996), p. 190.

12. *Fairplay*, July 3, 1975, p. 37; Slack, “Pawns in the Game,” p. 582; Turner, Windle, and Dresner, “North American Containerport Productivity,” p. 351; author’s interview with Mike Beritzhoff, Oakland, CA, January 25, 2005.

13. Boschken, *Strategic Design*, p. 200.

14. Hans J. Peters, “Private Sector Involvement in East and Southeast Asian Ports: An Overview of Contractual Arrangements,” *Infrastructure Notes*, World Bank, March 1995.

15. Pearson and Fossey, *World Deep-Sea Container Shipping*.

16. *Lloyd’s Shipping Economist*, January 1983, p. 10.

17. Ibid., p. 12 and March 1985, p. 4.

18. Daniel Machalaba, “McLean Bets that Jumbo Freighter Fleet Can Revive Industry,” *Wall Street Journal*, September 26, 1986; Ron Katims interview, COHP.

19. Broeze, *The Globalisation of the Oceans*, p. 95.

20. Ibid., p. 84; *Lloyd’s Shipping Economist*, April 1984, p. 7, and March 1986, p. 3; UNCTAD, *Review of Maritime Transport 1989*, p. 25; *JOC*, October 15, 1986.

21. Bruce Barnard, “Evergreen Set to Drop Felixstowe,” *JOC*, October 22, 1986; Machalaba, “McLean Bets”; Kuby and Reid, “Technological Change,” p. 279.

22. *Lloyd's Shipping Economist*, January 1987; Gibson and Donovan, *The Abandoned Ocean*, p. 218; Susan F. Rasky, "Bankruptcy Step Taken by McLean," *NYT*, November 25, 1986.

23. The bankruptcy filing, *in re McLean Industries, Inc.*, was in the Southern District of New York, case numbers 86-12238 through 86-12241. This paragraph draws on docket nos. 106, 107, 111, 133, and 163. On the vessel sale, see Daniel Machalaba, "Sea-Land Will Buy 12 Superfreighters Idled by U.S. Lines Inc. for $160 Million," *Wall Street Journal*, February 9, 1988.

24. Author's interview with Gerald Toomey, May 5, 1993; Daniel Machalaba, "Container Shipping's Inventor Plans to Start Florida–Puerto Rico Service," *Wall Street Journal*, January 31, 1992. For the views of a former U.S. Lines employee, see "McLean Doesn't Deserve Award," letter to the editor, *JOC*, September 16, 1992.

25. R. M. Katims, "Keynote Address: Terminal of the Future," in National Research Council, Transportation Research Board, *Facing the Challenge: The Intermodal Terminal of the Future* (Washington, DC, 1986), pp. 1–3.

## Chapter 13
## The Shippers' Revenge

1. Comment of Karl Heinz Sager cited in Broeze, *The Globalisation of the Oceans*, p. 41.

2. UNCTAD, *Review of Maritime Transport 1975*, p. 43.

3. *Fairplay*, July 15, 1971, pp. 47 and 53. UNCTAD's estimated shipping costs were:

Average Cost of Handling One Cubic Meter of Freight, 1970

| | *Capital Cost* | *Operating Cost* | *Cargo Handling* | *Total Cost* |
|---|---|---|---|---|
| Conventional ship | $2.30 | $3.81 | $17.00 | $23.11 |
| Containership | $2.50 | $2.47 | $ 5.90 | $10.87 |

*Source*: UNCTAD.

4. Matson Research Corp., *The Impact of Containerization*, pp. 40–41; *Fairplay*, February 1, 1968, p. 8.

5. OECD, "Ocean Freight Rates as Part of Total Transport Costs" (Paris, 1968), p. 24.

6. Antwerp data taken from Bremer Ausschuß für Wirtschaftsforschung, *Container Facilities*. Dart Container Line spent nearly $300,000 in 1973 on a computer to keep track of its 20,000 containers. *Fairplay*, April

5, 1973, p. 40. By 1974, U.S. Lines was spending $1.7 million a year to operate its computers; see *Fairplay*, April 4, 1974, p. 76.

7. Broeze, *The Globalisation of the Oceans*, pp. 55–56. Worldwide, the containerships entering the fleet in 1973 traveled at an average speed of 25 knots, compared with 20 knots or less for almost all breakbulk and containerships built before 1968. Wallin, "The Development, Economics, and Impact," p. 642. The 85 percent breakeven point is cited in U.S. Congress, Office of Technology Assessment, *An Assessment of Maritime Technology and Trade* (Washington, DC, 1983), p. 71. Three ship lines surveyed by J. E. Davies in 1980 reported that their fixed costs were between 53 and 65 percent of total costs, implying much lower breakeven points; "An Analysis of Cost and Supply Conditions in the Liner Shipping Industry," *Journal of Industrial Economics* 31, no. 4 (1983): 420.

8. *Fairplay*, February 4, 1971.

9. Sletmo and Williams, *Liner Conferences*, chap. 5; Benjamin Bridgman, "Energy Prices and the Expansion of World Trade," Working Paper, Louisiana State University, November 2003. Fuel cost as a share of operating costs are given in Office of Technology Assessment, *An Assessment of Maritime Technology and Trade*, p. 71. The International Monetary Fund cites the increase in market concentration in shipping following the introduction of containers as another reason for the failure of shipping rates to fall. However, it is not at all clear that pooling agreements and other anticompetitive practices succeeded in holding shipping rates above competitive levels for extended periods. International Monetary Fund, *World Economic Outlook*, September 2002, p. 116; Sjostrom, "Ocean Shipping Cartels," pp. 107–134.

10. *Fairplay*, July 15, 1974, p. 50. In principle, it should be possible to quantify transport-cost saving over time by comparing a country's imports under two different definitions, free on board (f.o.b.), which represents the value of merchandise at the point of export, and cost of insurance and freight (c.i.f.), which is the value at the point of import, including transport costs. In practice, however, the difference between c.i.f. and f.o.b. imports provides little guidance concerning freight-cost trends. The accuracy of the underlying data is questionable; if IMF figures are to be believed, insurance and freight accounted for a mere 1 percent of Switzerland's imports as long ago as 1960. Data for countries with large-scale trade in bulk products, such as coal and oil, may not reflect changes affecting manufactured goods. More problematic, the use of aggregate c.i.f. and f.o.b. data assumes that the composition and origin of imports have not changed over time. Scott L. Baier and Jeffrey H. Bergstrand, "The Growth of World Trade: Tariffs, Transport Costs, and Income Similarity," *Journal of International Economics* 53, no. 1 (2001): 1–27, show that for 16 wealthy nations, transport fell from 8.2 per-

cent to 4.3 percent of import value between 1958–60 and 1986–88, but the various factors cited above make this conclusion unpersuasive.

11. Indexes of "tramp" charter rates were compiled during the 1960s and 1970s by several sources, including the *Norwegian Shipping News* and the British Chamber of Shipping. The price for a single-voyage charter, adjusted for the capacity of the vessel, yields a vessel cost per ton shipped. Japanese shippers were the tramps' main customers. The tramp market was somnolent in the early 1970s, and it appears that most tramp charters involved bulk freight rather than breakbulk freight of the sort that would have been competitive with container shipping. *Fairplay*, July 1, 1971, p. 73.

12. The German Liner Index did not purport to measure rates on freight in other parts of the world. As UNCTAD pointed out, it did not fully reflect changes in rates or surcharges and was "rather narrowly based and greatly influenced by declining currency exchange ratios of the Deutschmark versus the United States dollar." See UNCTAD, *Review of Maritime Transport 1972–73*, p. 81, and *1984*, p. 42. It also appears that the index as published in the 1960s and 1970s did not clearly distinguish liner freight rates overall from container shipping rates. The index as published in the 1990s did make this distinction, revealing very different trends. The index for all liner rates, for example, fell from 101 in January 1994 to 96 in June 1997, whereas the subindex for container rates fell much more steeply, from 101 to 90, during the same period. See UNCTAD, *Review of Maritime Transport 1997*, p. 50. For more discussion of both indexes and an attempt to adjust them for inflation, see Hummels, "Have International Transportation Costs Declined?"

13. For the Hansen index, see *Fairplay*, January 15, 1981, p. 15.

14. Tursi, White, and McQuilkin, *Lost Empire*, p. 185.

15. UNCTAD's annual *Review of Maritime Transport* provides data on container shipping in developing countries; see also Pearson and Fossey, *World Deep-Sea Container Shipping*, p. 27. The containerized share of U.S. imports appears in the *Review* for 1974, p. 51.

16. Sletmo and Williams, *Liner Conferences*, p. 80.

17. According to Pacific Maritime Association data, base wages for longshoremen on the U.S. Pacific Coast nearly doubled, from $3.88 per hour in July 1966 to $7.52 in July 1976. Data available at www.pmanet.org. At U.S. North Atlantic ports, longshoremen who were entitled to four weeks' vacation and eleven paid holidays in 1966 were eligible for six weeks' vacation and thirteen paid holidays by the early 1970s. *Longshore News*, November 1969, p. 4A.

18. OECD, "Ocean Freight Rates as Part of Total Transport Costs," p. 31.

19. Hummels, "Have International Transportation Costs Declined?"; *Fairplay*, May 16, 1968, p. 49.

20. On New Zealand, see *Fairplay*, February 19, 1976, p. 3.

21. No accurate measures of insurance-rate changes are available. Insurers initially resisted lowering rates for container shipments, mainly with the reasoning that container shipping might lead to less frequent but larger losses if an entire container were stolen or damaged. In addition, a full container was usually handed off from one carrier to another without being opened, making it difficult to determine which carrier was responsible if damage did occur. *Fairplay*, September 2, 1971; Insurance Institute of London, "An Examination of the Changing Nature of Cargo Insurance Following the Introduction of Containers," January 1969. By 1973, an insurance expert was ready to admit that "[c]argoes carried in containers appear to be bringing improved claims experience." *Fairplay*, July 5, 1973, p. 55.

22. Marad, "Current Trends in Port Pricing" (Washington, DC, 1978), p. 19.

23. Real oil prices in dollar terms rose until 1981; see U.S. Department of Energy, *Annual Energy Review* (Washington, DC, 2003), Table 5.21. The German liner-freight index discussed above also began to decline in the late 1970s, after adjustment for inflation.

24. Pedro L. Marin and Richard Sicotte, "Exclusive Contracts and Market Power: Evidence from Ocean Shipping," Discussion Paper 2028, Centre for Economic Policy Research, June 2001; comment from J. G. Payne, vice chairman of Blue Star Line, in *Fairplay*, April 11, 1974, p. 7.

25. The former competitors involved in the North Atlantic Pool were American Export–Isbrandtsen Line, Belgian Line, Bristol City Line, Clarke Traffic Services, Cunard Line, French Line, Hamburg-American Line, Holland-America Line, North German Lloyd, Sea-Land Service, Seatrain Lines, Swedish American Lines, Swedish Transatlantic Lines, United States Lines, and Wallenius Line. On shippers' councils, see U.S. General Accounting Office, *Changes in Federal Maritime Regulation Can Increase Efficiency and Reduce Costs in the Ocean Liner Shipping Industry* (Washington, DC, 1982), chap. 5. UNCTAD encouraged the formation of regional shippers' councils, which were formed in Central America, East Africa, and Southeast Asia.

26. *Fairplay*, July 1, 1971; UNCTAD, *Review of Maritime Transport 1972–73*, p. 80, and *1975*, p. 44.

27. Office of Technology Assessment, *An Assessment of Maritime Technology and Trade*, p. 72.

28. U.S. General Accounting Office, *Centralized Department of Defense Management of Cargo Shipped in Containers Would Save Millions and Improve Service* (Washington, DC, 1977).

29. Author's telephone interview with Cliff Sayre, former vice president of transportation at DuPont, January 24, 1992.

30. According to Sayre, DuPont had more than fifty loyalty agreements and had relationships with more than three hundred individual ocean carriers in 1978.

31. Prior to containerization, Evergreen Line operated as a nonconference carrier on the Japan–Red Sea route, pricing 10 to 15 percent below conference rate. On the Japan-India route, however, Evergreen decided to join the conference after finding that Japanese steel mills would not use its services because they had loyalty agreements with the conference. See *Fairplay*, August 9, 1973, p. 60.

32. Broeze, *The Globalisation of the Oceans*, p. 65.

33. *Fairplay*, September 21, 1972, p. 11; November 23, 1972, p. 59; and June 28, 1973, p. 44; Eric Pace, "Freighters' Rate War Hurting U.S. Exporters," *NYT*, September 11, 1980; *Fairplay*, February 12, 1981, p. 9.

34. James C. Nelson, "The Economic Effects of Transport Deregulation in Australia," *Transport Journal* 16, no. 2 (1976): 48–71.

35. U.S. General Accounting Office, *Issues in Regulating Interstate Motor Carriers* (Washington, DC, 1980), p. 35.

36. Matson Research Corp., *The Impact of Containerization*, 2:64; U.S. General Accounting Office, *Combined Truck/Rail Transportation Service: Action Needed to Enhance Effectiveness* (Washington, DC, 1977). One company reported in 1978 that sending a trailer on a flatcar the 1,068 miles from Minneapolis to Atlanta cost $723; the cost of through trucking service for the same commodity was $693. See Frederick J. Beier and Stephen W. Frick, "The Limits of Piggyback: Light at the End of the Tunnel," *Transportation Journal* 18, no. 2 (1978): 17.

37. Iain Wallace, "Containerization at Canadian Ports," *Annals of the Association of American Geographers* 65, no. 3 (1976): 444; "The 'Minibridge' That Makes the ILA Boil," *Business Week*, May 19, 1975; General Accounting Office, *American Seaports—Changes Affecting Operations and Development* (Washington, DC, 1979); Lee Dembart, " 'Minibridge' Shipping Is Raising Costs and Costing Jobs in New York," *NYT*, February 27, 1977; Marad, "Current Trends in Port Pricing," p. 20.

38. Robert E. Gallamore, "Regulation and Innovation: Lessons from the American Railroad Industry," in *Essays in Transportation Economics and Policy: A Handbook in Honor of John R. Meyer*, ed. José A. Gómez-Ibañez, William B. Tye, and Clifford Winston (Washington, DC, 1999), p. 515. Number of contracts appears in Wayne K. Talley, "Wage Differentials of Intermodal Transportation Carriers and Ports: Deregulation versus Regulation," *Review of Network Economics* 3, no. 2 (2004): 209. Clifford Winston, Thomas M. Corsi, Curtis M. Grimm, and Carol A. Evans, *The Economic*

*Effects of Surface Freight Deregulation* (Washington, DC, 1990), p. 41, estimate the total saving from deregulation at $20 billion in 1988 dollars, with the loss to railroad and trucking workers estimated at $3 billion.

39. Gallamore, "Regulation and Innovation, p. 516; John F. Strauss, Jr., *The Burlington Northern: An Operational Chronology, 1970–1995*, chap. 6, available online at www.fobnr.org/bnstore/ch6.htm; Kuby and Reid, "Technological Change," p. 282. Paul Stephen Dempsey, "The Law of Intermodal Transportation: What It Was, What It Is, What It Should Be," *Transportation Law Journal* 27, no. 3 (2000), looks at the history of regulations governing intermodal freight.

40. Robert C. Waters, "The Military Sealift Command versus the U.S. Flag Liner Operators," *Transportation Journal* 28, no. 4 (1989): 30–31.

41. *Lloyd's Shipping Economist*, various issues; Hans J. Peters, "The Commercial Aspects of Freight Transport: Ocean Transport: Freight Rates and Tariffs," World Bank *Infrastructure Notes*, January 1991; author's interview with William Hubbard.

## Chapter 14
## Just in Time

1. Paul Lukas, "Mattel: Toy Story," *Fortune Small Business*, April 18, 2003; Holiday Dmitri, "Barbie's Taiwanese Homecoming," *Reason*, May 2005. For discussion of the toy industry's supply chains, see Francis Snyder, "Global Economic Networks and Global Legal Pluralism," European University Institute Working Paper Law No. 99/6, August 1999.

2. This description of just-in-time procedures is taken from G.J.R. Linge, "Just-in-Time: More or Less Flexible?" *Economic Geography* 67, no. 4 (1991): 316–332.

3. The counts, drawn from approximately a thousand business and management periodicals, are taken from Paul D. Larson and H. Barry Spraggins, "The American Railroad Industry: Twenty Years after Staggers," *Transportation Quarterly* 52, no. 2 (2000): 37; Robert C. Lieb and Robert A. Miller, "JIT and Corporate Transportation Requirements," *Transportation Journal* 27, no. 3 (1988): 5–10; author's interview with Cliff Sayre.

4. According to calculations based on the U.S. National Income and Product Accounts, private nonfarm inventories in 2004 averaged about $1.65 trillion, or about 13 percent of final sales. Through the early 1980s, the ratio was in the range of 22 to 25 percent. That 9 percentage point reduction measured against 2004 final sales of $12.2 trillion yields an annual saving approaching $1.1 trillion. An alternative measurement examines the average length of time goods are held in inventory by retailers, wholesalers, and manufacturers. Analyzed in this way, if inventories had

risen at the same rate as sales since the early 1980s, U.S. department and discount stores would have kept an additional $30 billion of stock on average during 2000, durable goods manufacturers would have held an additional $240 billion of inventories, manufacturers of nondurables would have had inventories about $40 billion higher than the actual number, and wholesale inventories might have been $30–$40 billion higher. This method yields a decline in average inventories relative to sales in these sectors of more than $400 billion. See U.S. Census Bureau, *Monthly Retail Trade Report*, and Hong Chen, Murray Z. Brank, and Owen Q. Wu, "U.S. Retail and Wholesale Inventory Performance from 1981 to 2003," Working Paper, University of British Columbia, 2005.

5. On earlier forms of globalization, see Kevin H. O'Rourke and Jeffrey G. Williamson, *Globalization and History: The Evolution of a Nineteenth-Century Atlantic Economy* (Cambridge, MA, 1999), and O'Rourke and Williamson, "When Did Globalization Begin?" Working Paper 7632, NBER, April 2000.

6. Robert Feenstra, "Integration of Trade and Disintegration of Production in the Global Economy," *Journal of Economic Perspectives* 12, no. 4 (1998); Rabach, "By Sea," p. 203.

7. David Hummels, "Toward a Geography of Trade Costs," mimeo, University of Chicago, January 1999; Will Martin and Vlad Manole, "China's Emergence as the Workshop of the World," Working Paper, World Bank, September 2003.

8. Ximena Clark, David Dollar, and Alejandro Micco, "Port Efficiency, Maritime Transport Costs, and Bilateral Trade," *Journal of Development Economics* 74, no. 3 (2004): 417–450.

9. Erie, *Globalizing L.A.*, p. 208; Miriam Dossal Panjwani, "Space as Determinant: Neighbourhoods, Clubs and Other Strategies of Survival," in Davies et al., *Dock Workers*, 2:759; Robin Carruthers, Jitendra N. Bajpai, and David Hummels, "Trade and Logistics: An East Asian Perspective," in *East Asia Integrates: A Trade Policy Agenda for Shared Growth* (Washington, DC, 2003), pp. 117–137.

10. David Hummels, "Time as a Trade Barrier," mimeo, Purdue University, July 2001.

11. Joel Mokyr, *The Gifts of Athena: Historical Origins of the Knowledge Economy* (Princeton, 2002), p. 232.

12. Clark, Dollar, and Micco, "Port Efficiency," p. 422; Nuno Limão and Anthony J. Venables, "Infrastructure, Geographical Disadvantage and Transport Costs," *World Bank Economic Review* 15, no. 3 (2001): 451–479; Robin Carruthers and Jitendra N. Bajpai, "Trends in Trade and Logistics: An East Asian Perspective," Working Paper No. 2, Transport Sector Unit, World Bank, 2002.

13. Clark, Dollar, and Micco, "Port Efficiency," p. 422.

14. Increase in shipping volume cited in Carruthers and Bajpai, "Trends in Trade and Logistics," p. 12; on Hamburg, see Dieter Läpple, "Les mutations des ports maritimes et leurs implications pour les dockers et les regions portuaires: l'exemple de Hambourg," in *Dockers de la Méditerranée*, p. 55.

15. Claude Comtois and Peter J. Rimmer, "China's Competitive Push for Global Trade," in Pinder and Slack, *Shipping and Ports*, pp. 40–61, offer an interesting discussion of the logic behind Chinese port development.

16. Clark, Dollar, and Micco, "Port Efficiency," p. 441.

17. Gibson and Donovan, *The Abandoned Ocean*, articulate in detail the connection between U.S. maritime policy and the decline of American shipping.

18. Chinitz, *Freight and the Metropolis*, pp. 161–162 and p. 100. For examples of such studies, see A. D. Little, "Containerisation on the North Atlantic," and Litton Systems, "Oceanborne Shipping."

19. Container data estimated from UNCTAD, *Review of Maritime Transport 2004*, pp. 73–76, and Institute for Shipping and Logistics, "Shipping Statistics and Market Review" (Bremen, 2004).

20. Jeffrey C. Mays, "Newark Sees Cash in Containers," *Star-Ledger*, February 4, 2004; Natural Resources Defense Council, "Harboring Pollution: The Dirty Truth about U.S. Ports" (New York, 2004); Deborah Schoch, "Pollution Task Force to Meet for Last Time on L.A. Port," *Los Angeles Times*, June 21, 2005.

21. Institute for Shipping and Logistics, "ISL Market Analysis 2005: World Merchant Fleet, OECD Shipping and Shipbuilding," viewed July 31, 2005 at http://www.isl.org/products_services/publications/pdf/comment_1-2-2005_short.pdf; Richard C. Roenbeck, "Containership Losses Due to Head-Sea Parametric Rolling: Implications for Cargo Insurers" (paper presented to International Union of Marine Insurers, 2003), available at http://www.iumi.com/Conferences/2003_sevilla/1609/RRoenbeck.pdf. Among the early proposals for offshore ports was one at Scapa Flow in the Orkney Islands, off the northern coast of Scotland. As envisioned, huge ships arriving from Asia or North America would make such a location their only port of call in northern Europe; the savings in ships' port time and sailing time were estimated to more than make up for the additional cost of transshipping the containers. See Scottish Executive, "Container Transshipment and Demand for Container Terminal Capacity in Scotland" (prepared by Transport Research Institute, Napier University, December 2003).

# Bibliography


### Archival Collections

Containerization Oral History Project, National Museum of American History, Smithsonian Institution, Washington, DC.

International Longshoremen's Association District 1 Files, 1954–56, Kheel Center, Catherwood Library, Cornell University, Ithaca, NY.

International Longshoremen's Association Files, Robert F. Wagner Labor Archives, New York University, New York, NY.

Mayor Abraham Beame Papers, New York Municipal Archives, New York, NY.

Mayor John V. Lindsay Papers, New York Municipal Archives, New York, NY.

Mayor Robert F. Wagner Papers, New York Municipal Archives, New York, NY.

National Archives and Records Administration, Modern Military Records Branch, College Park, MD.

Operational Archives Branch, Naval Historical Center, Washington, DC.

Penn Central Transportation Company Archives, Hagley Museum and Library, Wilmington, DE.

Port of New York Authority records (Doig Files), New Jersey State Archives, Trenton, NJ.

Regional Oral History Program, Bancroft Library, University of California, Berkeley, CA.

Robert B. Meyner Papers, New Jersey State Archives, Trenton, NJ.

Robert F. Wagner Labor Archives, New York University, New York, NY.

Robert F. Wagner Papers, LaGuardia and Wagner Archives, LaGuardia Community College, Queens, NY.

U.S. Army Materiel Command Historical Office, Fort Belvoir, VA.

Vernon H. Jensen papers, Kheel Center, Catherwood Library, Cornell University, Ithaca, NY.

## U.S. Government Documents

Interstate Commerce Commission. *Motor Carrier Cases.*

———. *Revenue and Traffic of Carriers by Water.* 1952–64.

———. *Reports.*

———. *Transport Statistics in the United States.* 1954–74.

Interstate Commerce Commission, Bureau of Transport Economics and Statistics. *Monthly Comment on Transportation Statistics.* 1945–April 1955.

———. *Transport Economics.* May 1955–December 1957.

Office of Technology Assessment. *An Assessment of Maritime Trade and Technology.* Washington, DC: U.S. Government Printing Office, 1983.

U.S. Army, Corps of Engineers. *Waterborne Commerce of the United States.* Annual.

U.S. Bureau of the Census. *Census of Manufactures.* Washington, DC: U.S. Government Printing Office, various years.

———. *County Business Patterns.* Washington, DC: U.S. Government Printing Office, various years.

———. *Historical Statistics of the United States.* 2 vols. Washington, DC: U.S. Government Printing Office, 1975.

———. *U.S. Census of Population and Housing.* Washington, DC: U.S. Government Printing Office, various years.

U.S. Congress, Joint Economic Committee. *Discriminatory Ocean Freight Rates and the Balance of Payments.* Washington, DC: U.S. Government Printing Office, 1964.

U.S. Department of Commerce, Bureau of Economic Analysis. *National Income and Product Accounts.*

U.S. Department of Commerce, Federal Maritime Board and Maritime Administration. *Annual Reports.*

U.S. Department of Commerce, Maritime Administration. "Shoreside Facilities for Trailership, Trainship, and Containership Services." Washington, DC: U.S. Department of Commerce, 1956.

U.S. Department of Transportation, Bureau of Transportation Statistics. *Transportation Statistics Annual Report.* Washington, DC: U.S. Department of Transportation, 1994–2004.

U.S. Department of Transportation, Maritime Administration. *United States Port Development Expenditure Report, 1946–1989.* Washington, DC: Office of Port and Intermodal Development, 1989.

U.S. Economic Stabilization Program, Pay Board. "East and Gulf Coast Longshore Contract." May 2, 1972.

U.S. General Accounting Office. *American Seaports—Changes Affecting Operations and Development.* Washington, DC: GAO 1979.

———. *Centralized Department of Defense Management of Cargo Shipped in Containers Would Save Millions and Improve Service*. Washington, DC: GAO, 1977.
———. *Changes in Federal Maritime Regulation Can Increase Efficiency and Reduce Costs in the Ocean Liner Shipping Industry*. Washington, DC: GAO, 1982.
———. *Combined Truck/Rail Transportation Service: Action Needed to Enhance Effectiveness*. Washington, DC: GAO, 1977.
———. *Issues in Regulating Interstate Motor Carriers*. Washington, DC: GAO, 1980.
U.S. House of Representatives, Committee on Merchant Marine and Fisheries. *Cargo Container Dimensions*. October 31 and November 1, 8, and 16, 1967. Washington, DC: U.S. Government Printing Office, 1968.
———. *Hearings on HR 8637, To Facilitate Private Financing of New Ship Construction*. April 9, 27, 28, 29, and 30, 1954.
———. *Study of Harbor Conditions in Los Angeles and Long Beach Harbor*. October 19–21, 1955 and July 16, 1956.
U.S. National Academy of Sciences. *Roll-On, Roll-Off Sea Transportation*. Washington, DC: U.S. Government Printing Office, 1957.
U.S. National Research Council. *Research Techniques in Marine Transportation*. Publication 720. Washington, DC: National Academy of Sciences, 1959.
U.S. National Research Council, Maritime Cargo Transportation Conference. *Cargo Ship Loading*. Publication 474. Washington, DC: National Academy of Sciences 1957.
———. *The SS Warrior*. Publication 339. Washington, DC: National Academy of Sciences, 1954.
———. *Transportation of Subsistence to NEAC*. Washington, DC: National Academy of Sciences 1956.
U.S. National Research Council, Transportation Research Board. *Facing the Challenge: The Intermodal Terminal of the Future*. Washington, DC: Transportation Research Board, 1986.

## Other Government Documents

Arthur D. Little, Inc. *Community Renewal Programming: A San Francisco Case Study*. New York, 1966.
Board of Inquiry into Waterfront Labor Conditions, New York 1951.
Booz-Allen & Hamilton. "General Administrative Survey of the Port of Seattle." January 20, 1958.
New York City Planning Commission. "The Port of New York: Proposals for Development." 1964.

New York City Planning Commission. "Redevelopment of Lower Manhattan East River Piers." September 1959.
———. *The Waterfront*. 1971.
New York State Department of Labor. "Employment Conditions in the Longshore Industry." *Industrial Bulletin* 31, no. 2 (1952).
———. *Population and Income Statistics*.
Port Authority of New York and New Jersey. *Foreign Trade 1976*.
Port of New York Authority (subsequently Port Authority of New York and New Jersey). *Annual Reports*.
———. *Container Shipping: Full Ahead*. 1967.
———. *Marine Terminal Survey of the New Jersey Waterfront*. February 10, 1949.
———. *Metropolitan Transportation 1980*.
———. *Outlook for Waterborne Commerce through the Port of New York*. 1948.
———. *The Port of New York*. 1952.
———. "Proposal for Development of the Municipally Owned Waterfront and Piers of New York City." February 10, 1948.
———. *Via—Port of New York*.
Port of Seattle, Marine Planning and Development Department. "Container Terminal Development Plan." October 1991.
Port of Seattle, Planning and Research Department. "A Conceptual Framework for the Physical Development of the Port of Seattle." April 1966.
Port of Singapore Authority. *Annual Report and Accounts*. Various years.
———. *A Review of the Past and a Look into the Future*. Singapore: Port of Singapore Authority, 1971.
Scottish Executive. "Container Transshipment and Demand for Container Terminal Capacity in Scotland." Transport Research Institute, Napier University, Edinburgh, December 2003.
Seattle Port Commission. "Container Terminals 1970–1975: A Development Strategy." November 1969.
———. "Report of the Marine Terminals Task Force to the Citizens' Port Committee." 1959.
———. "Review of Port Development and Financing." December 1968.
———. *Shipping Statistics Handbook*. 1963.
State of New York. *Record of the Public Hearing Held by Governor Thomas E. Dewey on the Recommendations of the New York State Crime Commission for Remedying Conditions on the Waterfront of the Port of New York*. June 8–9, 1953.
Tippetts-Abbett-McCarthy-Stratton. "Shoreside Facilities for Trailership, Trainship, and Containership Services." Washington, DC: U.S. Department of Commerce, Maritime Administration, 1956.

UK Department for Transport. "Recent Developments and Prospects at UK Container Ports." London, 2000.
———. *Transport Statistics Report: Maritime Statistics 2002*. London, 2003.
Waterfront Commission of New York Harbor. *Annual Reports*.

**International Agency Documents**

International Monetary Fund. *World Economic Outlook*. September 2002.
Organisation for Economic Co-operation and Development. "Ocean Freight Rates as Part of Total Transport Costs." Paris, 1968.
Pan American Union. "Recent Developments in the Use and Handling of Unitized Cargoes." Document UP/CIES-8 ES-CTPP-Doc. 12, Washington, DC, 1964.
United Nations Conference on Trade and Development. *Review of Maritime Transport*. Annual.
United Nations Department of Economic and Social Affairs, "An Examination of Some Aspects of the Unit-Load System of Cargo Shipments: Application to Developing Countries." 1966.
United Nations Economic and Social Commission for Asia and the Pacific. *Commercial Development of Regional Ports as Logistics Centres*. New York 2002.
World Bank. *East Asia Integrates: A Trade Policy Agenda for Shared Growth*. Washington, DC: World Bank, 2003.
World Trade Organization. *World Trade Report 2004* (Geneva, 2005).

**Private Reports and Documents**

A. D. Little. "Containerisation on the North Atlantic." London: National Ports Council, 1967.
AOTOS Awards, program, 1984.
Association of American Railroads. *Carloads of Revenue Freight Loaded*.
Containerization and Intermodal Institute, "Containerization: The First 25 Years." New York, 1981.
Downtown–Lower Manhattan Association. "Lower Manhattan." 1958.
First National City Bank. "The Port of New York: Challenge and Opportunity." 1967.
Insurance Institute of London, Advanced Study Group No. 188. "An Examination of the Changing Nature of Cargo Insurance following the Introduction of Containers." 1969.
International Cargo Handling Coordination Association. "Containerization Symposium Proceedings, New York City, June 15, 1955."
International Longshoremen's and Warehousemen's Union. "Report of the Officers to the Thirteenth Biennial Convention." April 6–7, 1959.

Litton Systems Inc. "Oceanborne Shipping: Demand and Technology Forecast." Report for U.S. Department of Transportation, Office of the Secretary, June 1968.

Matson Research Corporation. "The Impact of Containerization on the U.S. Economy." 2 vols. Report for U.S. Department of Commerce, September 1970.

McKinsey & Co. "Containerization: A 5-Year Balance Sheet." 1972.

———. "Containerization: The Key to Low Cost Transport." Report for British Transport Docks Board, June 1967.

McLean Industries Inc. *Annual Reports.*

Muncy, Dorothy. "Inventory of Port-Oriented Land: Baltimore Region." Report for Maryland Economic Development Commission, Arlington, VA, 1963.

National Ports Council. *Annual Digest of Port Statistics.*

———. *Container and Roll-On Port Statistics, Great Britain.*

New York Shipping Association. "Progress Report 1959."

Pacific Maritime Association and International Longshoremen's and Warehousemen's Union. "Memorandum of Agreement on Mechanization and Modernization." October 18, 1960.

Pan-Atlantic Steamship Corporation. "Summary of Post–World War II Coastwise Operations." Mimeo, n.d.

Schott, John G. *Piggyback and the Future of Freight Transportation.* Washington, DC: Public Affairs Institute, 1960.

———. *Progress in Piggyback and Containerization.* Washington, DC: Public Affairs Institute, 1960.

Tantlinger, K. W. "U.S. Containerization: From the Beginning through Standardization." Paper presented to the World Port Conference, Rotterdam, 1982.

Tobin, Austin J. "Transportation in the New York Metropolitan Region during the Next Twenty-five Years." New York: Regional Plan Association, 1954.

Woodruff, G. C. "The Container Car as the Solution of the Less Than Carload Lot Problem." Speech to Associated Industries of Massachusetts, October 23, 1929.

———. "Freight Container Service." Speech to Traffic Club of New York, March 25, 1930.

## Newspapers and Periodicals

*Army Logistician*
*Baltimore Sun*
*Brooklyn Eagle*

*Business Week*
*Civil Engineering*
*Containerisation International Yearbook*
*Containers: Bulletin of the International Container Bureau*
*Fairplay International Shipping Journal*
*Jane's Freight Containers*
*Journal of Commerce*
*Longshore News*
*Marine Engineering/Log*
*Newark Evening News*
*News and Observer* (Raleigh)
*New York*
*New York Herald Tribune*
*New York Times*
*New York World Telegram and Sun*
*Robesonian*
*Sealift*

## Interviews and Oral Histories

Bell, Peter. Interview by Debra Bernhardt, August 29, 1981. New Yorkers at Work Oral History Collection, Robert F. Wagner Labor Archive, New York University.

Boylston, John. Interview by Arthur Donovan and Andrew Gibson, December 7, 1998. Containerization Oral History Project, National Museum of American History, Smithsonian Institution, Washington, DC.

Bulcke, Germain. "Longshore Leader and ILWU-Pacific Maritime Association Arbitrator." Interviews by Estolv Ethan Ward, 1983. Berkeley: Regional Oral History Office, Bancroft Library, University of California, 1984.

Campbell, Robert N. Author's telephone interview, June 25, 1993.

Cox, John Parr. "Parr Terminal: Fifty Years of Industry on the Richmond Waterfront." Interviews by Judith K. Dunning, 1986. Berkeley: Regional Oral History Office, Bancroft Library, University of California, 1992.

Crowley, Thomas B. "Crowley Maritime Corporation: San Francisco Bay Tugboats to International Transportation Fleet." Interviews by Miriam Feingold Stein, 1973–75. Berkeley: Regional Oral History Office, Bancroft Library, University of California, 1983.

Cushing, Charles. Author's interviews, New York, NY, April 7, 1993, and June 2, 1993.

Czachowski, Bernard. Author's interview, New York, NY, January 24, 1992.

Francis, Amadeo. Author's telephone interview, April 28, 2005.

Gleason, Thomas W. Author's interview, New York, NY, September 29, 1992.

———. Interview by Debra Bernhardt, July 31, 1981. New Yorkers at Work Oral History Collection, Robert F. Wagner Labor Archive, New York University.

Goldblatt, Louis. "Working Class Leader in the ILWU, 1935–1977." Interviews by Estolv Ethan Ward, 1977–78. Berkeley: Regional Oral History Office, Bancroft Library, University of California, 1980.

Grey, Vincent. Author's telephone interview, May 1, 2005.

Hall, Earl. Author's telephone interview, May 12, 1993.

Harlander, Les. Author's telephone interview, November 2, 2004.

———. Interview by Arthur Donovan and Andrew Gibson, June 19, 1997. Containerization Oral History Project, National Museum of American History, Smithsonian Institution, Washington, DC.

Hayden, Frank. Author's telephone interview, June 29, 2004.

Healey, Richard. Author's telephone interview, January 9, 1994.

Hooper, Edwin B. "Reminiscences of Vice Admiral Edwin B. Hooper." Annapolis: U.S. Naval Institute, 1978.

Hubbard, William. Author's telephone interview, July 1, 1993.

Irvin, William D. "Reminiscences of Rear Admiral William D. Irvin." Annapolis: U.S. Naval Institute, 1980.

Katims, Ron. Author's interview, New York, NY, October 30, 1992.

———. Interview by Arthur Donovan and Andrew Gibson, August 15, 1997. Containerization Oral History Project, National Museum of American History, Smithsonian Institution, Washington, DC.

Morrison, Scott. Interview by Arthur Donovan and Andrew Gibson, March 27, 1997. Containerization Oral History Project, National Museum of American History, Smithsonian Institution, Washington, DC.

Nutter, Ben E. "The Port of Oakland: Modernization and Expansion of Shipping, Airport, and Real Estate Operations, 1957–1977." Interview by Ann Lage, 1991. Berkeley: Regional Oral History Office, Bancroft Library, University of California, 1994.

Ramage, Lawson P. "Reminiscences of Vice Admiral Lawson P. Ramage." Annapolis: U.S. Naval Institute, 1970.

Richardson, Paul. Author's interviews, Holmdel, NJ, January 14, 1992, and July 20, 1992.

Roger, Sidney. "A Liberal Journalist on the Air and on the Waterfront." Interviews by Julie Shearer, 1989–90. Berkeley: Regional Oral History Office, Bancroft Library, University of California, 1983.

Sayre, Cliff. Author's telephone interview, January 24, 1992.

Schmidt, Henry. "Secondary Leadership in the ILWU, 1933–1966." Interviews by Miriam F. Stein and Estolv Ethan Ward, 1974, 1975, 1981.

Regional Oral History Office, Bancroft Library, University of California, 1983.
Stickles, Milton J. Author's telephone interview, June 1, 2004.
St. Sure, J. Paul. "Some Comments on Employer Organizations and Collective Bargaining in Northern California since 1934." Interview by Corinne Gilb, 1957. Berkeley: Institute of Industrial Relations Oral History Project, University of California, 1957.
Tantlinger, Keith. Author's interview, San Diego, CA, January 3, 1993.
Tolan, David J. Interview by Debra Bernhardt, August 1, 1980. New Yorkers at Work Oral History Collection, Robert F. Wagner Labor Archive, New York University.
Toomey, Gerald. Author's interview, New York, NY, May 5, 1993.
Tozzoli, Guy F. Author's interview, New York, NY, January 13, 2004.
Van Dyk, Dena. Author's telephone interview, May 2, 1994.
Ward, William. ILWU Oral History Collection, viewed July 5, 2004, at http://www.ilwu.org/history/oral-histories/bill-ward.cfm.
Wriston, Walter. Author's interview, New York, NY, June 30, 1992.
Younger, Ken. Author's telephone interview, December 16, 1991.

### Dissertations and Theses

Fitzgerald, Donald. "A History of Containerization in the California Maritime Industry: The Case of San Francisco." Ph.D. diss., University of California at Santa Barbara, 1986.
Holcomb, Kenneth Johnson. "History, Description and Economic Analysis of Trailer-on-Flatcar (Piggyback) Transportation." Ph.D. diss., University of Arkansas, 1962.
Rabach, Eileen Rhea. "By Sea: The Port Nexus in the Global Commodity Network (The Case of the West Coast Ports)." Ph.D. diss., University of Southern California, 2002.
Rosenstein, Mark. "The Rise of Maritime Containerization in the Port of Oakland, 1950–1970." M.A. thesis, New York University, 2000.
Wallin, Theodore O. "The Development, Economics, and Impact of Technological Change in Transportation: The Case of Containerization." Ph.D. diss., Cornell University, 1974.
Winter, Jennifer Marie. "Thirty Years of Collective Bargaining: Joseph Paul St. Sure, Management Labor Negotiator 1902–1966," M.A. thesis, California State University at Sacramento, 1991.

### Books

Albion, Robert Greenhalgh. *The Rise of New York Port.* New York: Scribner, 1939. Reprint, 1971.

Arnesen, Eric. *Waterfront Workers of New Orleans: Race, Class, and Politics 1863–1923*. New York: Oxford University Press, 1991.

Aronowitz, Stanley. *From the Ashes of the Old: American Labor and America's Future*. Boston: Houghton Mifflin, 1998.

Arthur, W. Brian. *Increasing Returns and Path Dependence in the Economy*. Ann Arbor: University of Michigan Press, 1994.

Beckert, Sven. *The Monied Metropolis*. Cambridge, UK: Cambridge University Press, 1993.

Bernstein, Peter L. *Wedding of the Waters: The Erie Canal and the Making of a Great Nation*. New York: Norton, 2005.

Borruey, René. *Le port de Marseille: du dock au conteneur*. Marseilles: Chambre de commerce et d'industrie Marseille-Provence, 1994.

Boschken, Herman L. *Strategic Design and Organizational Change: Pacific Rim Seaports in Transition*. Tuscaloosa: University of Alabama Press, 1988.

Bremer Ausschuß für Wirtschaftsforschung. *Container Facilities and Traffic in 71 Ports of the World Midyear 1970*. Bremen: University of Bremen, 1971.

Broeze, Frank. *The Globalisation of the Oceans: Containerisation from the 1950s to the Present*. St. John's, NF, 2002.

Burke, Padraic. *A History of the Port of Seattle*. Seattle: Port of Seattle, 1976.

Cafruny, Alan W. *Ruling the Waves: The Political Economy of International Shipping*. Berkeley and Los Angeles: University of California Press, 1987.

Cannato, Vincent J. *John Lindsay and His Struggle to Save New York*. New York: Basic Books, 2001.

Caro, Robert A. *The Power Broker: Robert Moses and the Fall of New York*. New York: Knopf, 1974.

Chandler, Alfred D., Jr. *The Visible Hand: The Managerial Revolution in American Business*. Cambridge, MA: Harvard University Press, 1977.

Chinitz, Benjamin. *Freight and the Metropolis: The Impact of America's Transport Revolution on the New York Region*. Cambridge, MA: Harvard University Press, 1960.

Condit, Carl W. *The Port of New York*. 2 vols. Chicago: University of Chicago Press, 1980–81.

Cronon, William. *Nature's Metropolis: Chicago and the Great West*. New York: Norton, 1991.

Davies, Sam., et al., eds. *Dock Workers: International Explorations in Comparative Labour History*. 2 vols. Aldershot, UK: Ashgate Publishers, 2000.

DiFazio, William. *Longshoremen: Community and Resistance on the Brooklyn Waterfront*. South Hadley, MA: Bergin & Garvey, 1985.

*Dockers de la Méditerranée à la Mer du Nord*. Avignon: ÉDISUD, 1999.

Doig, Jameson W. *Empire on the Hudson: Entrepreneurial Vision and Political Power at the Port of New York Authority*. New York: Columbia University Press, 2001.

Duncan, Beverly, and Stanley Lieberson. *Metropolis and Region in Transition.* Beverly Hills: Sage Publications, 1970.

Elphick, Peter. *Liberty: The Ships That Won the War.* London: Chatham Publishing, 2001.

Erie, Steven P. *Globalizing L.A.: Trade, Infrastructure, and Regional Development.* Stanford: Stanford University Press, 2004.

Fairley, Lincoln. *Facing Mechanization: The West Coast Longshore Plan.* Los Angeles: University of California at Los Angeles Institute of Industrial Relations, 1979.

Finlay, William. *Work on the Waterfront: Worker Power and Technological Change in a West Coast Port.* Philadelphia: Temple University Press, 1988.

Fishlow, Albert. *American Railroads and the Transformation of the Antebellum Economy.* Cambridge, MA: Harvard University Press, 1965.

Flynn, Stephen E. *America the Vulnerable: How the U.S. Has Failed to Secure the Homeland and Protect Its People from Terror.* New York: Harper Collins, 2004.

Fogel, Robert William. *Railroads and American Economic Development.* Baltimore: Johns Hopkins University Press, 1964.

Freeman, Joshua. *Working-Class New York.* New York: New Press, 2000.

Fujita, Masahisa, Paul Krugman, and Anthony J. Venables. *The Spatial Economy: Cities, Regions, and International Trade.* Cambridge, MA: MIT Press, 1999.

Gibson, Andrew, and Arthur Donovan. *The Abandoned Ocean: A History of United States Maritime Policy.* Columbia: University of South Carolina Press, 2000.

Glaskowsky, Nicholas A., Jr., ed. *Management for Tomorrow.* Stanford, CA: Graduate School of Business, Stanford University, 1958.

Gómez-Ibañez, José A., William B. Tye, and Clifford Winston, eds. *Essays in Transportation Economics and Policy: A Handbook in Honor of John R. Meyer.* Washington, DC: Brookings Institution, 1999.

Griffin, John I. *The Port of New York.* New York: Arco Publishing, 1959.

Grunwald, Joseph, and Kenneth Flamm. *The Global Factory: Foreign Assembly in International Trade.* Washington, DC: Brookings Institution, 1985.

Hagel, Otto, and Louis Goldblatt. *Men and Machines: A Story about Longshoring on the West Coast Waterfront.* San Francisco: Pacific Maritime Association and International Longshoremen's and Warehousemen's Union, 1963.

Haig, Robert Murray. *Major Economic Factors in Metropolitan Growth and Arrangement.* New York: Regional Plan Association, 1927. Reprint, New York: Arno Press, 1974.

Hannes, Matt. *The Container Revolution.* Anaheim: Canterbury Press, 1996.

Hartman, Paul T. *Collective Bargaining and Productivity*. Berkeley and Los Angeles: University of California Press, 1969.

Heilbrun, James. *Urban Economics and Public Policy*. New York: St. Martin's, 1974.

Heiser, Joseph M. Jr. *A Soldier Supporting Soldiers*. Washington, DC: U.S. Army Center of Military History, 1991.

———. *Vietnam Studies: Logistic Support*. Washington, DC: Department of the Army, 1974.

Herod, Andrew. *Labor Geographies: Workers and the Landscapes of Capitalism*. New York: Guilford Press, 2001.

Hooper, Edwin B. *Mobility Support Endurance: A Story of Naval Operational Logistics in the Vietnam War, 1965–1968*. Washington, DC: Naval History Division, 1972.

Hoover, Edgar M., and Raymond Vernon. *Anatomy of a Metropolis*. Cambridge, MA: Harvard University Press, 1959.

Immer, John R. *Container Services of the Atlantic*, 2nd ed. Washington, DC: Work Saving International, 1970.

International Longshoremen's and Warehousemen's Union. *The ILWU Story: Two Decades of Militant Unionism*. San Francisco: ILWU, 1955.

Jensen, Vernon. *Hiring of Dock Workers and Employment Practices in the Ports of New York, Liverpool, London, Rotterdam, and Marseilles*. Cambridge, MA: Harvard University Press, 1964.

———. *Strife on the Waterfront*. Ithaca: Cornell University Press, 1974.

Kessner, Thomas. *Fiorello H. LaGuardia and the Making of Modern New York*. New York: McGraw-Hill, 1989.

Kimeldorf, Howard. *Reds or Rackets? The Making of Radical and Conservative Unions on the Waterfront*. Berkeley and Los Angeles: University of California Press, 1988.

Kirsh, Benjamin S. *Automation and Collective Bargaining*. New York: Central Book Co., 1964.

Kornhauser, Arthur, Robert Dublin, and Arthur M. Ross, eds. *Industrial Conflict*. New York: McGraw-Hill, 1954.

Kreps, Juanita M. *Automation and Employment*. New York: Holt, Reinhart & Winston, 1964.

Larrowe, Charles P. *Harry Bridges: The Rise and Fall of Radical Labor in the United States*. New York: Lawrence Hill, 1972.

——— *Shape Up and Hiring Hall*. Berkeley and Los Angeles: University of California Press, 1955.

Larson, John L. *Internal Improvement*. Chapel Hill: University of North Carolina Press, 2001.

Latimer, Jack. *Friendship among Equals*. Geneva: International Standards Organisation, 1997.

Lockwood, Rupert. *Ship to Shore: A History of Melbourne's Waterfront and its Union Struggles*. Sydney: Hale & Iremonger, 1990.

Magden, Ronald E. *The Working Longshoreman*. Tacoma: International Longshoremen's and Warehousemen's Union Local 23, 1996.

Magden, Ronald E., and A. D. Martinson. *The Working Waterfront: The Story of Tacoma's Ships and Men*. Tacoma: International Longshoremen's and Warehousemen's Union Local 23 and Port of Tacoma, 1982.

Maldonado, A. W. *Teodoro Moscoso and Puerto Rico's Operation Bootstrap*. Gainesville: University Press of Florida, 1997.

Marolda, Edward J., and Oscar P. Fitzgerald. *The United States Navy and the Vietnam Conflict*. Vol. 2, *From Military Assistance to Combat, 1959–1965*. Washington, DC: Naval Historical Center, 1986.

McDougall, Ian. *Voices of Leith Dockers*. Edinburgh: Mercat Press, 2001.

McNickle, Chris. *To Be Mayor of New York: Ethnic Politics in the City*. New York: Columbia University Press, 1993.

Minor, Woodruff. *Pacific Gateway: An Illustrated History of the Port of Oakland*. Oakland: Port of Oakland, 2000.

Mokyr, Joel. *The Gifts of Athena: Historical Origins of the Knowledge Economy*. Princeton: Princeton University Press, 2002.

Mollenkopf, John, and Manuel Castells, eds. *Dual City: Restructuring New York*. New York: Russell Sage Foundation, 1992.

Moses, Robert. *Public Works: A Dangerous Trade*. New York: McGraw-Hill, 1970.

Nelson, Bruce. *Divided We Stand: American Workers and the Struggle for Black Equality*. Princeton: Princeton University Press, 2001.

———. *Workers on the Waterfront: Seamen, Longshoremen, and Unionism in the 1930s*. Champaign: University of Illinois Press, 1990.

Niven, John. *American President Lines and Its Forebears, 1848–1984*. Newark: University of Delaware Press, 1987.

*North Carolina: A Guide to the Old North State*. Chapel Hill: University of North Carolina Press, 1939.

O'Rourke, Kevin H., and Jeffrey G. Williamson. *Globalization and History: The Evolution of a Nineteenth-Century Atlantic Economy*. Cambridge, MA: MIT Press, 1999.

Pacini, Alfred, and Dominique Pons. *Docker à Marseille*. Paris: Payot & Rivages, 1996.

Pearson, Roy, and John Fossey. *World Deep-Sea Container Shipping: A Geographical, Economic, and Statistical Analysis*. Liverpool: Marine Transport Centre, University of Liverpool, 1983.

Pelzman, Sam, and Clifford Winston, eds. *Deregulation of Network Industries: What's Next?* Washington, DC: AEI-Brookings Joint Center for Regulatory Studies, 2000.

Pilcher, William W. *The Portland Longshoremen: A Dispersed Urban Community.* New York: Norton, 1972.

Pinder, David, and Brian Slack, eds. *Shipping and Ports in the Twenty-first Century: Globalisation, Technological Change, and the Environment.* London: Routledge, 2004.

Pollak, Richard. *The Colombo Bay.* New York: Simon & Schuster, 2004.

Rath, Eric. *Container Systems.* New York, Wiley, 1973.

Ricardo, David. *The Principles of Political Economy and Taxation.* London, 1821. Reprint, New York, 1965.

Rosenberg, Daniel. *New Orleans Dockworkers: Race, Labor, and Unionism, 1892–1923.* Albany: State University of New York Press, 1988.

Rostow, W. W. *Stages of Economic Growth.* Cambridge, UK: Cambridge University Press, 1960.

Rubin, Lester. *The Negro in the Longshore Industry.* Philadelphia: Wharton School, 1974.

Ruppenthal, Karl M., ed. *Revolution in Transportation.* Stanford: Stanford University Graduate School of Business, 1960.

Sayre, Wallace S., and Herbert Kaufman. *Governing New York City: Politics in the Metropolis.* New York: Norton, 1960.

Schenker, Eric, and Harry C. Brockel, eds. *Port Planning and Development as Related to Problems of U.S. Ports and the U.S. Coastal Environment.* Cambridge, MD: Maritime Press of America, 1974.

Schwartz, Stephen. *Brotherhood of the Sea: A History of the Sailors' Union of the Pacific, 1885–1985.* Piscataway, NJ: Transaction Books, 1986.

Seligman, Ben B. *Most Notorious Victory: Man in an Age of Automation.* New York: Free Press, 1966.

Selvin, David F. *A Terrible Anger.* Detroit: Wayne State University Press, 1996.

Shields, Jerry. *The Invisible Billionaire: Daniel Ludwig.* Boston: Houghton Mifflin, 1986.

Simey, T. S., ed., *The Dock Worker: An Analysis of Conditions of Employment in the Port of Manchester.* Liverpool: Liverpool University Press, 1956.

Sletmo, Gunnar K., and Ernest W. Williams, Jr. *Liner Conferences in the Container Age: U.S. Policy at Sea.* New York: Macmillan Publishing, 1981.

Snyder-Grenier, Ellen M. *Brooklyn! An Illustrated History.* Philadelphia: Temple University Press, 1996.

Starr, Roger. *The Rise and Fall of New York City.* New York: Basic Books, 1985.

Tursi, Frank V., Susan E. White, and Steve McQuilkin. *Lost Empire: The Fall of R. J. Reynolds Tobacco Company.* Winston-Salem: Winston-Salem Journal, 2000.

Vernon, Raymond. *Metropolis 1985: An Interpretation of the Findings of the New York Metropolitan Region Study*. Cambridge, MA: Harvard University Press, 1960.

Whittaker, J. R. *Containerization*. Washington, DC: Hemisphere Publishing, 1975.

Williams, Ernest W. *The Regulation of Rail-Motor Rate Competition*. New York: Harper, 1958.

Wilson, David F. *Dockers: The Impact of Industrial Change*. London: Fontana, 1972.

Wilson, Rosalyn A. *Transportation in America*. Washington, DC: Eno Transportation Foundation, 2002.

Winston, Clifford, Thomas M. Corsi, Curtis M. Grimm, and Carol A. Evans. *The Economic Effects of Surface Freight Deregulation*. Washington, DC: Brookings Institution, 1990.

Worden, William L. *Cargoes: Matson's First Century in the Pacific*. Honolulu: University Press of Hawaii, 1981.

Zweig, Phillip L. *Wriston: Walter Wriston, Citibank, and the Rise and Fall of American Financial Supremacy*. New York: Crown Publishers, 1995.

### Articles and Working Papers

Anderson, James E., and Eric van Wincoop. "Trade Costs." *Journal of Economic Literature* 42 (September 2004): 691–751.

Baer, Werner. "Puerto Rico: An Evaluation of a Successful Development Program." *Quarterly Journal of Economics* 73, no. 4 (1959).

Beier, Frederick J., and Stephen W. Frick. "The Limits of Piggyback: Light at the End of the Tunnel." *Transportation Journal* 18, no. 2 (1978).

Bougheas, Spiros, Panicos O. Demetriades, and Edgar L. W. Morgenroth. "Infrastructure, Transport Costs, and Trade." *Journal of International Economics* 47 (1999).

Broda, Christian, and David E. Weinstein. "Globalization and the Gains from Variety." Working Paper 10314, NBER, February 2004.

Brynjolfsson, Erik, and Lorin M. Hitt. "Beyond Computation: Information Technology, Organizational Transformation, and Business Performance." *Journal of Economic Perspectives* 14, no. 4 (2000).

Carrere, Céline, and Maurice Schiff. "On the Geography of Trade: Distance Is Alive and Well." World Bank Policy Research Working Paper 3206, February 2004.

Carruthers, Robin, and Jitendra N. Bajpai. "Trends in Trade and Logistics: An East Asian Perspective." Working Paper No. 2, Transport Sector Unit, World Bank (2002).

Chen, Hong, Murray Z. Brank, and Owen Q. Wu. "U.S. Retail and Wholesale Inventory Performance from 1981 to 2003." Working Paper, University of British Columbia, 2005.

Clark, Ximena, David Dollar, and Alejandro Micco, "Port Efficiency, Maritime Transport Costs, and Bilateral Trade." *Journal of Development Economics* 74, no. 3 (2004): 417–450.

Coe, David, et al. "The Missing Globalization Puzzle." International Monetary Fund Working Paper WP/02/171, October 2002.

Connolly, D. J. "Social Repercussions of New Cargo Handling Methods in the Port of London." *International Labour Review* 105 (1972): 543–568.

Cushing, Charles R. "The Development of Cargo Ships in the United States and Canada in the Last Fifty Years." Manuscript, January 8, 1992.

David, Paul A. "The Dynamo and the Computer: An Historical Perspective on the Modern Productivity Paradox." *American Economic Review* 80 (1990): 355–361.

Davies, J. E. "An Analysis of Cost and Supply Conditions in the Liner Shipping Industry." *Journal of Industrial Economics* 31 (1983): 417–436.

Dempsey, Paul Stephen. "The Law of Intermodal Transportation: What It Was, What It Is, What It Should Be." *Transportation Law Journal* 27, no. 3 (2000).

Devine, Warren D., Jr. "From Shafts to Wires: Historical Perspective on Electrification." *Journal of Economic History* 43 (1983): 347–372.

Egedyi, Tineke M. "The Standardized Container: Gateway Technologies in Cargo Transportation." Working Paper, Delft University of Technology, 2000.

Farrell, Joseph, and Garth Saloner. "Installed Base and Compatibility: Innovation, Product Preannouncements, and Predation." *American Economic Review* 76, no. 5 (1986): 940–955.

Feenstra, Robert. "Integration of Trade and Disintegration of Production in the Global Economy." *Journal of Economic Perspectives*. 12, no. 4 (1998).

Fishlow, Albert. "Antebellum Regional Trade Reconsidered." *American Economic Review* (1965).

Fujita, Masahisa, and Tomoya Mori. "The Role of Ports in the Making of Major Cities: Self-Agglomeration and Hub-Effect." *Journal of Development Economics* 49 (1996).

Gilman, Roger H. "The Port, a Focal Point." *Transactions of the American Society of Civil Engineers*, 1958.

Glaeser, Edward L. "Reinventing Boston: 1640–2003." Working Paper No. 10166, National Bureau of Economic Research (2003).

Glaeser, Edward L., and Janet E. Kohlhase. "Cities, Regions, and the Decline of Transport Costs." Working Paper No. 9886, National Bureau of Economic Research (2003).

Godfrey, Brian J. "Restructuring and Decentralization in a World City." *Geographical Review* (thematic issue, *American Urban Geography*) 85 (1995).

Hall, John, and D. Caradog Jones. "Social Grading of Occupations." *British Journal of Sociology* 1 (1950): 31–55.

Harbeson, Robert W. "Recent Trends in the Regulation of Intermodal Rate Competition in Transportation." *Land Economics* 42, no. 3 (1966).

Harlander, L.A. "Container System Design Developments over Two Decades." *Marine Technology* 19 (1982): 364–376.

———. "Engineering Development of a Container System for the West Coast–Hawaiian Trade." *Transactions of the Society of Naval Architects and Marine Engineers* 68 (1960).

———. "Further Development of a Container System for the West Coast–Hawaiian Trade." *Transactions of the Society of Naval Architects and Marine Engineers* 69 (1961).

———. "The Role of the 24-Foot Container in Intermodal Transportation." Paper for American Standards Association MH-5 Sectional Committee, June 1966.

Helle, Horst Jürgen. "Der Hafenarbeiter zwischen Segelschiff und Vollbeschäftigung." *Economisch en Sociaal Tijdschrift* 19, no. 4 (1965).

Hoare, Anthony. "British Ports and Their Export Hinterlands: A Rapidly Changing Geography." *Geografiska Annaler, Series B. Human Geography* 68, no. 1 (1986).

Hopper, A. G., P. H. Judd, and G. Williams. "Cargo Handling and Its Effect on Dry Cargo Ship Design." *Quarterly Transactions of the Royal Institution of Naval Architects* 106, no. 2 (1964).

Hummels, David. "Have International Transportation Costs Declined?" Working Paper, University of Chicago Graduate School of Business, 1999.

———. "Time as a Trade Barrier." Mimeo, Purdue University, 2001.

———. "Towards a Geography of Trade Costs." Mimeo, University of Chicago, 1999.

Isard, Caroline, and Walter Isard. "Economic Implications of Aircraft." *Quarterly Journal of Economics* 59 (1945): 145–169.

Jorgenson, Dale W., and Kevin J. Stiroh. "Information Technology and Growth." *American Economic Review* 89, no. 2 (1999): 109–115.

Katz, Michael L., and Carl Shapiro. "Systems Competition and Network Effects." *Journal of Economic Perspectives* 8, no. 2 (1994): 93–115.

Kelley, Robin D. G. " 'We Are Not What We Seem': Rethinking Black Working-Class Opposition in the Jim Crow South." *Journal of American History* 80, no. 1 (1993).

Kerr, Clark, and Lloyd Fisher. "Conflict on the Waterfront." *Atlantic* 184, no. 3 (1949).

King, Robert C., George M. Adams, and G. Lloyd Wilson, "The Freight Container as a Contribution to Efficiency in Transportation." *Annals of the American Academy of Political and Social Science* 187 (1936): 27–36.

Konishi, Hideo. "Formation of Hub Cities: Transportation Cost Advantage and Population Agglomeration." *Journal of Urban Economics* 48 (2000).

Kovarsky, Irving. "State Piggyback Statutes and Federalism." *Industrial and Labor Relations Review* 18, no. 1 (1964).

Krausse, Gerald H. "The Urban Coast in Singapore: Uses and Management." *Asian Journal of Public Administration* 5, no. 1 (June 1983): 33–67.

Krugman, Paul. "Growing World Trade: Causes and Consequences." *Brookings Papers on Economic Activity* 1995, no. 1 (1995): 327–377.

———. "Increasing Returns and Economic Geography." *Journal of Political Economy* 99, no. 3 (1991): 483–499.

Krugman, Paul, and Anthony J. Venables. "Globalization and the Inequality of Nations." *Quarterly Journal of Economics* 110, no. 4 (1995): 857–880.

Kuby, Michael, and Neil Reid, "Technological Change and the Concentration of the U.S. General Cargo Port System: 1970–88." *Economic Geography* 68, no. 3 (1993).

Kujovich, Mary Yeager. "The Refrigerator Car and the Growth of the American Dressed Beef Industry." *Business History Review* 44 (1970): 460–482.

Lamoreaux, Naomi R., Daniel M. G. Raff, and Peter Temin. "Beyond Markets and Hierarchies: Toward a New Synthesis of American Business History." *American Historical Review* 108 (2003).

Larson, Paul D., and H. Barry Spraggins. "The American Railroad Industry: Twenty Years after Staggers." *Transportation Quarterly* 52, no. 2 (2000): 5–9.

Lieb, Robert C., and Robert A. Miller. "JIT and Corporate Transportation Requirements." *Transportation Journal* 27, no. 3 (1988): 5–10.

Liebowitz, S. J., and Stephen E. Margolis. "Network Externality: An Uncommon Tragedy." *Journal of Economic Perspectives* 8, no. 2 (1994): 133–150.

Limão Nuno and Anthony J. Venables. "Infrastructure, Geographical Disadvantage and Transport Costs." *World Bank Economic Review* 15, no. 3 (2001): 451–479.

Linge, G.J.R. "Just-in-Time: More or Less Flexible?" *Economic Geography* 67, no. 4 (1991): 316–332.

MacMillan, Douglas C., and T. B. Westfall. "Competitive General Cargo Ships." *Transactions of the Society of Naval Architects and Marine Engineers* 68 (1960).

Manning, Seaton Wesley. "Negro Trade Unionists in Boston." *Social Forces* 17, no. 2 (1938).

Marin, Pedro L., and Richard Sicotte. "Exclusive Contracts and Market Power: Evidence from Ocean Shipping." Discussion Paper 2028, Centre for Economic Policy Research, June 2001.

Martin, Will, and Vlad Manole. "China's Emergence as the Workshop of the World." Working Paper, World Bank, 2003.

McCarney, Bernard J. "Oligopoly Theory and Intermodal Transport Price Competition: Some Empirical Findings." *Land Economics* 46, no. 4 (1970).

McDonald, Lucille. "Alaska Steam: A Pictorial History of the Alaska Steamship Company." *Alaska Geographic* 11, no. 4 (1984).

McLean, Malcolm P. "Opportunity Begins at Home." *American Magazine* 149 (May 1950).

Miller, Raymond Charles. "The Dockworker Subculture and Some Problems in Cross-Cultural and Cross-Time Generalizations." *Comparative Studies in Society and History* 11. no. 3 (1969).

Mills, Edwin S., and Luan Sendé. "Inner Cities." *Journal of Economic Literature* 35 (1997).

Mills, Herb. "The San Francisco Waterfront—Labor/Management Relations: On the Ships and Docks. Part One: 'The Good Old Days.' " Berkeley: University of California Institute for the Study of Social Change, 1978.

———. "The San Francisco Waterfront—Labor/Management Relations: On the Ships and Docks. Part Two: Modern Longshore Operations." Berkeley: University of California Institute for the Study of Social Change, 1978.

Mokyr, Joel. "Technological Inertia in Economic History." *Journal of Economic History* 52 (1992): 325–338.

Morton, Alexander Lyall. "Intermodal Competition for the Intercity Transport of Manufactures." *Land Economics* 48, no. 4 (1972).

Nelson, James C. "The Economic Effects of Transport Deregulation in Australia." *Transportation Journal* 16, no. 2 (1976): 48–71.

North, Douglass. "Ocean Freight Rates and Economic Development 1750–1913." *Journal of Economic History* 18 (1958): 537–555.

Oliner, Stephen D., and Daniel E. Sichel. "The Resurgence of Growth in the Late 1990s: Is Information Technology the Story?" *Journal of Economic Perspectives* 14, no. 4 (2000): 3–22.

O'Rourke, Kevin H., and Jeffrey G. Williamson. "When Did Globalization Begin?" Working Paper 7632, National Bureau of Economic Research, 2000.

Overman, Henry G., and L. Alan Winters. "The Geography of UK International Trade." Working Paper CEPDP0606, Centre for Economic Performance, London, January 2004.

Pauling, Norman G. "Some Neglected Areas of Research on the Effects of Automation and Technological Change on Workers." *Journal of Business* 37, no. 3 (1964): 261–273.

Puffert, Douglas J. "Path Dependence in Spatial Networks: The Standardization of Railway Track Gauge." *Explorations in Economic History* 39 (2002): 282–314.

———. "The Standardization of Track Gauge on North American Railways, 1830–1890." *Journal of Economic History* 60, no. 4 (2000): 933–960.

Ripley, William Z. "A Peculiar Eight Hour Problem." *Quarterly Journal of Economics* 33, no. 3 (1919).

Romer, Paul M. "Why, Indeed, in America? Theory, History, and the Origins of Modern Economic Growth." Working Paper 5443, NBER, January 1996.

Rosenberg, Nathan. "On Technological Expectations." *Economic Journal* 86, no. 343 (1976): 523–535.

Ross, Philip. "Waterfront Labor Response to Technological Change: A Tale of Two Unions." *Labor Law Journal* 21, no. 7 (1970).

Rubin, Marilyn, Ilene Wagner, and Pearl Kamer. "Industrial Migration: A Case Study of Destination by City-Suburban Origin within the New York Metropolitan Area." *Journal of the American Real Estate and Urban Economics Association* 6 (1978).

Ryon, Roderick N. "An Ambiguous Legacy: Baltimore Blacks and the CIO, 1936–1941." *Journal of Negro History.* 65, no. 1 (1980).

Sasuly, Richard. "Why They Stick to the ILA." *Monthly Review* (January 1956).

Sjostrom, William. "Ocean Shipping Cartels: A Survey." *Review of Network Economics* 3, no. 2 (2004).

Slack, Brian. "Pawns in the Game: Ports in a Global Transportation System." *Growth and Change* 24, no. 4 (1993).

Slesinger, Reuben E. "The Pace of Automation: An American View." *Journal of Industrial Economics* 6, no. 3 (1958): 241–261.

Solow, Robert. "Technical Change and the Aggregate Production Function." *Review of Economics and Statistics* 39, no. 2 (1957): 65–94.

Stocker, H. E. "Cargo Handling and Stowage." Society of Naval Architects and Marine Engineers, November 1933.

Summers, Clyde W. "Admission Policies of Labor Unions." *Quarterly Journal of Economics* 61, no. 1 (1946).
Sung, Nai-Ching, and Michael C. Bunamo. "Competition for Handling U.S. Foreign Trade Cargoes: The Port of New York's Experience." *Economic Geography* 49, no. 2 (1973): 156–162.
Taft, Philip. "The Responses of the Bakers, Longshoremen and Teamsters to Public Exposure." *Quarterly Journal of Economics* 74, no. 3 (1960).
Todd, Daniel. "The Interplay of Trade, Regional and Technical Factors in the Evolution of a Port System: The Case of Taiwan." *Geografiska Annaler, Series B. Human Geography* 75, no. 1 (1993).
Tozzoli, Anthony J. "Containerization and Its Impact on Port Development." *Journal of the Waterways, Harbors and Coastal Engineering Division, Proceedings of the American Society of Civil Engineers* 98, no. WW3 (1972).
Turnbull, Peter. "Contesting Globalization on the Waterfront." *Politics and Society* 28, no. 3 (2000).
"Uniform Containerization of Freight: Early Steps in the Evolution of an Idea." *Business History Review* 63, no. 1 (1969): 84.
Vamplew, Wray. "Railways and the Transformation of the Scottish Economy." *Economic History Review* 24 (1971).
Wallace, Michael, and Arne L. Kalleberg. "Industrial Transformation and the Decline of Craft: The Decomposition of Skill in the Printing Industry, 1931–1978." *American Sociological Review* 47, no. 3 (1982): 307–324.
Waters, Robert C. "The Military Sealift Command versus the U.S. Flag Liner Operators." *Transport Journal* 28, no. 4 (1989): 28–34.
Weldon, Foster L. "Cargo Containerization in the West Coast–Hawaiian Trade." *Operations Research* 6 (1958).

# Index